Biofuels and Bioenergy

Biofuels and Bioenergy

Edited by John Love and John A. Bryant
Biosciences, College of Life and Environmental Sciences,
University of Exeter, UK

This edition first published 2017 © 2017 by John Wiley and Sons Ltd

Registered Office
John Wiley & Sons Ltd, The Atrium, Southern Gate, Chichester, West Sussex, PO19 8SQ, UK

Editorial Offices
9600 Garsington Road, Oxford, OX4 2DQ, UK
The Atrium, Southern Gate, Chichester, West Sussex, PO19 8SQ, UK

For details of our global editorial offices, for customer services and for information about how to apply for permission to reuse the copyright material in this book please see our website at www.wiley.com/wiley-blackwell.

The right of John Love and John A. Bryant to be identified as the author of the editorial material in this work has been asserted in accordance with the UK Copyright, Designs and Patents Act 1988.

Library of Congress Cataloging-in-Publication Data

Names: Love, John and Bryant, John A.
Title: Biofuels and bioenergy / [edited by] John Love and John A. Bryant.
Description: Chichester, West Sussex : John Wiley & Sons, Inc., 2017. |
    Includes bibliographical references and index.
Identifiers: LCCN 2016059291 (print) | LCCN 2017000214 (ebook) | ISBN 9781118350560 (cloth) |
    ISBN 9781118350546 (pdf) | ISBN 9781118350539 (epub)
Subjects: LCSH: Biomass energy. | Algal biofuels.
Classification: LCC TP339 .B5394 2017 (print) | LCC TP339 (ebook) | DDC 662/.88–dc23
LC record available at https://lccn.loc.gov/2016059291

A catalogue record for this book is available from the British Library.

Wiley also publishes its books in a variety of electronic formats. Some content that appears in print may not be available in electronic books.

Cover image: Courtesy of John Love
Anaerobic Digestion of food or agricultural waste and sewage is a tried and tested method for producing renewable methane, or "Biogas", which can either be used directly or to make electricity. The residues from the process are then used as fertiliser to grow crops. The cover image is of an anaerobic digestor plant at Bygrave near Baldock, England, that is owned and run by Biogen. Every year, the plant transforms 45,000 tonnes of food waste into enough electricity to power about 4,500 homes.

Cover design by Wiley

Set in 10/12pt Warnock by SPi Global, Pondicherry, India
Printed and bound in Malaysia by Vivar Printing Sdn Bhd
10 9 8 7 6 5 4 3 2 1

Contents

List of Contributors

Jessica Adams
University of Aberystwyth, Wales, UK

Michael J. Allen
Plymouth Marine Laboratory,
Plymouth, UK

John Bombardiere
West Virginia State University, USA

Leah M. Brown
University of Georgia, Athens, USA

John A. Bryant
University of Exeter, Exeter, UK

Christopher J. Chuck
University of Bath, Bath, UK

Lionel Clarke
BionerG Ltd, Chester, UK

John Clifton-Brown
University of Aberystwyth, Wales, UK

Charlotte Cook
University of Exeter, Exeter, UK

John C. Cushman
University of Nevada, Reno, USA

Chappandra Dayananda
Central Food Technological Research
Institute, Mysore, India

Joy Doran-Peterson
University of Georgia, Athens, USA

Stephen C. Fry
University of Edinburgh, Edinburgh, UK

Astley Hastings
University of Aberdeen, Scotland, UK

Leyla T. Hathwaik
University of Nevada, Reno, USA

Gary M. Hawkins
University of Georgia, Athens, USA

Thomas P. Howard
University of Newcastle, Newcastle-on-
Tyne, UK

Christopher J. Howe
University of Cambridge, Cambridge, UK

Steve Hughes
University of Exeter, Exeter, UK

C.B. Jamieson
World Agroforestry Centre, Laguna,
Philippines and
Next Generation, Hertfordshire, UK

Rhodri W. Jenkins
University of Bath, Bath, UK

R.D. Lasco
World Agroforestry Centre, Laguna,
Philippines

David J. Lea-Smith
University of Cambridge,
Cambridge, UK

Alessandro Marco Lizzul
University College London,
London, UK

John Love
University of Exeter, Exeter, UK

J.M. Lynch
University of Surrey, Guildford, UK

Jon McCalmont
University of Aberystwyth, Wales, UK

E.T. Rasco
PhilRice, Munoz, Philippines

Lisa A. Sargeant
University of Bath, Bath, UK

David A. Stafford
Enviro-Control Ltd., Devon, UK

Richard K. Tennant
University of Exeter, Exeter, UK

Preface

A century ago, petroleum – what we call oil – was just an obscure commodity; today it is almost as vital to human existence as water.
James Buchan, Political commentator and author

The use of vegetable oils for engine fuels may seem insignificant today. But such oils may become in course of time as important as petroleum and the coal tar products of the present time.
Rudolf Diesel, engineer and inventor of the compression engine (quotation dates from 1912)

Most people have difficulty coming to grips with the sheer enormity of energy consumption.
Rex Tillerson, Civil engineer, businessman and President/CEO of the Exxon-Mobil Corporation

We have to rethink our whole energy approach, which is hard to do because we're so dependent on oil, not just for fuel but also plastic ... We have to think quite carefully about using oil and its derivatives, because it's not going to be around forever.
Margaret Atwood, Author, literary critic and environmental activist

We can no longer allow America's dependence on foreign oil to compromise our energy security. Instead, we must invest in inventing new ways to power our cars and our economy. I'll put my faith in American science and ingenuity any day before I depend on Saudi Arabia.
Senator John Kerry, US Secretary of State, 2013–2016

There is an urgent need to stop subsidizing the fossil fuel industry, dramatically reduce wasted energy, and significantly shift our power supplies from oil, coal, and natural gas to wind, solar, geothermal, and other renewable energy sources.
Bill McKibben, Author, educator and environmental activist

These quotations provide a nice series of snapshots. The world is energy-hungry and increasingly so. The need for fuels for transport makes up a large proportion of that hunger. That need is largely met by petroleum (literally 'rock-oil'), 70% of which is used for transport by road, air or sea, but there are issues related to its continued use and availability (even if concerns about 'peak oil' – the moment when global oil production reaches its maximum – have declined somewhat). There are concerns that some developed countries' need for oil may make them economic or moral hostages to countries that are oil producers. And above all is the realisation that burning fossil fuels (principally coal, natural gas and oil) is the major contributor to anthropogenic climate change. There is thus a drive to develop renewable sources of energy, sources of energy that do not involve burning fossil fuels. And we have to say, as is evident from Chapter 1, there has been very good progress in generation of electricity via environmental energy sources. Electricity can of course be used to power some forms of transport. but it still glaringly obvious that transport is very dependent on oil (and to a lesser extent, coal and gas), and will continue to be dependent on liquid fuels well into the future. So we come to this book, which deals with current areas of research aimed at finding ways of using renewable biological resources to provide fuels mainly for transport, research which becomes ever more urgent as the reality of climate change becomes more apparent.

Biofuels, therefore, should be viewed in the context of sustainability, either as alternatives to reduce petroleum use during the transition to other forms of transport, energy or primary materials, or as a way to mitigate climate change. The use of petroleum distillate in mass transport did not happen overnight (indeed petrol was once considered a waste product of oil refining); likewise, biofuels are at the very early stages of development. Biofuels research is intense, with new options being imagined and solutions being proposed almost weekly. Every new technology explores a previously unimagined design landscape. The issue with biofuels is that technical developments are heavily constrained by existing infrastructure, land use and global commerce in commodities. As yet, we cannot pick the future biofuel 'winners', but any biofuel solution (and there may be several) must be responsive to a number of criteria, including cost, technical feasibility, efficiency, reliability, sustainability and, arguably most difficult of all, our lifestyle expectations.

We are bound to say that the book has been a long time coming. It is actually several years since a conversation in an Oxford coffee shop between JAB and Rachel Wade of Wiley-Blackwell led to the idea of an edited text on Biofuels. It took a long time to recruit authors and even then, other factors outside the control of editors and publisher, led to further delays and loss of some of the planned and contracted chapters. Against this background, we are especially grateful to those authors who have remained with the project and provided the excellent and interesting range of chapters presented here. A number of them have been remarkably patient as they waited for news of further progress after submitting their chapters. We are also grateful to colleagues who have given us their time in discussion and/or provided diagrams and figures for us to use. JAB expresses special thanks to Dr David Stafford of Enviro-Control Ltd, Professor Jim Lynch of the University of Surrey and Professor Steve Hughes of the University of Exeter for their long-term friendship, support and readiness to share their knowledge and expertise; also to environmental engineer, Rachel Oates of the Lee Abbey Community, Devon, for her knowledgeable enthusiasm and readiness to talk about environmental energy sources, especially 'micro-hydro' (see Chapter 1). JL is especially indebted to Professor Rob Lee of Royal Dutch Shell, his 'partner in slime' for the past 15 years and to Drs Mike Goosey and Jeremy Shears, both directors of Shell Biodomain, for their positive and supportive vision of open innovation between academia and industry to solve global problems. Heartfelt thanks also to all of the current and past members of the Exeter Microbial Biofuels Group for their talent, commitment, professionalism and humour, to our numerous collaborators in academia and in industry, and to the BBSRC for supporting our research. Our programme would not be possible without the direct support of the University of Exeter. JL is personally grateful to our Vice Chancellor Professor Sir Steve Smith and to our Deputy Vice-Chancellor for research, Professor Nick Talbot, FRS (formerly his head of department) for their genuine and sustained interest, and to his all colleagues in professional services, notably Linda Peka and Caroline Hampson, for their immeasurable patience in underpinning our research.

Finally, we want to say a big 'Thank You' for the patience of our publishers at Wiley-Blackwell and especially those closely involved with this book, Rachel Wade at Oxford who helped to initiate the project, Fiona Seymour at Chichester who was for a long time 'our' editor, Audrie Tan at Singapore who took over from Fiona as the last few chapters came in and Vinodhini Mathiyalagan together with Shummy Metilda who supervised the book's production. It would have been so easy for these individuals and for Wiley-Blackwell themselves to abandon the book as we sought yet another re-scheduling. We are very grateful that they stuck with us.

John Love
John A. Bryant
Exeter, August 2015

List of Abbreviations

AAT	ATP-ADP translocase
ACC	Acetyl-CoA carboxylase
ACL	ATP-citrate lyase
Ack	Acetate kinase
ACP	Acyl carrier protein
ACS	Acyl-CoA synthetase
AD	Anaerobic digestion
ADH	Alcohol dehydrogenase
AdhE2	Butaraldehyde/butanol dehydrogenase
AFEX	Ammonia fibre expansion
ADO	Aldehyde decarbonylase
ALR	Airlift reactor
AS	ATP synthase
ATJ	Alcohol to jet
AtoAD	Acetoacetyl-CoA transferase
ATP-CL	ATP-citrate lyase
B100	100% biodiesel
BBSRC	Biotechnology and Biological Sciences Research Council
bbl	Barrels
Bcd	Butyrl-CoA dehydrogenase
BCKD	Branched chain ketoacid dehydrogense
BDC	Buoyant density centrifugation
BDGC	Buoyant density gradient centrifugation
BIS	Bisbolene synthase
BOD	Biological oxygen demand
BTL	Bio-to-liquids
CAR	Carboxylic acid reductase
Ccr	Cronotyl-CoA reductase
CCX	Chicago Climate Exchange
CFPP	Cold Filter Plugging Point
CH	Catalytic hydrothermolysis
CI	Cetane index
C/N	Carbon to nitrogen ratio
CN	Cetane number
CP	Cloud point

CRP   Cyclic AMP (cAMP) receptor protein
CSL   Cellulose synthase-like
CtfAB   Coenzyme A transferase, A and B subunits
DAG   Diacylglycerol
DAGAT   Diacyglycerol acyltransferase
DDG   Dried distillers grain
DDGS   Dried distillers grains with solubles
DEFRA   Department of Food, Rural Affairs and Agriculture
DIC   Dicarboxylate carrier
DM   Dry matter
DMAPP   Dimethylallyll diphosphate
DMSO   Dimethyl sulfoxide
DoE   Department of Energy
DPF   Diesel particulate filter
DSHC   Direct sugar to hydrocarbons
E85   85% ethanol
EBI   Energy Biosciences Institute
ECM   Extracellular matrix
EJ   Exajoules
ELA   Extremely low acid
EMP   Embden-Meyerhof-Parnas
EPA   Environmental Protection Agency
EPIC   Environmental Policy Integrated Climate
ER   Endoplasmic reticulum
EU   European Union
FA   Fatty acid
FAAE   Fatty acid alkyl esters
FACS   Fluorescence activated cell sorting
fadD   Fatty acyl-CoA synthetase
FAEE   Fatty-acid ethyl ester
FAME   Fatty acid methyl ester
FAO   Food and Energy Organisation
FAR   Fatty acid reductase
FAS   Fatty-acid synthase
FCM   Flow cytometry
FPP   Farneseyl pyrophosphate
FS   Farnesene synthase
G-3-P   Glycerol-3-phosphate
GAT   G-3-P acyltransferase
GC   Gas chromatography
GDSL   Gly-Asp-Ser-Leu
GGPP   Geranylgeranyl pyrophosphate
GHG   Greenhouse gases
GIS   Geographic Information System
GM   Genetically modified
GPP   Geranyl pyrophosphate
GTL   Gas-to-liquids

GTME	Global transcriptional machinery engineering
GW	Gigawatts
Hbd	3-hydroxybutyrl-CoA dehydrogenase
HCCI	Homogenous charge compression ignition
HDCJ	Hydrotreated depolymerised cellulosic jet
HEFA	Hydroprocessed esters and fatty acids
HGA	Homogalacturonan
HMF	Hydroxymethylfurfural
HRAP	High rate algal pond
HT	Hydrotreatment
IFES	Integrated food-energy systems
IFQC	International Fuel Quality Centre
iGEM	Internationally Genetically Engineered Machines
LA	Lysophosphatic acid
ILUC	Indirect Land Use Change/integrated land use change
IOU	Investor-owned utilities
IPP	Isopentenyl diphosphate
IPPC	Inter-Governmental Panel on Climate Change
ispS	Isoprene synthase
KDC	Ketoacid decarboxylase
kW	Kilowatts
LB	Lipid bodies
LCA	Life cycle analysis/Life Cycle Assessment
Ldh	L-lactate dehydrogenase
LED	Light emitting diode
LNG	Liquefied Natural Gas
LNS	Light natural sandwich
lpa	Lysophosphatidic acid
LPAAT	Lysophosphatidic acid acyltransferase
LS	Limonene synthase
LUC	Land Use Change
MAG	2-monoacylglycerol
MDH	Malate dehydrogenase
ME1	Malic enzyme
MEP	2-C-methyl-D-erythritol-4-phosphate
MEV	Mevalonate
MIT	Massachusetts Institute of Technology
MLG	Mixed-linkage glucan
MSW	Municipal solid waste
MW	Megawatts
MXE	MLG:xyloglucan endotransglucosylase
NDP	An *N*-galacturonoyl amide
NGO	Non-Governmental Organisation
NMR	Nuclear magnetic resonance
NO_x	Nitrogen oxides
OECD	Organisation for Economic Cooperation and Development
OEM	Optical Emission Spectroscopy

OPEC	Organization of the Petroleum Exporting Countries
PA	Phosphatidic acid
PAP	Phosphatic acid phosphohydrolase
PBR	Photobioreactor
PC	Pyruvate carboxylase
Pdc	Thiamine pyrophosphate
PDH	Pyruvate dehydrogenase
PHB	Polyhydroxybutyrate
PME	Pectin methylesterase
PP	Pour point
PPi	Pyrophosphate
ppm	Parts per million
PPP	Pentose phosphate pathway
PS	Pinene synthase
PT	Pyruvate transporter
Pta	Phosphate acetyltransferase
PV	Photovoltaic
PVC	Polyvinyl chloride
PVP	Polyvinylpyrrolidone
QTL	Quantitative trait locus/loci
R&D	Research and Development
REACH	Registration, Evaluation and Authorisation of Chemicals
REC	Renewable Energy Certificate
RED	Renewable Energy Directive
REDD+	Reduction of Emissions Due to Deforestation and Forest Degradation
RFS	Renewable Fuel Standards
RME	Rape-seed methyl ester
RG	Rhamnogalacturonan
RPS	Renewables Portfolio Standard
RVP	Reid Vapour Pressure
S-AM	S-adenosyl methonine transferase
SI	Spark-ignition
SCO	Single-cell oil
SOC	Soil organic carbon
SSL	Squalene synthase-like
TAG	Triacylglycerol
TBA	Tertiary butanol
TCA	Tricarboxylic acid
TCL	Thermo-chemical liquefaction
Te	Thioesterase
TFA	Trifluoroacetic acid
Thl	Thiolase
TIC	Tricarboxylate carrier
TPP	Thiamine pyrophosphate
TW	Terawatts
ULS	Ultra-low sulphur
USDA	US Department of Agriculture

USDoE	US Department of Energy
USPS	US Postal Service
UV	Ultra-violet
WRAP	Waste Resources Action Programme
XEG	Xyloglucan endo-glucanase
XET	Xyloglucan endotransglucosylase
XTH	Xyloglucan endotransglucosylase/hydrolase

1

Biofuels: The Back Story

John A. Bryant and John Love

College of Life and Environmental Sciences, University of Exeter, Exeter, UK

Summary

This chapter looks at the history of the use of fossil and non-fossil fuels and of environmental energy sources from the earliest phases of human society right up to the present day. Factors, especially climate change, which affect the use of particular fuels are discussed. The chapter ends with an overview of biofuels, thus setting the scene for the rest of the book.

1.1   Introduction

The earliest recorded use of the word *biofuel* was in 1970 when it was defined as 'a fuel (such as wood or ethanol) composed of or produced from biological raw materials.' Use of the term gradually became more frequent but it is only in last 15 years or so that it has entered into everyday speech. The definition has also widened: the *Oxford Dictionary* On-line now simply states 'a fuel derived immediately from living matter.' This clearly covers much more than wood and ethanol; the range will be apparent from the chapters in this book. The purpose of this chapter is to provide the context for, and to discuss the reasons behind, this increased interest in biofuels. It is an unfolding story of human ingenuity and inventiveness in the search for sources of light and heat and of energy for industry, transport, commerce and domestic appliances. It is a fascinating story that sets the scene for the rest of the book.

1.2   Some History

1.2.1   Wood and Charcoal

Although the first recorded use of the word was relatively recent, the use of biofuels actually goes back much further. Biological materials have been used as energy

Biofuels and Bioenergy, First Edition. Edited by John Love and John A. Bryant.
© 2017 John Wiley & Sons Ltd. Published 2017 by John Wiley & Sons Ltd.

sources throughout human existence; indeed it is likely that the Neanderthals had discovered fire and the use of wood as a fuel. On a small local scale, burning of wood as a fuel may be regarded as having a very small 'ecological footprint', especially since, reflecting our modern concerns, it releases only recently fixed CO_2 into the atmosphere.

Pyrolysis of wood in the absence of air produces charcoal, a form of carbon that burns at a higher temperature than wood and can thus be used in metal smelting. The use of charcoal as a fuel dates back at least 6,000 years (and probably longer). Initially it was confined to Egypt and what is now known as the Middle East. Its use soon spread across Europe so that by the Middle Ages, charcoal production was very widespread and resulted in extensive deforestation over large areas. It thus had an ecological/environmental impact that today we would regard at least as undesirable.

With the invention of a method for making coke from coal (*i.e.* a fossil fuel), charcoal production declined dramatically, especially from 1900 onwards (although one of us can remember seeing charcoal burners in woods in Surrey in the middle years of the 20th century). Today the use of charcoal as a fuel[1] in developed countries is largely confined to domestic barbecues. However, across the world, wood and charcoal are still the mostly widely used fuels. This includes the use of wood-burning stoves in people's homes and wood-burning power stations, often regarded as environmentally friendly because, as noted before, it is recently fixed CO_2 that is released. This CO_2 release may be further mitigated by the planting of replacement trees in managed forestry systems. However, there is no universal agreement on this; some think that growing wood just for burning is not wise when the wood could have so many other uses[2]. Furthermore, in many parts of the world where emissions are less stringently controlled, burning of wood often causes serious smoke pollution and damaging effects on human health.

1.2.2 Dung as Fuel

Evidence for use of dried animal dung as fuel dates back about 9,000 years to Neolithic communities in which cattle, sheep, goats and pigs had been domesticated. It is still used today in many less-developed countries. There is also evidence for use by Native Americans of dung from wild bison in the prairies where wood fuel was very scarce or non-existent. There is undoubtedly today support for increased use of dung as fuel, both in what we might call traditional or semi-traditional methods and by anaerobic digestion (see Chapter 3).

1.2.3 Oils and Fats

The use of natural oils for lighting dates back to about 15,000 years. Most ancient oil lamps ran on plant oils. Thus the lamps referred to in the Old and New Testaments of the Bible and in the Qur'an were fuelled with olive oil. Both plant and animal oils were also used for lighting in ancient Egypt, dating back to about 3000 BC: rushlights, precursors to candles, were made by dipping rolled-up papyrus into oil or into melted

---

1 Powdered charcoal has several non-fuel uses, including absorption of gases and purification of liquids.
2 http://www.usewoodwisely.co.uk/

beeswax or melted animal fat. The Romans are generally credited with invention of the true candle containing a wick that ran through the length of a cylinder of bees' wax (other solid animal fats may also be used).

Another animal-based biofuel is whale oil, which was used for lighting from the 17th until the second half of the 19th century, when it was finally displaced by kerosene and by coal gas (see also Sections 1.3.2 and 1.3.3). It was noted that 'sperm oil' (from the head of the sperm whale) gave a much cleaner and less odoriferous flame than whale blubber oil and this was one of the factors that led to intensive hunting of sperm whales in the 18th and 19th centuries. Thus, in the period between 1770 and 1775, the north-eastern United States produced about 7.16 million litres of sperm oil per year. At least 6,400 sperm whales would have been killed annually to supply this amount of oil. Hunting at this intensity continued until the second half of the 19th century and of course was not confined to US-based whaling fleets. It is estimated that the world population of sperm whales declined by about 235,000 in the 18th century alone and it seems very likely that they would have been hunted to extinction[3] had petroleum oil and oil-based products not displaced sperm oil as fuels of choice.

1.2.4 Peat

The last traditional biological fuel we wish to consider is peat. This occurs in the wetter areas of the world, covering between 2% and 3% of the global land area, and consists of compressed and partly rotted remains of plants, especially *Sphagnum* moss. It may thus be regarded as being part way to forming lignite, a form of coal. Peat is cut from the bog in slices, known in Ireland and Scotland as turves (singular *turf*), which are left to dry before being burned as fuel. One of the problems with peat is that it takes a long time to form, growing at a mean rate of 1 mm/year in a typical peat wetland. It is thus regarded as a semi-renewable fuel. However, in many areas, the rate of exploitation far exceeds the rate of re-growth, resulting in denudation of the peat bog and increased run-off of water, leading to flooding.

Peat is a less efficient fuel than coal and natural gas, which means that per unit of energy, peat releases twice as much CO_2 as natural gas and 15% more than coal. This difference is only partly mitigated by the slow renewal of peat fuel (as mentioned above). Large-scale peat fires, sometimes initiated by lightning strikes and sometimes by illegal 'slash and burn' activities, in addition to releasing large amounts of CO_2 into the air, also cause very extensive particulate pollution. One of us was working in Singapore during the notorious 1997 Southeast Asia haze, caused by illegal burning of forest trees and subsequent out-of-control peat fires in Indonesia. Visibility was very poor, the air smelt of smoke and we were advised not to exercise outside. Right across the region there were deleterious effects on human and animal health. Similar hazes have occurred several times since 1997, as exemplified in Figure 1.1.

Peat may be regarded as being on its way to becoming a fossil fuel. However, the true fossil fuels are coal (including lignite), oil and natural gas. We thus move on to discuss the history of their use.

---

3 See also Hoare P (2008).

Figure 1.1 Haze over Singapore 2013.

1.3   Fossil Fuels

1.3.1   Coal

On Caerphilly Common, between Cardiff and Caerphilly in South Wales, are some shallow depressions and some small mounds. These are the remains of bell pits and their associated spoil tips, providing evidence of coal mining in the area dating from the 14th century. However, use of coal as a fuel for heating, cooking and even smelting metals actually goes back several thousand years. Its use was recorded in China at around 1000 BC and in Ancient Greece at around 350 BC. In Britain, surface or outcrop coal has been used since the Bronze Age (2000–3000 BC). In Roman times, houses and baths were heated by burning coal and a brazier of coal was kept permanently alight in the Temple of Minerva in Aquae Sulis (now known as Bath). The Romans also used coal for smelting iron.

However, it was not until the end of the 18th century that coal mining became really organised. Those simple bell pits at Caerphilly Common were tapping into the enormous South Wales coalfield and it was coal mined from this field that fuelled the Industrial Revolution in that part of Britain. Indeed, the Industrial Revolution led to a large increase in the demand for coal, which was mined in nearly all of the countries in which industry was increasing. Coal mining in the UK has nearly ended now, not because stocks have run out but because it has become uneconomical to mine it. Nevertheless, coal is still mined in many other countries and known global reserves will last for centuries, even taking into account the acceleration in use in countries like India and especially China. Consumption of coal currently runs at nearly 8×10^9 tonnes per year, with China being responsible for just over 50% of that total. Combustion of that amount of coal, assuming that it is all carbon, releases 22.88×10^9 tonnes of CO_2 into the atmosphere.

1.3.2 Petroleum Oil

As with coal, the use of petroleum oil as a fuel goes back much further than we might suppose. Oil, pitch and asphalt were all known in Babylon and Persia about 4,000 years ago and were used for lighting, heating and in building work. The first recorded drilling for oil occurred in China in the 4th century AD and oil was first distilled to make lighter products such as kerosene in Persia in the 9th century. From there the practice of distillation spread through the Arab world and eventually reached Spain, *via* Morocco, in the 12th century. Deposits of oil and/or asphalt were also known from many locations in Europe, including France, Greece and Romania and some of these have been worked until the second half of the 20th century. The first recorded oil refinery was built in Russia in 1745, producing kerosene, mainly for lighting of churches, monasteries and the homes of aristocrats. Oil and associated products were also known from the New World. In 1595, Sir Walter Raleigh described the asphalt lake in Trinidad and from the beginning of the 17th century onwards, oil springs and other types of oil deposits were discovered in many places in North America, including the now-notorious tar sands in Canada.

It was a Scotsman, James Young, who is often credited with starting the modern oil and petrol industry. In 1847, he refined petroleum that was seeping from a coal seam in a mine in Derbyshire (in the English Midlands) and obtained both a light oil suitable for lighting (as in the earlier Russian refinery) and a heavier oil suitable for lubrication. Within a few years he had also shown that oil could be obtained by a simple chemical treatment of coal and in 1851 set up a factory at Bathgate (in Scotland) for refining oil from coal. So, from around that time there was a flurry of activity of establishing oil wells and refineries in known oil fields all over the world. In the 1860s, the most productive refinery was in Baku[4], then in Russia but now in Azerbaijan, which produced 90% of the oil being used in the world, and this despite the rapid growth of the industry in North America.

However, it was the internal combustion engine that catalysed our modern dependence on oil. The first such engines, built around the middle of the 19th century, used mixtures of various flammable gasses as fuels. Later in the same century, engines were invented that used refined oil (gasoline/petrol) or heavier oil ('diesel') as fuel. These were ideal fuels for the cars and other motor vehicles that were being developed in Europe and North America. The growth in car production and ownership since the early 20th century has been phenomenal and continues apace as countries like India, China and Brazil undergo rapid economic growth. At the beginning of 2014, the global *daily* use of oil was 14 billion (14×10^9) litres and transport accounted for 25% of global CO_2 emissions (see Section 1.4).

---

4 Thus Edmund de Waal, in his chronicle of the very wealthy Russian-Jewish Ephrussi dynasty, *The Hare with Amber Eyes*, writes of 'new possibilities in oil in Baku and gold near Lake Baikal' (p. 31).

1.3.3 Natural Gas

In the late 19th century and for over half of the 20th century, much of the gas used as fuel in Europe, including Britain and North America, was coal gas produced by the destructive distillation of coal or as a by-product of producing coke. However, the existence of another fossil fuel, natural gas, had been known since about 500 BC in China, where it was even transported in pipes made from large bamboo canes. The world's first industrial-scale extraction plant was built in 1825 at Fredonia on the shores of Lake Erie in New York State but for many years, natural gas was known mainly from flaring-off of the gas associated with oil deposits. Nevertheless, commercial and industrial use of natural gas increased from around the middle of the 20th century, so that the fuel now makes a major contribution to the world's energy budget. At the time of writing, global consumption of natural gas is about 3.4 trillion (3.4×10^{12}) litres per year; at this rate of consumption we have about 250 years-worth of known recoverable reserves.

A significant proportion of the known reserves are in the form of shale gas, which is released by a procedure known as hydraulic fracturing ('fracking'). For some, this is a controversial procedure and it has been claimed to pollute ground waters and even cause earthquakes[5]. Thus, at the time of writing, the very word 'fracking' elicits a fierce and angry response from some groups of environmental campaigners who, in the UK and the USA, have attempted to block new fracking sites, even though some fracking 'wells' have been running safely for several years. Indeed, fracking has been practised in the USA since 1940 and in the UK since about 1982[6]. On a more positive note, natural gas is the 'cleanest' of the fossil fuels. The output of CO_2 per unit of energy is 29% less than with oil and 44% less than with coal (see Section 1.4). Indeed, in 2013, output of CO_2 in the USA was at its lowest for 20 years, despite increased energy production. Some of this is attributable to increased use of solar and wind power (see Section 1.4), but the major part of the reduction is due to increased use of shale gas with a concomitant reduction in the use of oil and coal. Furthermore, burning natural gas releases much lower amounts of sulphur dioxide and nitric oxides than either oil or coal.

1.4 Fossil Fuels and Carbon Dioxide

1.4.1 The Club of Rome

In 1972, the Club of Rome, a global 'think-tank', founded in 1968 and describing itself as 'a group of world citizens, sharing a common concern for the future of humanity' published *Limits to Growth*. This dealt with the difficulties of sustaining economic and social development in a world of finite resources. Amongst several pessimistic scenarios was a prediction that at some point in the early years of the 21st century, the ability to extract oil would not keep up with demand. This

---

5 Actually, in some rock formations, fracking can cause very minor, harmless earthquakes: see Royal Society/Royal Academy of Engineering (2012).
6 Fracking has been widely used in the off-shore oil and gas fields in the North Sea since the 1970s.

prediction has become known as 'peak oil'. Now, we need to say that a subsequent report, *Mankind at the Turning Point* written by Eduard Pestel and Mihajlo Mesarovic, published in 1974 and based on a very much wider analysis than *Limits to Growth*, presented a more optimistic prognosis. In particular, it noted that many of the factors affecting the future of humankind on our planet are within human control. Thus it was suggested that with appropriate use of science, technology, architecture and so on, environmental and economic disasters are preventable or avoidable. But there was one major factor that did not enter the reckoning in either of these reports, and that was global climate change[7].

1.4.2 Climate Change

It may well be that, at least in terms of fuel availability, the Club of Rome was too pessimistic. The concept of peak oil has remained with us over the years, but the actual year in which this will occur has crept slowly backwards. The trend in oil extraction continues upwards and some commentators are now suggesting that peak oil will be never be reached. However, since the world's oil deposits are in fact finite, peak oil will certainly be reached at some time, albeit many years into the future. Even then, the very extensive known reserves of coal mean that oil production from coal (as described earlier) will continue to be possible for very many years. Oil shortage on its own then will not drive a reduction in oil usage[8]. It is climate change that is now the main driver for reduction in the use of all fossil fuels.

We need to be clear that without 'greenhouse gases' the Earth would be a giant snowball. These gases, especially water vapour, CO_2 and methane (CH_4), prevent some of the Sun's heat energy from being lost into space, thus warming the Earth's surface and making life as we know it possible. However, since the start of the Industrial Revolution, the concentration and balance of the greenhouse gases has changed. Every year, the CO_2 concentration in the atmosphere peaks in May, and in May 2013, it reached 400 ppm. Not only does that represent an increase of 42% on the pre-industrial baseline of 280 ppm, but it is also the highest concentration that the Earth's atmosphere has contained for at least 800,000 years and possibly since the Pliocene era, over 3 million years ago. This increase in the atmospheric concentration of CO_2 is attributed to the hugely increased rate of burning of fossil fuels in the industrial era. Fuels that took millions of years to be laid down are being burned in a tiny fraction of that time.

As we have already noted, CO_2 is a greenhouse gas and it was as long ago as 1938 that a paper was published that drew attention to the likely warming effects of the increasing concentration of CO_2.[9]

Awareness of the problem slowly grew; by the 1970s, meteorologists and atmospheric physicists were concerned about global warming, a concern that had spread to the wider scientific community by the early 1980s. The World Climate Research Programme was set up in 1979, followed in 1988 by the establishment of the Inter-Governmental Panel on Climate Change (IPCC). In a series of reports starting in 1990,

---

7 We note that climate change has been a major feature of the Club's more recent reports.
8 Although we also note that the oil price crisis of 1973 was a driver for seeking alternative power sources.
9 See Callendar (1938).

the IPCC has shown with increasing clarity and certainty that the increased atmospheric CO_2 concentration is caused by human activity and that this is causing global warming.

So, how much warming might we expect? As discussed by Robert Kunzig of the *National Geographic* magazine, when the atmospheric concentration of CO_2 was at 400 ppm in the Pliocene era, the Earth was two to three degrees (C) warmer than it is now, in the late stages of a long 'greenhouse epoch'. Horses and camels lived in the high Arctic and sea levels were around 10 m higher, at a level that would flood many major cities around the world today. That does not mean that temperatures will rise this much in the current era; several other factors also affect the Earth's temperature. Nevertheless, it does seem very likely that there will a further increase on top of the currently experienced (as of early 2015), mean global temperature that is 0.7 °C higher than the pre-industrial mean.

There has been much discussion about how much CO_2 and therefore how great an average temperature rise can be accommodated. Comparison with earlier geological eras is not necessarily helpful. First, because, as already noted, CO_2 concentration is not the only driver of temperature and second, because we are now living in an era in which over 7 billion humans inhabit the planet. Furthermore, based on geological and geochemical evidence, there has been no period in which CO_2 concentration has risen so rapidly. Some climate scientists are now discussing the possibility of tipping points, points at which either the temperature or the CO_2 concentration or both, lead to an event which then accelerates the change. Such possible events include the melting of permafrost, leading to release of large amounts of previously trapped methane, which is 20 times more effective a greenhouse gas than CO_2.

Because of all these uncertainties there has been a strong and widely supported suggestion that we should attempt to reduce the atmospheric CO_2 to 350 ppm, the value that was last seen in about 1988. At this concentration, the global temperature is likely to stabilise at about 0.5 °C above the previous running mean, provided that no serious tipping points have occurred. A target of 350 ppm will be very difficult to hit. Even if we stopped burning fossil fuels today, it will take about 500 years for the CO_2 concentration to drop far enough. In the face of such difficulties, there is an international effort (supported by national governments with varying degrees of enthusiasm) to limit the temperature rise to 2 °C (probably equivalent to a CO_2 concentration of about 450 ppm). Even this target is not easy; furthermore, even if it is achieved, there will still be a good deal of disruption across the globe, not least because of changing weather patterns and rising sea levels. Indeed, the March 2014 report of the IPCC presents a gloomy picture. We are already seeing the effects of 'irreversible' changes in the climate, with increased frequency of extreme weather events, flooding, ecological problems and reductions in crop yield.

One final comment is needed. In the geological history of the Earth, temperatures have at times been significantly higher than they are now or even than they are projected to be. The planet survived. So, when we hear people talking about saving the planet, what is mainly meant, as we see from the recent IPPC reports, is maintaining the planet in a state where it can continue to support the human population (and, if there is an interest in the environment for its own sake, also to support the current range of living organisms).

1.5   Alternative Energy Sources

1.5.1   Introduction

If the rise in mean global temperature is to be kept to 2 °C or lower, it will take a good deal of hard work on an integrated global scale. Fossil fuels currently provide about 83% of the world's energy (but only 66% in the USA) and this proportion needs to drop very dramatically. Even if the proportion drops, it is predicted that the world's energy demands will have increased by around 50% between 2010 and 2035, with largest increases being seen in China and India. A decreased proportion may still mean a larger amount. So, where fossil fuels continue to be used, carbon-capture techniques, currently in their 'technological infancy', will need to be deployed.

One obvious way to reduce the use of fossil fuels is to turn to alternative energy sources. Thus, the European Union, in its 2009 Renewable Energy Directive, determined that the Union should obtain 20% of its energy from renewable sources. This includes both fuels for combustion and means of generating electricity.

1.5.2   Environmental Energy Sources

By 'environmental' we mean using the 'forces of nature' to generate energy for human use. Actually, utilisation of some forms of environmental energy dates back many centuries. Windmills for example, have been used to grind cereal grains for at least 1,500 years and possibly longer[10]; their first use in Europe dates back to the late 12th century. Use of wind to propel boats goes back much further, probably to about 3000 BC in ancient Egypt; by 500 BC, two-masted cargo ships were regularly plying the Mediterranean trade routes.

By the end of the 19th century, windmills were in wide use in northern Europe, exemplified by Denmark where there are 2,500 windmills, mostly used in mills and for pumping. Up to 6 million were installed on farms in the mid-west of the USA, where their main use was to power irrigation pumps.

Water wheels (or water mills), with applications in irrigation and as a power source, were used by the Greeks (and were later copied by the Romans) as far back as the 3rd century BC. They were certainly known in Europe by the 6th century AD and at the time when William the Conqueror's assistants compiled the Doomsday Book, there were at least 6,000 water mills, including tide mills, in England alone. Overall, in pre-industrial Europe, water mills outnumbered windmills by at least two to one.

In more modern times, the water wheel was mostly replaced in the Industrial Revolution by the water turbine and in the 19th century, water power began to be used both in small local hydro-electric plants and in more large-scale generation of electricity in hydro-electric power stations. An example of the former was at Lynmouth in North Devon, southwest England. The small hydro-electric plant, built in 1890, harnessed the power of the River Lyn which flowed rapidly from Exmoor, a high rainfall area. It remained working until 1952, at which time it was destroyed by severe floods.

---

10 Indeed, in the 1st century ad, the Greek engineer, Heron of Alexander built a device that used wind power to drive simple machinery, including a pipe organ.

Increased use of hydro-power has continued right through to the 21st century; in the USA for example, hydro-electric power generation contributes to about 7% of the energy budget. However, large-scale hydro-electric power generation has its downside in that the trapping of water behind a dam causes loss of land and in many cases, displacement of people.

Some water wheels are still in operation in several countries and increasing numbers of organisations with access to appropriate water courses are installing modern water turbines. Thus, at a conference centre in North Devon, southwest England, a water turbine provides an average of 30% of the electricity used at the centre (Figure 1.2). In the same area, several villages have built or are planning to build their own hydro-electric stations. Across the globe, small-scale hydro-electric generation accounts for about 20% of the total hydro-electric output.

There has also been a renewal of interest in the power of the tides: the first modern tidal power station was built on the estuary of the Rance (between St Malo and Dinard in Brittany, northwest France) and opened in 1966. For several years, 'La Rance' remained the sole example of tidal power generation and even when other tidal power stations began to be built, the Rance power station remained the largest (with a capacity of 240 MW) until it was 'overtaken' by Sihwa Lake tidal power station (254 MW) in South Korea. None of the other six tidal power stations in the world (as at the end of 2013) has a generating capacity greater than 3.2 MW, although several very large ones are planned for South Korea and Russia. In fact, there are many sites over the world where the tidal range would merit construction of a tidal power station. However, this would destroy the inter-tidal zone which in many places, for example in South Wales and in southwest England, are important wild-life habitats – an example of clashing environmental ethics priorities.

There is also extensive interest in harnessing the power of waves but as yet, much of the work on this is only at the experimental stage. The world's first commercial wave-power plant, commissioned in 2000, was a very small one[11], situated on the shore of Islay, an island off the west coast of Scotland and generating only 0.5 MW. Since then only four other wave-driven power plants have been built (as in early 2014), the largest of which, in Orkney, Scotland, generates 2.4 MW. Experimental rigs are working in several places in Europe and North America and it is likely that more commercial wave-power plants will be built in the next ten years.

As with water power, the use of wind as a power source has also had a renaissance in modern times. Once again, a Scotsman was a key pioneer in this: in 1887, Professor James Blyth built a windmill that charged accumulators (effectively large rechargeable batteries), which then supplied the electricity to light his holiday cottage. This cottage, at Marykirk in eastern Scotland, was the first house in the world to use electricity generated by wind power. Windmills for electricity generation were also built in the USA in the later years of the 19th century, but it was in Denmark that the most rapid progress was made. In 1895, the Danish scientist Paul la Cour modified an earlier-designed wind turbine to produce enough electricity to light a whole village. From then, wind turbines were built all over Denmark as part of a programme to decentralise the generation of electricity; the largest of these turbines generated 25 kW. The rest of the world quickly

---

11 The 'Islay Limpet'.

(a)

(b)

(c)

Figure 1.2 (a) Hydro-electric generating station at Lee Abbey Conference Centre, Devon. Even a small station like this can generate a significant proportion of the electricity used by up to 120 delegates plus 80 resident members of the community. This picture shows the turbine house and tail race through which the water is exiting. (b) Self-cleaning Coanda screen at the head of the high pressure inlet pipe in the Lee Abbey system. (c) The generator in the turbine house at the Lee Abbey conference centre. Thanks to Lee Abbey, Devon, for supplying these photos.

followed with the use of wind turbines to generate electricity in places distant from any centralised supply. This included remote regions of the USA, Australia, parts of Africa and even Antarctica.

The first wind turbine of the modern type that we are now familiar with was built at Yalta in Russia in 1931 and ran for about ten years. This generated 100 kW but an experimental rig, set up in 1941 in Vermont, USA was capable of generating 1.25 MW; however, it only ran for 1,100 hours before a blade collapsed. Meanwhile, developments continued in Denmark; in 1957, Johannes Juul built and installed a 200 kW wind turbine[12] that fed alternating current directly into the grid. The turbine incorporated a number of technical innovations which were adopted by other manufacturers as wind power became more widely used. The term 'Danish design' is still used in the industry.

However, it was the oil price crisis of 1973 that provided a strong impetus for the exploitation of alternative energy sources in general and of wind power in particular. Even though oil prices declined again in the 1980s, other concerns such as energy security and climate change continued to drive the development of the use of renewable energy. The visual impact of this will be familiar to our readers. There are large turbines installed in windy places all over the world, sometimes individually or in small groups, but often in large arrays known as wind-farms (Figure 1.3) Some of these installations are at sea ('off-shore'), including the recently opened London Array located off the Kent coast. The maximum power output achieved so far by an individual turbine (early 2014) is 8 MW, although most turbines generate between

Figure 1.3 Wind Farm at Bears' Down, Cornwall, UK. Photograph by Ron Stutt; reproduced under the Creative Commons Attribution ShareAlike Licence.

---

12 The famous Gedser wind turbine.

2 and 5 MW. The largest wind farm in the world, the Alta Wind Energy Centre in California, has a total generating capacity of 1,320 MW (1.32 GW).

At the time of writing, the use of wind power across the world is growing by 30% per year. Focusing on the UK as a specific example, overall mean power generation rose 35-fold between 2000 and 2013 (300 MW to 10.5 GW). At the end of 2013, wind power generated about 6% of the electricity used in the UK. However, in windy periods, that percentage can be much higher. For example, in December 2013, wind power generated approximately 10% of the UK's electricity demand. In the week beginning 26 December, 13% of Britain's total electricity needs were met by wind power[13], with a record of 17% being set on 21 December. In 2016, for a limited time, Scotland was able to generate all its electricity requirement using renewable sources. It is figures like these that add weight to the view that the UK is one of the most favourable countries in Europe for wind power.

More off-shore and land-based wind farms are either under construction or are being planned. However, it is here that we meet another of those environmental ethical tensions. It is stating the obvious that wind turbines are most efficient in windy places, but many of those windy places are in areas of wilderness and/or rugged natural beauty. In a small country like the UK, there are few areas of wilderness and thus they are very precious. For this reason, some environmental campaigners, while remaining very concerned about climate change, actually oppose many of the plans to build more land-based wind farms.

This brings us to the last of the major environmental energy sources, the Sun. While the UK may be the most favourable place in Europe for wind power, the same cannot be said about sunshine. Despite this, modern methods for capturing the Sun's energy mean that the UK can generate considerable amounts of energy by this route and of course this is even more so for the sunnier regions of the world, including southern Europe.

Like other forms of environmental energy, use of the Sun's power first occurred many centuries ago. Mirrors and lenses were used to concentrate the Sun's rays to start fires (including the lighting of lamps) as long ago as the 7th century BC; in ancient Greece, the Olympic torch was lit in this way. There is even a legend, actually not verified, that Archimedes set fire to Roman ships by using mirrors to focus the Sun's rays. The Romans themselves were among the first to use glass in windows, which allowed them to exploit passive solar gain in the heating of bath houses. In the UK, glass became widely used in homes in the 16th and 17th centuries and passive solar gain was (and still is today) used to heat orangeries and greenhouses (hence the term 'greenhouse effect').

The more technological approaches to using solar power did not begin until the late 19th century. The first solar powered water heater was installed in the USA in 1896 and the technology slowly gained in popularity from the 1920s onwards. In both the USA and across the 'developed' world, there was an increase in uptake from the 1960s onwards and especially after the oil crisis of 1973. Modern systems have very efficient heat exchangers and provide hot water both for the tapped supply and for central heating. Solar water heating systems are especially popular in China,

---

13 It was indeed a very windy week; across the UK, thousands of trees were blown down, including over 200 in the National Trust's estate at Killerton in Devon, UK (and two in the garden of one of the present authors)!

which currently accounts for over 80% of the world's new solar heating installations. The huge array at Kramer Junction in California (built in 1986) is also essentially a hot water system: mirrors focus the Sun's rays onto the heat exchangers, leading to the production of steam. This is used to drive turbines which generate electricity.

However, the most rapid recent growth in the use of solar power has been in the more direct generation of electricity. In 1839, a very young Edmund Becquerel (who would later share a Nobel Prize with Pierre and Marie Curie for the discovery of radioactivity) demonstrated the photovoltaic (PV) effect: illumination of some but not all metals causes the generation of an electric current. Einstein later called this the photoelectric effect and worked out the theoretical basis for its occurrence. Increasing awareness of the types of material that exhibited the effect and continuing sophistication and therefore efficiency of the experimental devices employing the effect, eventually led to its use in generating electricity on a practical scale. Since 1958, satellites have been powered by PV cells; in the 1960s, some small hand-held devices were powered by PV cells (now commonplace in things like calculators) and by the 1970s larger pieces of equipment and machines could be solar powered. The first applications in providing electricity for buildings were also in the 1970s.

By the 1980s, it became apparent that PV technology has enormous potential for widespread generation of electricity. It had proved its worth in pilot studies on all the world's continents. In 1981, a solar-powered plane, piloted by Paul McCready, flew across the English Channel; in 1982, a one-MW PV power station was opened in California and in the same year, Hans Holstrup drove a solar-powered car across Australia. Since then, there has been immense progress in developing ever lighter arrays, in increasing the energy conversion efficiencies and increasing the efficiency at lower light intensities. There are PV power stations all over the world, the largest two of which (as in early 2014) are both in California; each generates 500 MW. A 700 MW array is under construction in Arizona, USA. There has also been, in the 21st century, increasing uptake by domestic users. In the UK and in some other countries, this has doubtless been accelerated by the 'feed-in' tariffs: electricity in excess of the consumer's use is fed back into the grid, attracting a payment to the consumer. Thus, in the UK, it is becoming common for farmers to cover a whole field with an array of large PV panels (Figure 1.4).

By the end of 2013, installed capacity of PV arrays, from small domestic roof panels to vast solar parks, exceeded 100 GW, of which the UK contributed about 2.5 GW, but these numbers are growing daily. This increased uptake, combined with the greater conversion efficiency of modern PV materials and their ability to utilise lower light intensities, means that the generating capacity from PV installations is growing faster than the generating capacities of either wind or water power, although in terms of actual totals, PV remains in third place. However, the potential for use of solar power is huge and we have barely scratched the surface of what is possible.

While water, wind and sun are undoubtedly the major sources of environmental energy, we must also mention geothermal energy using the heat from the Earth. This may involve a direct use of hot springs and hot water in underground aquifers, as happens for example in Iceland, New Zealand and many other countries where such supplies are readily available. However, geothermal energy (which may include dry sources such as 'hot rocks') is increasingly used to generate electricity, first done

Figure 1.4 Array of PV solar collectors installed by the Scottish dairy and food company, Mackie's, at their farm in Aberdeenshire. With 7,000 panels, it is the largest array in Scotland, covering 4 hectares and generating 1.8 MW of electricity. When running at full capacity, it will reduce CO_2 emissions by about 850 tonnes/year.

on a commercial scale in Italy in 1911[14] and eventually, from 1958 onwards, in New Zealand and then in several other countries. By 2013, total generating capacity across the globe was 10.7 GW, with five countries – El Salvador, Kenya, The Philippines, Iceland and Costa Rica generating more than 15% of their electricity from this source.

Finally, there are ground-source heat pumps. Technically, the source of energy is not geothermal because pumps exploit the more or less stable temperatures (with some variation according to latitude) a few metres below the soil surface, temperatures that arise from the Sun heating the Earth's surface. Temperatures at a few metres depth are usually higher than ambient in winter and lower in summer, meaning that the heat pumps may be used for summer cooling as well as winter heating. There are at least 2 million installed over the world and in Finland all new houses built since 2006 have had ground-source heat pumps installed.

1.5.3 Nuclear Power

Another major energy source is nuclear fission. It is not technically a renewable energy source in that fuel supplies are finite (but far from being exhausted), but it is a 'clean' energy source in that it does not use fossil fuels and does not release CO_2 into the atmosphere. In the Second World War, nuclear fission had been used in the bombs that destroyed Nagasaki and Hiroshima, but after the war there was a strong motivation to use 'atoms for peace' (as in President Eisenhower's speech to the United Nations in

---

14 Following an experiment in 1904, in which geothermally generated electricity powered four light bulbs.

1953). The world's first commercial nuclear power station, at Windscale (now known as Sellafield) in the UK, was commissioned in 1956 and from then, use of nuclear power increased rapidly across the industrial nations until the mid-1980s. However, there had always been some opposition, first because of the accumulation of very long-lived radioactive waste for which there has been (and still is) no universally acceptable solution. Second, there have always been background concerns about safety and these fears had been exacerbated by an accident at Three Mile Island in Pennsylvania, USA in 1979. Anxiety was then further stoked up by a major accident at Chernobyl, Ukraine in April 1986. From that time there was a major slowdown in the building of new nuclear power stations. However, there were some exceptions. In the UK, the Sizewell B power station was built in the late 1980s. France in particular continued to invest in nuclear power and by early 2014, obtained 75% of its electricity from its 58 nuclear power stations. This compares to an average of about 12% across the rest of the world. A further 15% of France's electricity comes from hydro-power (including tidal – see earlier), which means that per unit of electricity generated, France has by far the lowest CO_2 output. It also produces the cheapest electricity in Europe and thus can export to other nations, including the UK.

In the 21st century, with the increased awareness of climate change, there has been a revival of interest in nuclear power, at least as a supplement to the growing use of renewable energy sources. Indeed, the French model tells us that a combination of nuclear and environmental energy can be very effective in lowering CO_2 emissions. Thus, France continues to build nuclear power stations, about 25 new stations are under construction in China and in the UK a power station is under construction at Hinkley Point. In the USA, which is actually the world's largest producer of electricity from nuclear power (30% of the global total; 19% of its own electricity), there had been a long pause since the Three Mile Island accident. However, by early 2014, five new nuclear power stations were under construction and licences had been granted for six more.

Against this increased activity in nuclear power generation we need to set reactions to the disaster at Fukushima in Japan. On 11 March 2011, there was a huge earthquake off the coast of Japan. The Fukushima-Daiichi power station had been built to withstand earthquakes and indeed it did so. What it did not withstand was the enormous tsunami that followed the earthquake. The power station was badly damaged and three of the six reactors went into meltdown[15]. It was a very serious accident and catalysed extensive opposition to nuclear power, both amongst the public and at government level in several countries. Although it did not halt the planning of the new power stations that are now being built in France, the UK and the USA, the nuclear power industry was brought to a sharp halt in other countries. In Japan itself, all the other 48 nuclear power plants were shut down in September 2013 and some remained shut down to April 2014. Since these nuclear power stations previously provided 30% of Japan's electricity, this has made a major 'dent' in supplies. In Germany, all of the old-style nuclear power plants were rapidly shut down and all of the newer ones will be shut down by 2022. Plans to build new nuclear power stations were halted in Bahrain, Kuwait, Malaysia and the Philippines while in

---

15 For further details, see archive at http://fukushimaupdate.com/

China, plans were put temporarily on hold but were re-instated in mid-2012. Since nuclear power is regarded as having a role to play in reducing CO_2 emissions (as has clearly happened in France), these negative responses make the adoption of renewable energy sources even more urgent.

1.5.4   Hydrogen

Although each of the foregoing energy sources has a role to play in reducing the use of fossil fuels, none directly produces fuel for combustion. It is here that we encounter the largest difficulty in reducing our reliance on fossil fuels. For example, as stated earlier in this chapter, in Spring 2014, the global use of petroleum oil was 14 billion litres per day[16] Much of this is used to fuel transport of various types. Now it is true that some transport systems use other sources of power such as nuclear power (as in submarines) and electricity, as with many trains and trams and, increasingly, some cars. Despite this, we will need combustible fuels for many years to come. Hydrogen gas has been seen by some as playing a significant role in dealing with this problem. It is certainly true that hydrogen is a renewable fuel: it is made mainly from water and turns back to water when burned; it is thus a very clean fuel. It is also true that vehicles can be adapted to use it either in a fuel cell to produce electricity or in an internal combustion engine (as with conventional fuels).

The first hydrogen-fuelled vehicles were made in Germany in the late 19th century and by the start of the Second World War, thousands of hydrogen-fuelled lorries and buses (and some submarines) were in use in that country. Today, Germany is still a world leader in hydrogen-fuel technology and, combined with its expertise in automotive innovation, may yet produce hydrogen-powered vehicles for the mass market. Hydrogen and hydrogen fuel cells were amongst the propulsion systems used in space rockets from the 1960s onwards and the US Navy developed some fuel-cell-driven submarines. However, it took the oil crisis of 1973 to generate a wider interest in the potential for the domestic market. The first vehicles to use hydrogen fuel cells were built in the 1990s, although the technology had been invented in the 1830s and 1840s in Switzerland and England[17], and from that time there has been parallel development in Europe, Japan, Canada, Australia and the USA of both hydrogen-fuelled vehicles and fuel-cell-driven vehicles.

In 1998, Iceland announced that it was working towards a hydrogen economy, freeing the country from dependence on fossil fuels by 2040. The first hydrogen-fuelled buses were introduced in 2003 and the first cars in 2007 (admittedly not many); there is also one passenger ferry that uses hydrogen. Although the project is behind schedule, it is understood to still be part of government policy. Indeed, Iceland is an ideal country for such a policy. Producing hydrogen, either by release from organic compounds or more usually by the electrolysis of water, uses energy and by the laws of thermodynamics, less energy is obtained by burning the fuel than went into producing it. It is thus Iceland's valuable resources of renewable energy – geothermal, hydroelectric and wind – that make this feasible. In a similar way, hydrogen production in the USA generally uses electricity generated by solar or wind power.

---

16 Ireland (2014).
17 Sir William Grove is known as the Father of the Fuel Cell.

1.6 Biofuels

And so we return to the theme with which we started the chapter. As we noted, 'traditional' biofuels have been in use for many centuries. We have also seen that there is now a worldwide imperative to move away from fossil fuels; biofuels clearly have a place in the mix of alternative energy sources. However, for all the excellence of many alternative energy sources, there is still a major shortfall in availability of combustible fuels that may be used for direct propulsion of, for example, planes and road vehicles, in contrast to indirect, in which fuel combustion is used to generate electricity. It is a major task to meet this need and biofuels will certainly have a role but it will not be easy. We have already mentioned the amount of oil that is consumed daily, but Tim Ireland illustrates the problem even more vividly: imagine a line of cars, say 500 m long. Based on productivity of biofuel crops, it would take an area of 500 m × 8 km to produce enough fuel for them. Thinking specifically of the UK, devoting all the available arable land that is currently used for conventional crops to biofuels, would provide only a tiny fraction of the nation's liquid fuel needs.

It is with these problems in mind that we start to look at the modern era in biofuel generation. This was initiated by two very different technologies, namely the use of ethyl alcohol as fuel and the generation of methane by anaerobic digestion. Use of alcohol as a fuel, either on its own, or mixed with petrol (gasoline), has occurred on a small scale since the end of the 19th century. However, it entered the modern era in Brazil in the early 1970s, where a decision was made to use sugar from the extensive sugar-cane plantations to make alcohol. In 1976, it was made compulsory for petrol to contain between 10 and 22% by volume of anhydrous alcohol; in 2003, the range was tightened up to 20–25%. For several years, Brazil was the world's biggest producer and biggest exporter of fuel-alcohol, although they have recently been overtaken in both areas by the USA. Between them, Brazil and the USA produce about 85% of the world's fuel alcohol, but it is in Brazil where alcohol makes a very significant contribution to the fuel economy. Further benefit is obtained from the waste sugar-cane material (bagasse), which is used to generate heat or power, the latter contributing to the 85% of electricity generated from renewable resources.

There is more on alcohol production in Chapters 5 and 8. For the present though, we now briefly consider anaerobic digestion. It has been known since at least the 17th century that rotting biological material produces a flammable gas, and the identification of the gas as methane arising from microbial metabolism was made in the 19th century. By the end of that century, the city of Exeter in the UK had installed a series of septic tanks that utilised anaerobic methanogenic bacteria to treat its waste water; the methane was burned to provide heating and lighting at the treatment works. A true anaerobic digester was built in India in 1897, using human waste as fuel to generate methane to light a leprosy hospital. These two types of application were developed across the world in a piecemeal fashion until the 1970s when two factors, the oil-price crisis of 1973 and, in many developed countries, tighter pollution controls, catalysed further research and development on 'biogas' generation. Today, anaerobic digesters are a relatively common sight at sewage treatment works and in places where biological waste is generated (see front cover). These may include farms of various types, breweries and food-processing factories (see Chapter 3). Some digesters are set up as central facilities where output from several farms, or waste food from schools, restaurants and supermarkets, can be digested. In most cases, the methane is used to drive electricity generators but some motor vehicles have been modified to run on methane, thereby helping in a small way with the transport fuels problem mentioned above.

Focusing specifically on liquid fuels, one more type is classed as first generation, namely fuel from plant lipids, that is, biodiesel. Like ethanol derived from fermentation of sucrose, it is mainly produced from crops that are normally used for food production. This raises the food *vs* fuel issue, which is discussed in several chapters in this book. What we need to see say here is that even if, in a country such as the UK, all the land currently used for growth of food crops was used instead for biofuel production, we would still be a very long way from being able to fulfil our liquid fuel requirements. As will become apparent in reading the rest of this book, progress is already being made on 'second–generation' and 'third-generation' fuels, with a look forward even to the fourth generation. So read on!

Selected References and Suggestions for Further Reading

Callendar, G.S. (1938) The artificial production of carbon dioxide and its influence on temperature. *Quarterly Journal of the Royal Meteorological Society*, **64**, 223–240.

Hoare, P. (2008) *Leviathan or The Whale*. Fourth Estate, London.

Inter-Governmental Panel on Climate Change (2014). *Climate Change, 2014*. IPCC, Geneva.

Ireland, T. (2014) Algal biofuel: in bloom or dead in the water? *The Biologist*, **61**, 20–23.

Janardhan, V. and Fesmire, B. (2010) *Energy Explained*. vol. 1, *Conventional Energy*; vol. 2, *Alternative Energy*. Rowman & Littlefield, Lanham MD, USA.

Lynas, M. (2011) *The God Species: How the Planet Can Survive the Age of Humans*. Fourth Estate, London.

Moran, E.F. (2006) *People and Nature*. Blackwell, Oxford.

Royal Society/Royal Academy of Engineering (2012) *Shale Gas Extraction in the UK: a Review of Hydraulic Fracturing*. Royal Society, London.

2

Biofuels in Operation

Lionel Clarke

BioenerG Ltd, Chester, UK

Summary

Commercial production and marketing of biofuels demands careful attention to many important issues, not least sustainability within the supply chain. In this chapter, we focus specifically on factors that shape and constrain the functional operation of biofuels within transport systems. At any particular time, hundreds of different fuel grades are in use around the world, and this mix is constantly evolving. To design biofuels for the future, it is necessary to understand the factors that currently shape and constrain fuel specifications and how shifting market supply and demand may impact design parameters. Both chemical and physical properties play an important role in determining functional performance, subject to an ever-shifting range of engineering advances, commercial challenges, regulatory constraints and consumer expectations.

2.1 Fuels for Transport

Mobility of people and movement of goods – whether on land, by air or sea – is both a driver and indicator of economic development and improving living standards. Mobility means access to markets, to jobs, to education, to economic opportunity and to extended communities. Mobility enables social interaction, global trade and economic development. In developing economies, increased economic activity leads to growing income *per capita* and, as populations and standards of living rise, so demand for supporting personal transportation increases.

   Globalisation of the internal combustion engine has been key to the mobility revolution and may be considered one of the defining technological developments of the 20th century. The number of vehicles on the road worldwide broke through the 1 billion mark in 2010[1], having doubled within the previous 25 years, and could reach 2 billion by 2050. Throughout the past century, keeping vehicles moving has depended almost

---

1 Ward's report: http://wardsauto.com/ar/world_vehicle_population_110815

Biofuels and Bioenergy, First Edition. Edited by John Love and John A. Bryant.
© 2017 John Wiley & Sons Ltd. Published 2017 by John Wiley & Sons Ltd.

entirely on the refining of crude oil to produce fuels in the form of gasoline and diesel. Around 76% of transport energy now goes on road (and land-based off-road) applications and a further 2% on rail. Aviation presents another major demand for liquid fossil fuels. Air traffic has been growing at between 4% and 5% per year over the past decade in terms of both passenger-kilometres and airfreight tonne-kilometres. Jet fuel currently comprises around 11% of global transport energy consumption. Marine transport has been growing steadily at around 2% to 3% per year, and also comprises around 11% global transport energy today[2].

Supplying fuel to the transport sector as a whole accounts for well over half of all crude oil use. In energy terms, this is equivalent to just under 100 Exajoules (EJ; 10^{18}, or 1 quintillion) of energy per year; just over 3 Terawatts (TW; 10^{12}, or 1 trillion) of power; just over 2 billion (10^9) tonnes oil equivalent per year; or just under 50 million (10^6) barrels (bbl) of oil per day. Many different units may be used, but they all convey the same message – worldwide demand for transport fuel is extremely high and continues to grow, posing considerable challenges for the development of viable alternatives that can be delivered at a meaningful scale. Over the next 25 years, demand for liquid fuels is projected to increase more rapidly in the transportation sector than in any other end-use sector, with most of the growth projected among the developing non-OECD nations, whilst consumption among the developed OECD nations is expected to remain relatively flat or to decline[3].

Attempts to substitute established fossil fuels with biofuels in the future must take into account the broad range of factors that have shaped combustion engines and fuel demand to date and generated the myriad of fuel qualities that currently exist around the world. In this chapter we consider the factors that shape fuel design today, how these may evolve in coming decades and the implications and opportunities for current biofuels and future biofuel development.

Different factors influence fuel design (Figure 2.1). The essential fuel properties that define fitness for purpose and are required to ensure engines operate reliably are at the core of fuel development. These properties, however, evolve and are defined in national or regional fuel specifications that are constantly reviewed and updated.

Fuel specifications exist alongside a range of associated standards, targets and regulations pertaining to health, safety and the environment. For example, in Europe, regulations for the Registration, Evaluation and Authorisation of Chemicals (REACH) came into force in 2007[4]. REACH requires a rigorous registration process for the market introduction of 'new chemicals' to market, in a system that defines 'existing chemicals' as those already in use in the market by 1981. Emissions and Fuel Consumption regulations have been introduced around the world since the early 1970s. These and other related legislation translate into national procedures such as vehicle homologation tests established to ensure that emissions, safety and road-worthiness of vehicles meet required standards before they can be released to the market. Regulations apply not only to vehicle design but also to their operation over time – regular vehicle test procedures may be applied to ensure that catalysts and other emission controlling equipment remain effective for their operating lifetime. Regulations apply to fuel handling, storage and distribution, addressing safety to individuals and the environment. Further

---

2 New Lens Scenarios, a shift in perspective for a world in transition; Royal Dutch Shell (2013).
3 IEA International Energy Outlook (2013).
4 http://ev.europa.eu/environmnet/chemicals/reach/reach_intro.htm

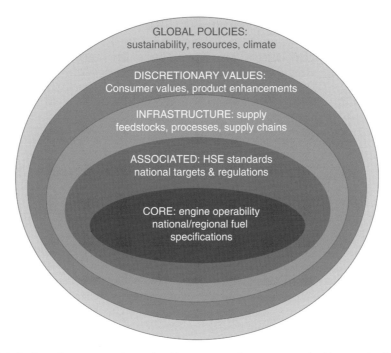

Figure 2.1 Fuel requirements are determined by numerous factors, categorised here under five broad headings, each evolving with time and impacting the nature of fuels required now and in the future.

regulations for classification and labelling are currently under consideration. The technical and commercial suitability of any new biofuel component must take into account this full array of associated requirements.

The fuel supply infrastructure has been established and optimised over a century for conventional fossil fuels. Not all biofuels will be fully compatible with existing engines, infrastructure and even health and safety regulations. Delivering both sufficient quantity and consistent quality are also major challenges facing the introduction of novel biofuels, involving the establishment of suitable conversion processes and potentially new or modified supply, handling and distribution facilities.

Delivering a quality fuel to the consumer is not exclusively a matter of appropriate chemistry and engineering; in addition to the considerations listed above, consumer attitudes and values can also play a very important role. For example, am I satisfied with the performance and fuel economy I get from my vehicle? Am I concerned about my carbon footprint? Or am I concerned about exposure to vapours during the process of refilling my fuel tank? Alternative fuel grades and differentiated fuel qualities help satisfy individual consumer needs and concerns, providing discretionary values that may vary between locations and cultures.

Global policies further shape and drive fuel requirements – for example, mounting concern over climate change, declining resources, and increasing focus on sustainability of the transport system as a whole. A more sustainable future will require both cleaner energy and more efficient use of energy[5]. Of particular relevance to the

―――――
5 Sustainability Report 2012, Royal Dutch Shell.

transport sector is the emerging awareness of CO_2 emissions, and the CO_2 equivalent emissions of other greenhouse gases (GHGs) and their potential impact on climate change. Today, the transport sector accounts for just under a quarter of energy-related CO_2 emissions, with around 17% of global emissions from the use of fossil fuel coming from road transport. Without significant developments, this surging demand for mobility could see transport CO_2 emissions increase by up to 80% by 2050. Although this increase in CO_2 emissions may be partially offset by improvements in operating efficiencies and from a variety of alternative fuel options including biofuels, hydrogen, electricity and natural gas, designed to generate lower overall emissions, in practice, the actual split between the various fuel options will vary from one market to another, depending on supply, demand and other local factors. The type and availability of feedstock, the biomass from which biofuels are derived, may be further enabled or constrained by international trading agreements, since some markets may become net exporters and others net importers of biofuels or feedstocks.

To date, three main drivers – energy security, domestic economy and global GHG emissions, have strongly influenced national biofuels policies. At present, over 65 countries have or are considering biofuel policies, with an increasing number focusing more on the GHG credentials of biofuels than on volume targets alone. Challenges and drivers have changed considerably in recent decades and it is reasonable to assume they will continue to evolve throughout the 21st century. It is only by recognising this breadth of interconnected issues as a whole, and considering long-term trends at each level, that meaningful assessments of future fuel requirements, the need for innovative biofuel solutions, and likely pace of development can be made.

2.2 Future Trends in Fuels Requirements and Technology

The future is notoriously uncertain. An approach that can help thinking about the possible span of future operating environments and associated policies is to consider contrasting yet internally consistent global scenarios, a process first developed by Shell more than 40 years ago, and refined and regularly updated ever since. The most recent set of 'New Lens Scenarios' contrasts a world in which *status quo* power is locked-in (the 'Mountains' scenario), aligning interests to unlock resources steadily and cautiously, and an alternative world in which power is devolved and market forces assume greater prominence (the 'Oceans' scenario). These scenarios generate internally self-consistent, yet divergent, frameworks within which future demand for biofuels and bioenergy may be estimated[6].

In 2010, supply levels of biofuels were around 2.5 million EJ per year. In a more stable and structured world, as reflected in the 'Mountains' scenario, biofuel supply could more than triple to reach 8.6 EJ/year by 2030 and 13.5 EJ/year by 2060. In the contrasting 'Oceans' scenario, an initial failure to impose global energy constraints arising from the more fluid political structures associated with this scenario would slow the initial pace at which biofuels would take longer to develop, doubling to around 5.5 EJ/year by 2030, but the consequently more critical need and greater urgency to develop local mitigation strategies that may arise, as the full impact of climate change becomes apparent, could subsequently lead to a significantly greater demand (as much as 25.6 EJ/year)

---

6 New Lens Scenarios, a shift in perspective for a world in transition; Royal Dutch Shell (2013).

by 2060. For comparison, the US EIA expectation for biofuels supply by 2040 is close to 6.0 EJ/year (2.8 m bbl/day), consistent with the lower end of the Oceans scenario[7]. Relative to the current 100 EJ/yr required for global transportation, it can be reasonably expected that biofuels will play a vital role for the foreseeable future, but only as part of a wide range of low-carbon solutions that will also be needed.

Changing consumer attitudes and other discretionary values are also difficult to anticipate over the very long term, but it may be worth noting that concerns over local emissions and their potential impact on private health arising initially in the USA during the 1960s have resulted in very significant changes to transport fuels and engines. For more than half a century since, emissions regulations responding to such interests have continued to evolve, driving down tailpipe emissions *via* an ongoing succession of increasingly stringent engineering initiatives and fuel property constraints. In contrast, public awareness of CO_2 emissions and personal "carbon footprint" issues has begun to have a discernible impact on consumer choice only within the past decade, so it is too early to determine how this increasing awareness may translate into biofuel design parameters in years to come.

There is a growing body of opinion that climate change poses not only a major long-term threat, but that, due to the longevity of GHGs already accumulated in the atmosphere, delay in responding to the threat could render future solutions increasingly challenging and costly[8]. This introduces a need for urgent solutions that is incompatible with the timescales for delivering radical changes to the transport infrastructure. Biofuels provide an important option – indeed one of the only options – that can introduce lower carbon solutions rapidly through existing supply infrastructures to the broad body of existing vehicles and engines.

A major factor determining the changing biofuels landscape will be the balance between what can be produced and what the vehicle parc (*i.e.* the totality of operating vehicles in a given market) can absorb.

A fundamental constraint is that the easiest and most cost-effective biofuels to produce today, ethanol from the fermentation of sugars and plant oil- or animal fat-derived biodiesels, so-called 'first generation biofuels', are not fully compatible with current engine designs that have been optimised over a century to work with hydrocarbons distilled from crude oil or petroleum. Either future engines must adapt better to biofuels or *vice versa*. In practice, both approaches are being, and will continue to be, pursued as advanced drivetrains and advanced biofuels continue to progress. In the meantime, readily available first-generation biofuels can be blended into the fuel pool within strict blend limits that ensure the resulting blended fuels remain fully compatible with reliable engine operation.

In simple terms, progress may result from any or all of:

1) developing more suitable feedstocks;
2) more efficient new conversion processes to produce more cost-effective and compatible biofuels;
3) better adapted engines; and
4) clear incentives and constraints within the end market (Figure 2.2).

---

7 IEA International Energy Outlook (2013), slide 12.
8 Stern, N (2006).

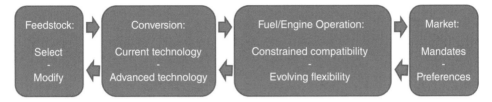

Figure 2.2 The future biofuels landscape will evolve over time in response to technological advances in feedstocks, conversion processes, fuel/engine operations and market demand. Opportunities will arise from innovative developments, whilst critical constraints within the system generate inertia.

2.3   Engines and Fuels – Progress *vs* Inertia

Consolidation of low-cost mass production and reliable operation in the initial decades of the 20th century laid the foundations for widespread adoption of the internal combustion engine. Research and development in engine technology, fuels and lubricants have combined to make modern engines remarkably efficient and reliable. A focus on high quality control and substantial investment into robust and reliable fuel supply infrastructures has helped advance the global adoption of the internal combustion engine. Despite numerous and continuing attempts to develop alternative powertrains, the basic principles of the internal combustion engine and its two contrasting principles of spark-ignition and compression ignition, have remained remarkably unchanged for more than a century, although at a detailed operational level it has progressed significantly and continues to do so, not least in efficient fuel delivery and emissions management systems.

The majority of road vehicles in numerical terms (~70%) may be classed as passenger and light-duty transport, the remaining 30% being commercial. Whereas the spark-ignition engine running on gasoline ('petrol') is the generally preferred powertrain for passenger and light-duty vehicles in the USA, in Europe a higher preference for light-duty diesels based on compression-ignition has emerged in recent decades. Worldwide consumption of diesel fuel is broadly comparable to gasoline, due not least to the higher payloads and distances travelled by commercial road transport[9] and to the many off-road uses of diesel ranging from agriculture to construction and power.

Another critical success factor for the prevailing transport system is the fact that liquid fuels not only provide a very convenient source of propulsion energy but also comprise a high density energy storage medium, providing low on-board weight and enabling rapid refilling. For aviation, the benefits of high energy density liquid fuels are even more pronounced, especially due to the relationship between total payload and fuel consumption during take-off. For these and related reasons, demand for liquid fuels is expected to continue to increase for decades to come. Figure 2.3 illustrates possible future trends, albeit subject to the caveats on the reliability of future projections given earlier. Conventional gasoline use is likely to diminish proportionately over time, as it is easier to displace light duty use with biofuels, alternative powertrains and improving

---

9 International Transport Forum (2012): http://www.internationaltransportforum.org/Pub/pdf/12Outlook.pdf

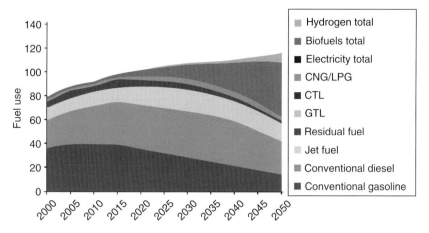

Figure 2.3 Fossil fuels may dominate until at least 2050. The proportion of conventional diesel is likely to increase relative to gasoline. The relative proportion of alternatives post-2030 is subject to many evolving technological and other external factors so is less predictable.

efficiency, whilst demand for diesel and jet fuel is likely to remain longer, subject to the development of suitable biofuel blending components.

The process of continuing to improve the operational reliability of increasingly sophisticated internal combustion engines has been enabled by increasingly stringent quality controls on fuel. Standardisation has been key to reliability and affordability. Over many years, fuel quality has consistently improved in terms of characteristics such as deposit-forming tendency, and engine designers have in turn been able to exploit the opportunities this affords, for example increasing efficiency through features such as higher temperatures and pressures, finer injector nozzles and precise injection timings, and increasing the longevity of emission control systems. This ongoing symbiotic relationship between fuel and engine development continues to refresh the potential to design performance-enhancing fuels.

To some extent, such optimisation and refinement has been achieved at the expense of increased sensitivity to fuel quality and intolerance to the introduction of new fuel components. On the other hand, greater flexibility has been enabled in recent years *via* the incorporation of on-board computers, which process a wide variety of feedback signals to help control and adapt the fuel delivery and combustion processes, and also by the inclusion of new materials such as improved elastomeric seals and through the provision of increasingly advanced lubricants.

However, market fuels must service all vehicles, not only the newest. Vehicles typically remain operational in the market for at least 15 years, in some cases much longer. New fuels must remain compatible with materials and operational limitations of earlier generation vehicles. This imposes a further constraint on the rate at which radically new fuel components may be introduced.

The fundamental distinction between gasoline and diesel, and the corresponding spark-ignition (SI) and compression-ignition (CI) engines, shapes the available fuel options. In an SI engine, fuel and air are fully mixed before the spark initiates combustion. Avoiding premature combustion (knock) is a key factor for operation, achieved by using gasoline and a relatively low compression ratio. Higher octane means greater resistance to knock and the potential to improve engine efficiency. Conversely, in a CI engine, air is first compressed under a relatively higher compression ratio, diesel fuel is injected near maximum compression and it burns as it mixes with the air. If there is a significant delay in the onset of ignition, the subsequent combustion of larger accumulations of fuel can cause rough and noisy operation. Fuels with higher cetane numbers have a shorter ignition delay – the resulting knock is less intense, and the engine runs more quietly and smoothly.

In addition to this reciprocity between octane and cetane, another major distinguishing factor between gasoline and diesel is the flash point, which relates to safe storage and handling. Safety in fuel storage and handling is a critical factor – the vapour space above gasoline storage is maintained as too rich to permit the mixture to burn ('high flash point') due to its relatively high volatility; whilst for diesel, the low volatility and corresponding low flash point preserve safety during storage, across the range of temperatures at which the mixture may be stored. It is essential to avoid creating a flammable atmosphere during storage, which could risk an explosion. Whereas the high volatility of ethanol gives it compatibility with gasoline storage systems, this introduces incompatibility with diesel storage systems. If ethanol is mixed with diesel, it must be handled under gasoline-like high flash point conditions.

2.4 Engine Constraints, Fuel Specifications and Enhanced Performance

Engines must be able to perform effectively under a wide range of conditions, whether starting from cold in winter or operating fully warmed-up in summer. Engine performance encompasses everything from idling and accelerating under start-stop urban driving through to long-distance cruising. Specifications on volatility, density and octane or cetane help ensure a minimum standard of performance for a given engine design. To ensure that performance does not degrade unacceptably over the long term and ultimately to avoid engine failure, further specifications seek to constrain the formation of combustion deposits, degradation of seals, and corrosion or failure of metal components. Further specifications address health, safety and environmental issues. For example, as noted above, the longevity of effective emission control systems is a legal requirement, so contaminants that could poison exhaust gas catalysts or oxygen sensors must also be controlled. Specifications vary from country to country, as appropriate to locally significant issues, but draw upon a broadly common set of parameters relating to common needs of fuel handling and engine performance.

Virtually all fuel specifications derive from observed performance in an engine. Conventional fossil fuels comprise hundreds of different molecular types that occur in a limitless range of subtly different combinations. Their combustion involves many thousands of chemical reactions. In the past (and to a large extend still today), empiricism provided the only realistic way to determine performance characteristics, and it forms the basis for devising tests that are both robust enough and sufficiently reproducible yet cost-effective to underpin the target objectives. This has important consequences for the introduction of alternative fuels comprising novel components.

Octane, for example, is an empirical measure of gasoline anti-knock performance based on comparison in a standard engine against a mixture of two reference fuels selected in 1929[10]. Although there is a vast body of data and understanding relating the composition of conventional fuels to this historical measure, components interact non-linearly, so it remains difficult to predict accurately the consequences of introducing any particular novel component chemistry, especially if significantly changing the balance of components outside the historically established range, as with novel oxygenates. Aromatics and branched alkanes tend to have higher octane than straight-chain alkanes, but performance ultimately depends on the total mix of components present.

Determining the cetane number (CN) of a novel diesel blend is equally challenging and must be determined empirically. Because of the complexity and cost of measuring cetane number, an approximate value, the cetane index (CI), may be computed using an algorithm based on boiling point and density, but this cannot be relied upon for more novel components or for complex blends outside the original correlation set. The relationship between octane (or cetane) and desired engine performance under real-world operating conditions must also be determined empirically. The operation of a modern vehicle engine can be radically different from that of a standard octane or cetane engine.

A significant consequence of the multi-component nature of both gasoline and diesel is their broad vaporisation-temperature profile, or 'distillation curve'. Maintaining the optimal ratio of fuel vapour to air is a critical determinant of efficient combustion, power generation and emissions. In broad terms, the more volatile components have a strong influence on starting, components with mid-range volatility tending to influence acceleration whilst the least volatile components play a bigger role during cruising. Oxygen sensors and other feedback mechanisms help modern injector systems to compensate for different fuel volatility profiles and densities, but there are inevitable consequences resulting from marked deviations from the original design space.

Post-combustion particulates and nitrogen oxides (NO_x) are important targets for diesel emissions reduction technology, with maximum emissions determined by a succession of increasingly stringent regulations, as set out over the years in the USA, EU and Japan. Advanced emissions control systems have been developed and incorporated in modern vehicles. Increasingly tight specifications on fuel sulphur content facilitate the initial and ongoing operation of such devices. Biofuels tend to have very low sulphur content, which can be clearly advantageous, but it is important that other trace materials that may also be present in the biofuel do not interfere with the operation of emissions control devices.

The use of waste products such as used cooking oils introduces the possibility of a range of contaminants, such as cooking salt (sodium chloride) being present in the feedstock which, if not effectively eliminated from the final product, could impact emissions and emission control equipment. Pure biodiesel specifications include limits on phosphorus (typically 4 or 10 ppm, max., depending on the prevailing engine regulations in each country), group I alkalis (K and Na – max. 5 ppm) and group II metals (Mg, Ca – max. 5 ppm).

For (pure) ethanol feedstock, national specifications may include solvent-washed gum (5 mg/100 ml), chloride (sometimes but not always specified as inorganic chloride), which currently can vary from 1 ppm max. to 40 ppm max., copper (generally max. 0.1 ppm), sulphur – ranging from 10 ppm max. in Europe to as high as 2,000 ppm max. in some developing markets and phosphorus (varying between 0.15 and 0.5 mg/l).

---

10 Owen *et al.* (2005).

Table 2.1 Ethanol specifications in EU and US. Note these are subject to change and many other countries have their own specifications. Within the US, California has its own denatured ethanol specifications, 120% of which differ slightly from those in this chart. The EU specifications cover all 28 member states and also apply to European-based EFTA countries (Iceland, Liechtenstein, Norway and Switzerland).

Ethanol			EU EN 15376:2001	US ASTM D4806-12
Implementation year			Aug-11	Jul-11
ethanol	wt%	min	98.7	92.1
water	wt%	max	0.3	1
chloride, inorganic	ppm	max	6	10
chloride	ppm	max	–	–
copper	ppm	max	0.1	0.1
methanol	vol%	max	1	0.5
C3-C5 alcohols	ppm	max	20.00	–
non-volatile matter	g/100ml	max	0.01	–
gum (solvent washed)	mg/100 ml	max	–	5
denaturant	vol%	min-max	–	1.96–5.0
sulphur	ppm	max	10	30
pH	(pHe)	min-max	–	6.5–9
electrical conductivity	uS/m	max	2.5	–
acidity	wt%	max	–	0.007
acetic acid	wt%	max	0.007	–
phosphorus	g/l		0.00015	–

Tables 2.1 and 2.2 give a list of representative EU and US specifications for ethanol and biodiesel as at the time of writing. However, these specifications are constantly evolving. It is important to check the latest worldwide fuel specifications as, for example, summarised on an annual basis by the International Fuel Quality Centre (IFQC)[11], but original regulatory documents should always be sourced for definitive detail.

As noted earlier, national and regional specifications have evolved mainly to help the performance of advancing engine technologies meet national policies or local requirements. Such specifications are necessarily broadly applicable and relatively simple to perform, to permit consistent and cost-effective measurement and monitoring. Because technology is constantly advancing, meeting an existing specification with a radically different fuel composition does not guarantee optimal or trouble-free performance

---

11 International Fuel Quality Centre, www.ifqc.org

Table 2.2 Biodiesel specifications in EU and US. Note these are subject to change and many other countries have their own specifications. The EU specifications cover all 28 member states and also apply to European-based EFTA countries (Iceland, Liechtenstein, Norway and Switzerland).

			EU	US
			EN 14214:2012	ASTM D 6751-12
Biodiesel			2012	Nov-12
cetane number		min	51	47
ester content	wt%	min	96.5	–
iodine number	g/100 ml	max	120	–
sulphur	ppm	max	10	15
density@ 15C	kg/m^3	min-max	860–900	15
viscosity@40C	cSt	min-max	3.50–5.00	1.9–6.0
distillation T90	C	max	–	360
Flash Pt	C	min	101	93
CCR 100%	wt%	max	–	0.05
CFPP	C	max	5 to −44	0.05
Cloud Pt	C	max	–	report
water and sediment	vol%	max	–	0.05
water	mg/kg	max	500	0.05
sulphated ash	wt%	max	0.02	0.02
acid value	mg KOH/g	max	0.5	0.5
methanol	wt%	max	0.2	0.2
ethanol		max	–	0.2
monoglycerides	wt%	max	0.7	0.4
diglycerides	wt%	max	0.2	0.4
triglycerides	wt%	max	0.2	0.4
free glycerol	wt%	max	0.02	0.02
total glycerol	wt%	max	0.25	0.24
phosphorus	ppm	max	4.0	10
alkali Group 1 (Na,K)	ppm	max	5.0	5.0
metals, group II (Ca, Mg)	ppm	max	5.0	5.0
oxidation stability @ 110C	hour	min	8.0	3.0

across the full range of modern vehicles that may ultimately use the fuel in the market. In practice, extensive engine and fleet testing is also required to establish fitness for purpose of the retail fuel blend.

Shifting priorities occasionally demand solutions that interfere with previously established fuel-engine balances. The introduction of 'ultra-low sulphur diesel' within the past decade, to help ensure the longevity of emission control hardware performance, risked catastrophic failure of certain in-line fuel-pumps that were reliant on the presence of a low level of sulphur in the fuel to reduce friction. Lubricity additives are now routinely used when needed to avoid fuel pump failures when operating on 'ultra-low sulphur' fuels. The capacity for certain biofuels to confer lubricity with near-zero sulphur fuels can be advantageous. Hence, biofuels may be perceived as fuel additives as well as fuels in their own right.

The ability to enhance fuel performance *via* additives is not limited to avoiding component degradation and failure. It should be noted that a number of performance-related requirements such as cold-start, cetane, injector cleanliness or lubricity may be met, not only through the properties of the base fuel, but through the inclusion of generic or proprietary additives. Leading fuel companies, such as Shell, may further ensure or enhance fuel quality by supplementing the national specifications with a wide range of more detailed tests of their own that are adapted to specific conditions and advanced technology needs. In this way they are able to deliver statistically significant improvements over the norm, such as greater power, engine protection or fuel economy.

The relationship between biofuels, specifications and engine operation must be determined on a case-by-case basis. In the following section, we illustrate these inter-relationships by initially considering opportunities and constraints with first-generation biofuels – ethanol and biodiesel – and then considering fuel quality issues associated with a selection of other biofuel options.

2.5 Biofuels – Implications and Opportunities

2.5.1 Introduction

To date, the production of biofuels has been strongly driven by the urgent need to deliver products to market and meeting biofuel mandates using available feedstocks and conversion technologies. This practical requirement essentially defines 'first generation' biofuels, which, as noted previously, are predominantly ethanol derived from simple, fermentable sugars and biodiesel derived from plant oils. In either case, the resulting fuel product is an oxygenated hydrocarbon that differs in physical and chemical properties from the hydrocarbons typical of crude-oil derived fuel and hence has limited compatibility with the prevailing vehicle parc. With the passage of time, new engines entering the market are being designed to tolerate higher concentrations of first-generation biocomponents, but relaxing the corresponding fuel specifications remains limited by the longevity of older engines in the vehicle parc. If biofuels are to be supplied in sufficient quantities to deliver significant low-carbon benefits or energy security, they must be produced in a form that can be used by the large majority of vehicles in the marketplace, and/or within suitably adapted vehicles supported by dedicated supply and distribution infrastructures.

2.5.2 Ethanol

Ethanol is currently produced by fermenting sugar (and to a much lesser extent, plant biomass: Chapter 5) and is added predominantly to gasoline or petrol. Ethanol disrupts hydrogen bonding in mixtures and forms azeotropes with lower molecular weight hydrocarbons in the fuel blend. This raises the Reid Vapour Pressure (RVP) and front-end volatility (specified either as E70 – meaning the percentage evaporated at 70°C as used in European specifications – or as T50 – meaning the temperature at which 50% of the fuel is evaporated, as used in US specifications), but does not necessarily aid cold starting, which for hydrocarbon fuels tends to correlate with a combination of RVP and front-end volatility. The RVP and volatility of conventional gasoline is carefully controlled to deliver good operability, and simply adding ethanol to standard gasoline would cause the total mixture to be too volatile for reliable operation across the range. More volatile components need to be removed from the base gasoline at the refinery to create a 'volatility hole' into which the ethanol can be blended, so that the resulting blend does not fall outside the target volatility range. This is called an 'RBOB' (Reformulated Blendstock for Oxygen Blending) gasoline. The quality of the final blend depends on the synergistic qualities of both the ethanol and the base gasoline, such that the combination meets, as a minimum quality standard, the regulatory properties for local market application. This further constrains ethanol blends to specific concentration bands (for RBOB in the US, this is currently assumed to be 10% by volume).

Unlike hydrocarbon fuels, which are hydrophobic and will tend to shed water present in more than a few hundred ppm water, ethanol is highly hygroscopic and may pick up water from within the supply chain, such as accumulated at the bottom of storage vessels or fuel tanks. In the three-phase ethanol/water/gasoline system, phase separation may then occur, separating out the water-ethanol phase, which acts as a corrosive medium or breaks down deposits and carries them through the fuel delivery system to impact performance. Retail site fuel storage tanks are always carefully dried before introducing ethanol for the first time. To ensure that this does not increase corrosion, the water content of ethanol for fuel blending is limited to between 0.3 wt% in some parts of the world including the EU, and up to 1 wt% in other parts including the USA. An exception is the hydrous ethanol grade, unique to Brazil and dedicated Brazilian vehicles, which contains up to 4.9% water and is only used in specially designed vehicles.

Being a single component, pure ethanol alone is not generally suitable as a fuel – it does not have sufficient volatility to start an engine, even in a tropical climate. Limiting (anhydrous) ethanol content to 85% leaves sufficient volatility from the gasoline fraction to assist starting without the need for a separate fuel tank. Most cars running on 'hydrous ethanol' (97% ethanol) in Brazil are equipped with a separate 1-litre gasoline store, which is drawn from to aid starting. However, in 2009, a new generation of flex fuel vehicles was launched in Brazil, which warms the ethanol fuel during starting, permitting cold start down to −5°C without the need for the extra gasoline fuel tank.

2.5.3 Biodiesel

Triglycerides are natural energy stores occurring as plant oils and animal fats and are the starting point for most biodiesel in current use (see Chapter 6). Triacylglycerol (TAG) comprises a three-carbon glycerol backbone to which individual long-chain fatty acids are attached covalently *via* ester linkages. Used neat, plant oils are poor fuel components. The viscosities of plant oils can be ten times higher than that of diesel fuel. The densities of neat plant oils also tend to be high, cetane numbers low, and they may contain relatively high concentrations of water and trace metals. Blending neat plant oils in engines can lead over time to many operating problems, ranging from filter blocking and engine stall to increased frictional wear and emissions.

Problems associated with neat oils can be significantly reduced *via* a process of trans-esterification with methanol. Transesterification, carried out in the presence of alkali hydroxides as a catalyst, yields fatty acid methyl ester (FAME) and free glycerol, and is a relatively energy-efficient process. Essentially reversible, transesterification is driven by adding an excess of the methanol, resulting in two readily separable phases – the ester phase above and the glycerol phase below. Benefits resulting from transesterification of triglycerides to FAMEs include improved viscosity, density and cetane number. The properties of FAME-based biodiesels are mainly determined by the chain length and saturation of these fatty acids and so their performance characteristics are closely linked to the nature of the original plant oil or animal fat.

Fatty acid chain lengths associated with any particular biodiesel feedstock may lie anywhere between C6 (6 carbons) and C22 (22 carbons), but are most commonly 16 or 18 carbons long. This is comparable to typical diesel ranges of C8–C25. Saturation (the amount of hydrogen associated with the carbons that make up the fatty acid chain) is another important determinant of performance; chains typically range from unsaturated (classified here as, for example, C18:0 for a chain length of 18 carbons and zero unsaturated bonds, also known as 'stearate') to possessing one, two or three unsaturated bonds (*e.g.* C18:1 (oleate), C18:2 (linoleate) and C18:3 (linolineate)).

In general, cetane number increases with increasing chain length and saturation. Wide variations in measured CN values can be found in the literature for given chain lengths and saturation, but oxidation of the FAME can also increase the CN, and may at least partly account for these widely differing measurements[12].

Saturated long-chain methyl esters have generally higher melting points, which translate into relatively poor cold temperature flow characteristics. The corresponding performance-related measure is the Cold Filter Plugging Point (CFPP), although cloud point (CP) and pour point (PP), are also sometimes still applied. Specification limits are set nationally to reflect local climate. Unsaturated methyl esters with their lower melting points are relatively more suitable for use in cold weather climates. Various strategies are applied to improve cold temperature flow characteristics, including the use of additives. One approach is to separate out the heavier, unsaturated, portions before blending ('winterisation'), but fundamental properties of the ester chain length and saturation are dominant and it is clearly better if possible to use suitably selected plant oils to start with.

Oxidation of FAMEs can result in organic acids and polymer deposits in the engine. The result of oxidative fuel degradation or 'ageing' during storage and handling may potentially cause filter plugging and poor injection control. Auto-oxidation rates are related to the number of unsaturated bonds and their position in the chain. The presence of *bis*-allylic carbon sites (as occurs for example at the C-11 carbon position in linoleic acid as a consequence of double bonds at $\Delta 9$ and $\Delta 12$) significantly increases

---

12 Bamgboye and Hansen (2008).

susceptibility to oxidation[13]. Relative to an oxidation rate of 1 for oleates, linoleates have been reported to oxidise at a rate of 41, whilst for linolenates with two bis-allylic positions at C-11 and C14, the relative oxidation rate was observed to be 98[14]. Consequently, oxidation stability can be substantially degraded by the presence of even small quantities of such compounds in biodiesel blends.

A simple test of saturation uses the absorptive properties of the unsaturated bonds to remove iodine and hence its colour from a solution. A high iodine number indicates a higher degree of unsaturation, and a maximum iodine number (typically 120 g/100 ml) is often specified for biodiesel. Although the Iodine method indicates the number of double bonds present in biodiesel, it does not distinguish between bond locations and is therefore a weak predictor of oxidation stability. Finding a suitable method for this purpose took a number of years, but recently a version of the Rancimat method has been adopted as a basis for specifications. Using this method, in Europe for example, stability must be maintained for a minimum of 8 hours whilst air is bubbled through the biodiesel held at 110°C. In other regions, the requirement is generally lower – 6 hours stability or 3 hours (as in the USA – ASTM D 6751-12) under similar test conditions. Other specifications, such as maximum acid value (typically set at 0.5 mg KOH/g max.) may also help limit the consequences of oxidation, including corrosivity.

Being derived from different plant species, different biodiesel feedstocks possess their own unique chain lengths/saturation 'fingerprint'. Figure 2.4 illustrates typical profiles of some common and alternative lipid-based feedstock types (see also Chapter 6). From this, it may be seen why rapeseed oil (*Brassica napus*; Canola), with its predominant C18:1 composition and minimal unsaturated components, is relatively well suited to cooler European conditions, whereas palm oil (*Elaeis spp.*), with a predominance of unsaturated C16:0 components has relatively poor cold flow properties and is thus generally less well suited for colder European conditions.

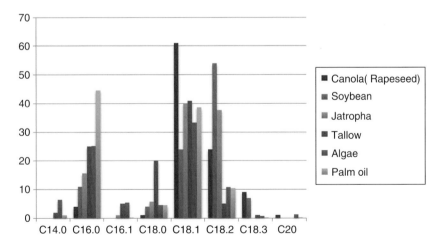

Figure 2.4 Illustrative chain length and saturation profiles of various biodiesel types.

13 Knothe (2005).
14 Frankel (1998).

The presence of components with longer chain lengths than typical diesel can be problematic, as these represent relatively non-volatile components that may fail to combust before impacting and condensing on the piston walls. This may be exacerbated for example in more advanced engine systems, where late fuel injection is required for diesel particulate filter (DPF) regeneration. Components condensed in this way can make their way down past the piston rings into the sump, causing crankcase oil dilution and gradual deterioration in its lubricating properties. For such reasons, a maximum blend volume of FAME is often set. In Europe, for general road use, the maximum FAME blend currently stands at 7% per volume, but in other countries, or for particular uses under specifically monitored conditions such as urban bus fleets, the permitted blend volume can be as high at 50%, or delivered as a 'pure' 100% fuel.

Although the process of transesterification is simple in principle, in practice there are many processing options that can impact the suitability of the final product for transport fuel use. For example, the most commonly-used catalysts are alkaline, which yield rapid methanolysis under relatively mild conditions. The reaction of the triacylglycerol with short-chain alcohol results in the desired fatty acid alkyl ester plus glycerol, which are almost completely immiscible and hence easy to separate. Stepwise methanolysis and extraction of one chain at a time is efficient, but can leave residual monoglycerides, diglycerides and triglygerides in addition to free fatty acids in the final fuel component. Their potential impact on performance is such that each and all are constrained by specifications (to a max. 0.8, 0.2, 0.2 and 0.02 wt% respectively and a max. 0.25 wt% on total glycerol).

Free fatty acids also present in the feedstock are readily converted into soaps and can interfere with the separation of the glycerols. Some feedstocks are naturally low in free fatty acids, but others, such as palm oil can be much higher. An upper limit for the free fatty acid content to avoid processing issues has been established to be around 3%[15]. Either the feedstock must be pretreated to reduce the free fatty acids, for example using an excess of the alcohol to esterify the fatty acids to a combination of esters and water, or acid-based catalysis must be used which requires more severe reaction conditions and generally produces a quantity of unwanted side products[16]. A constraint on acid value (max. 0.5 mg KOH/g) helps to limit potential consequences. Although large-scale processing plants are generally well equipped to effectively control the resulting fuel quality, this can be more challenging for smaller, simpler, processing plants.

The processing of waste oils (such as frying and cooking oils) as noted above, poses significant additional challenges, due to the potential for feedstock contamination including oxidation products and coloured materials. Even when effectively pre-cleaned and converted, it should be noted that the resulting fuel ester may be dominated by the chain-length and saturation properties of oils more suited to cooking than to meeting the stringent local requirements of transport fuels in colder climates.

In summary, achieving the right balance between cetane, cold-flow, oxidative stability and other properties using FAMEs, requires careful matching to local conditions. Many factors must be taken into account, and can only be assessed effectively when finally blended.

----

15 Ahn et al. (1995).
16 Mittelbach (2009).

Standard vehicles and storage/supply systems have been gradually adapted to tolerate the particular requirements of first-generation biofuels through the introduction of improved materials, fuel delivery systems and engine mapping. For example, standard elastomer seals in older cars were not specifically designed to accommodate blends containing ethanol, but any resulting effects such as elastomer swell can be kept within engineering tolerances provided the ethanol concentration is limited. Modernisation of the engine fleet has allowed the concentration of ethanol in gasoline to steadily increase from typically 5% per volume initially, to 10% or more per volume in some markets, the upper limit forming what is often referred to as a 'blend wall'.

More extensive engine-fuel adaptations enable high ethanol (typically up to 85% ethanol – E85), or up to 100% biodiesel (B100), but these require the provision of dedicated fuel supply systems. Such step-out options are only commercially feasible if mandated or heavily incentivised across a market, or can be supplied to dedicated local fleets, for example to urban bus fleets or *via* commercial heavy duty supply depots. Dedicated commercial fleet supply systems permit novel fuel-engine combinations to be considered that would not otherwise be practical within the general market, for example using di-ethyl ether or di-methyl ether. For widespread market applications, however, the development of advanced biofuels compatible with the current and future evolving vehicle parc has greatest potential. The Brazilian market must be viewed as a special case, based on a market-wide restructuring introduced from 1974 ('proalcool' initiative), which supports up to 24% ethanol blends and 97% hydrated ethanol. Attempting to emulate this approach is not a practical option for other developed markets today. For widespread market applications, the development of advanced biofuels compatible with the current and future evolving vehicle parc has greatest potential.

2.6 Advanced Biofuels as Alternatives to Ethanol and FAME

Given the constraints on blending volumes resulting from the imperfect match between first-generation biofuels ethanol and FAME and conventional gasoline and diesel respectively, and many other considerations including sustainability of feedstocks, considerable attention has been given to alternative processes and biofuel chemistries. In this book, the term 'first generation' applies, as it does in industry, to biofuels derived from fermentable sugars of plant oils. These sugars and oils derive predominantly from comestible plants. Here also, the term 'second generation' biofuels is taken to describe biofuels produced from the fermentation of non-food sugars, derived from treated lignocellulose. 'Advanced' or 'third generation' biofuels is used to describe biofuels derived from natural microbes and algae, and 'fourth generation' biofuels are those generated by synthetic biology. However, it must be noted that, in the fuels industry, such terminologies are not used. Indeed, there is no universally adopted definition of 'second generation' or 'advanced' biofuels. In the industry, the terminology is variously applied to novel fuel products, conversion processes or feedstocks – in practice, anything beyond the boundaries of first-generation biofuels as currently practised. What the different fuel products that may result are, and how suitable they will be for future transportation applications, is an area of intense research, globally.

To facilitate their adoption, the ideal end-point of alternative biofuel component developments is to behave as similarly as possible to existing fossil fuels. These are generally termed 'drop-in' fuels. Conventional fossil fuels in the market comprise a complex range of components, each contributing towards certain properties around which engines and fuels have been optimised over many years of co-development. It is therefore technologically challenging to generate bio-derived fuels that precisely meet the 'drop-in' requirement. There is also the commercial challenge of making such fuels cost-competitive, when the underlying need is to remove all oxygen from the original carbohydrate or lipid-based feedstocks as necessary to produce pure hydrocarbon streams, and to upgrade as needed to confer a suitable balance between straight-chain, branched and ring structures in the final blend.

A minor issue relates to nomenclature for advanced biofuels: The term 'biodiesel' was originally applied to fatty acid alkyl esters transesterified with methanol (*i.e.* FAMEs), and is generally applied with this specific form in mind, including the setting of formal specifications. However, many other bio-derived diesel chemistries can be envisaged. Either the term biodiesel must additionally be applied to these other classes of bio-derived diesel, or another term such as 'alternative bio-derived diesel' must be used. There is currently no commonly agreed approach. For the remainder of this chapter and to avoid confusion, we shall refer to each bio-derived diesel *via* a description of its specific process.

One approach to convert lipid-based feedstocks to hydrocarbons is by hydrotreatment (HT). In HT, the organic feedstock is reacted with hydrogen over a catalyst at elevated temperature and pressure to produce n-alkanes. The process is similar to that routinely used in refineries for the desulphurisation of diesel streams to meet sulphur specifications, but the more reactive nature of the oxygenated feedstock results in greater levels of carbon monoxide, carbon dioxide and other co-products including water. These secondary products limit the amount of bio-feedstock that can be co-processed in a conventional refinery unit, above which concentration a dedicated processing unit is required. The n-alkanes produced by HT are compatible with fossil diesel streams, with generally better cold-flow properties, density and higher cetane numbers than their FAME equivalents. For cold climates however, iso-alkanes are preferable, which can be achieved in principle by isomerising the n-alkane output from the hydrotreater.

Using gasification to convert ligno-cellulosic (*i.e.* wood or cellulose from plants) feedstock to syngas and then applying Fischer-Tropsch synthesis and hydrocracking (a form of bio-to-liquids (BTL) process) also enables the production of 'ultra-clean' bio-hydrocarbons for road transport. With virtually zero sulphur and aromatics and high cetane, this can be a very attractive blending component, helping to trim sulphur and aromatic levels in the final fuel, and delivering benefits for emissions control and other engine performance parameters. Gas-to-liquids (GTL) products, such as produced by Shell from its world-scale plant in Qatar ('Pearl'), are already widely used in the market to produced higher-quality premium diesel and more recently trialled in 50:50 blends with jet fuel[17]. If a BTL process could produce a similar range of product types, similar fuel performance benefits could apply.

Opportunities also exist to produce components that may not fulfil the 'drop-in' criterion, but could provide either specific performance advantages or cost benefits

---

17 http://www.shell.com/global/future-energy/natural-gas/gtl/products.html

compared to other biofuels. An important distinction exists between components that are essentially indistinguishable from components currently registered for use in fuels and those that are sufficiently different or novel to require additional regulatory approval.

One of the more energy efficient routes from ligno-cellulose to fuel grade components is *via* acid hydrolysis to levulinic acid[18]. Subsequent hydrogenation and esterification can be applied to produce ethyl valerate or pentenoate, potentially suitable as gasoline blending components, or pentyl valerate, potentially suitable as a diesel blending component up to 15 vol%. Despite the technical suitability of such components, they fall outside current regulatory approval systems. These and similar such approaches would either need to be taken through full approval process to gain full market access, or be restricted to niche applications.

Adopting a more holistic approach to the succession of stages from biomass to marketplace fuels, as outlined in Figure 2.2, introduces further opportunities for biofuel development.

Recent advances in synthetic biology are rapidly improving the ability to modify pathways in yeasts, bacteria and other micro-organisms as needed to produce commercially viable titers of potential fuel components. This can offer more direct and potentially efficient routes from sunlight to fuel than using natural plant sugars or oils and post-converting to suitable components. In some cases, toxicity of the target component to the host organism may provide a limitation on naturally occurring production that can be readily overcome by suitable separations engineering. Integrating routes from suitable energy crops to fuels *via* modified biological pathways and engineering processes can be envisaged as the basis for integrated bioprocessing.

The lipid pathway to biodiesel relies heavily on the properties of the fatty acids in the TAG to determine the performance characteristics of the final fuel. The potentially greater availability of ligno-cellulosic sugars as a feedstock raises the question as to whether ethanol could be used in the transesterification step to produce fatty-acid ethyl ester (FAEE) rather than FAME. FAEEs will generally have better cold flow performance and improved oxidation stability than their FAME equivalent, but higher viscosities and the poorer economics of using finished ethanol rather than methanol mean that FAEEs have not progressed beyond the laboratory to date. The conversion of ligno-cellulosic sugars directly to FAEE *via* modified bacteria or yeast has recently been demonstrated[19], and could provide a possible basis for an integrated bioprocessing scheme[20].

Regulations permit butanol to be blended in gasoline subject to constraints. Compared to ethanol, butanol has the advantages of greater volumetric energy density, lower oxygen content and lower water solubility, but the straight chain n-butanol form has significantly lower octane. Octane number tends to be higher with branching so that iso-butanol, with an octane of 102, represents a more attractive option. By comparison, tertiary butanol (TBA) has a high melting point, around 25°C, which would initially appear to make it very unsuitable, but in fact it has some history as a gasoline range extender, having been used as a co-solvent with methanol in

18 Lange *et al.* (2010).
19 Steen *et al.* (2010).
20 Bokinsky *et al.* (2011).

concentrations of up to 2%/2% methanol/TBA for the period in Germany from 1968. Pathways to many other potentially fuel-grade alcohols with varying degrees of branching can also be envisaged. EPA approval for iso-butanol blends in the USA have been gained by Gevo[21], which recently began commercial production using a modified yeast combined with effective separations technology to minimise toxicity issues.

The isoprenoid pathway in plants and algae provides an alternative to alcohols and fatty acids as a route to biofuels. Isoprenoids (or terpenes) comprise a broad family of molecules that have fuel potential, ranging from pinene to biabolene[22]. By modifying the relevant isoprene-related metabolic pathways in yeast, a fermentation route for large-scale production of farnasene has already been developed by Amyris (trademarked as Biofene™). Farnasene is a C15 branched hydrocarbon, with two isomers (six stereoisomers) found in nature. As a C15 hydrocarbon, it can be blended with road diesel.

As a further example of emerging microbial routes, it has been shown that *E. coli* bacteria can be modified to produce alkanes and alkenes across a range of relevant chain lengths from C8 to C18 and with varying degrees of branching. The ability to produce hydrocarbons tailored to meet market fuel blending requirements, as described earlier in this chapter, can greatly facilitate the delivery options for such biofuels to market.[23]

2.7   Biofuels for Aviation; 'Biojet'

Whereas ground transport fuels vary from region to region as described above, commercial fuels for aviation meet uniform international specifications, so that planes can fly and be refuelled anywhere in the world and always be assured of a consistent quality of fuel. The main international standard for jet fuel is called Jet A-1, as defined in ASTM specification D1655. Jet A-1 has a freezing point maximum of −47°C. This should not be confused with another grade, Jet A, which has a slightly higher freezing point limit and is generally restricted to applications in North America. Specialist grades also exist, for example for military uses. Test flights of a bio-jet blend by the US Air Force and Army have taken place[24,25].

There are few alternatives to the jet engine (gas turbine) for commercial flight, so biofuels are an important option in the quest to make air flight more sustainable. As D1655 applies specifically to crude-oil, a new specification, D7566, has been developed to support the inclusion of renewable fuels into the aviation sector. Renewable fuels produced using specified processes can now be blended up to 50% with conventional commercial and military jet fuel. Approved processes, and the specification of renewable fuel components produced using each process, are defined in annexes to D7566. Annex 1 refers to the production of jet fuel produced using Fischer-Tropsch synthesis methods. This applies to 'syn-crude' derived from coal, natural gas or biomass. Annex 2 addresses hydroprocessed esters and fatty acids (HEFA) from bio-oils. As specified in

---

21 http://domesticfuel.com/2013/08/26/senator-michael-bennet-visits-gevo/
22 Kung *et al.* (2012).
23 Howard *et al.* (2013).
24 http://www.biofuelsdigest.com/bdigest/2012/07/03/air-force-tests-gevos-isobutanol-based-biojet/
25 http://www.hydrocarbonprocessing.com/Article/3292494/US-Army-tests-isobutanol-bio-jet-fuel-for-helicopters.html

ASTM D7566-11, 'Specification for Aviation Turbine Fuel Containing Synthesized Hydrocarbons', this approach was approved on 1 July 2011. These are generally referred to as bio-SPK (synthetic paraffinic kerosenes). Other annexes are envisaged in future, to address other processes, including alcohol to jet (ATJ), direct sugar to hydrocarbons (DSHC), hydrotreated depolymerised cellulosic jet (HDCJ), catalytic hydrothermolysis (CH), and catalytic conversion of sugar[26].

Reference to current fuel specifications[27] is required to have a full understanding of the chemical and physical requirements of the blend component, but this is complex and evolving, so will not be attempted here. The final two digits refer to the year of approval or latest revision – ASTM D7566-13 refers to the revision made available in 2013. However, fuel specification is only part of the requirement. A range of fit-for-purpose properties (*e.g.* hydrocarbon chemistry (carbon number, type and distribution)) and trace materials; numerous bulk physical and performance properties, electrical properties, ground handling and safety properties (*e.g.* effect on clay filtration); and compatibility (*e.g.* with other approved additives and with engine and airframe seals, coatings and metallics) must also be met, and then a series of rig and engine tests for turbine, fuel and combuster system compatibility must be successfully negotiated, finally leading to OEM (original engine manufacturer) approval testing.

The final 'biojet' fuel needs to meet strict drop-in criteria, and this requirement opens up opportunities for the choice of biomass feedstock and the conversion process. One approach is to start with bio-oils with similar compositional ranges to typical jet fuels. Although most bio-oils fall within the C16–C18 range (see Figure 2.5), a number of plants produce oils with chain lengths and basic properties closer to the kerosene/jet

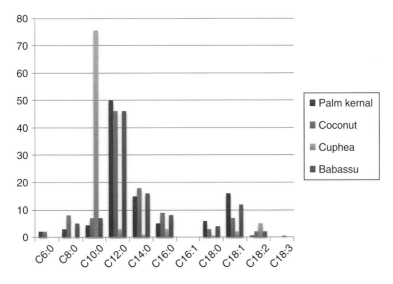

Figure 2.5 A sub-set of plant oils yields chain lengths dominated by 10 to 14 carbons.

26 Nate Brown (2012)

27 ASTM D7566 specification: http://www.astm.org/Standards/D7566.htm

fuel cut than diesel, as illustrated in Figure 2.5. The corresponding chain length range suitable for aviation gas turbines is C8-C17, mainly lying between C10 and C14. Aromatics at around 20% vol. may also be present. Jet A-1 specifies a maximum of 25% aromatics, so this is broadly within range.

Transesterification to FAME is not suitable to meet jet fuel requirements. Apart from anything else, it should be clear from the above that the predominance of saturated components introduces significant challenges to meet cold flow specifications, such as the freezing point max. of −47°C. Rather, hydrotreating plant oils to paraffins is required, followed by hydrocracking and separation to produce the finished jet fuel product, which will be predominantly in the form of iso-paraffins.

One of the earliest demonstration flights was carried out in 2008, using a 20% blend derived from coconut (*Cocos nucifera*) and babussa (*Attalea speciosa*) – consistent with the C-chain length logic described above – to fuel one engine of a Boeing 747 by Virgin Atlantic. This test-flight was not viewed as a genuinely commercial approach, but served the purpose of demonstrating that biojet could indeed be viewed as achievable for the airline industry.

Most subsequent demonstrations and test flights have focused on identifying more sustainable, available and economic feedstocks, and relying on the bioprocessing to deliver suitable chain lengths, rather than relying on plants with suitable oil properties. Tests using bio-oils derived from jatropha (*Jatropha spp.*) by New Zealand airlines and from camelina or false flax (*Camelina sativa*) by Japan Airlines were carried out in 2008/2009. In the same year, an algae-based biojet was developed by Solazyme[28] that could meet the D7566 specification. More recently, a wider range of feedstocks is being considered by different groups, ranging from algae and animal fats[29] to Ethiopian mustard (*Brassica carniata*; a form of canola more suited to more arid regions)[30], sugar cane (*Saccharum officinarum*)[31] and municipal waste. Syn-gas processes delivering products that meet D7566 Annex 1 are being considered for approval, while others such as butanol to jet[32] await future specification annexes.

2.8   Impact of Future Trends in Engine Design on Retail Biofuels

Fuels and engines have evolved together for over a century, and it is reasonable to expect this process to continue in the future. The biofuels being developed and tested today need to anticipate engine demands in coming decades. So how might the requirements for biofuel quality change? With the exception of the Brazilian market and certain fleet operations, the inclusion of biofuel components has mostly occurred within the past decade. This time frame is less than the turnover rate of older design vehicles in the market, which has actually constrained the rate at which first-generation biofuels could

---

28 http://solazyme.com/fuels
29 http://www.aliphajet.com/technology.html
30 http://agrisoma.com/images/pdfs/CarinataProductionGuide.pdf
31 http://biomassmagazine.com/articles/9598/amyris-to-enter-partnership-to-supply-biojet-to-gol-airlines
32 http://renewables.seenews.com/news/albemarle-to-use-cobalts-bio-n-butanol-in-navy-backed-biojet-fuel-project-262590

be assimilated. Over the past decade, a steadily increasing capacity to accommodate higher concentrations of biofuels has occurred as more compatible engines enter the market, not least in response to policy requirements to achieve lower net GHG emissions. In the meantime, increasing performance expectations and requirements, for example, to increase the longevity of emissions control equipment and duration between lubricant changes, plus targets to improve fuel economy, continues to increase demands on fuel quality.

Trends in engine operation include downsizing and greater efficiency, higher fuel line pressures, stop-start operations, increasingly sensitive emission control equipment and many other innovations. On-board computing capacity linked to sensors enables on the one hand greater flexibility to variable fuel types and on the other hand an ability to respond positively to higher quality fuels. Safety factors dictate that fuel must either be supplied in a gasoline-like 'over-rich' or diesel-like 'over-lean' vapour mode, but the greater control afforded by computer control and improving engineering precision does permit the concept of new and improved engine operating conditions that blur this historical distinction between modes of combustion. The underlying principle, known as homogenous charge compression ignition (HCCI), seeks to combine the efficiency benefits of current CI engines (*via* high compression ratios) and the cleanliness of current SI engines (ability to use three-way catalysts for tailpipe emissions control). In this HCCI concept, fuel and air are premixed as in the SI engine, but heat release occurs by auto-ignition – the same phenomenon that causes combustion in CI engines and (undesirable) knock in SI engines. However, controlling this is very difficult. A practical derivative of this, using partially-premixed combustion enables some of these benefits to be achieved under more manageable part-load conditions, and is already starting to be applied in more advanced heavy-duty engines. The impact of developments such as these, aided by increasingly sophisticated computer-aided engine design, is that the optimum fuel properties for such modified combustion processes, and hence the biofuels on sale at service stations, could continue to evolve for many years to come.

2.9   Conclusion

In this chapter, we have focused on the production and distribution of biofuels to the consumer. It is clear that not all biofuels are the same, and it is critically important that they are properly quality assured. The design of transport fuel must take a wide range of physical, chemical and other factors into account to deliver the same assurance of quality currently enjoyed by crude oil derived fuels. Engines have evolved to provide a highly efficient and cost-effective mechanism to convert fuel energy into propulsion, but the cost of this efficiency is a high degree of sensitivity to deviations from the prevailing norm. It is important to keep in mind that fuel requirements will continue to evolve and this will have shifting implications for the optimal design of biofuels in the long term.

Selected References and Suggestions for Further Reading

Ahn, E., Koncar, M., Mittelbach, M. and Marr, R. (1995) A low-waste process for the production of biodiesel. *Separation Science and Technology*, **30**, 2021–2033.

Bamgboye, A.I. and Hansen, A.C. (2008) *International Agrophysics*, **22**, 21–29.

Bokinsky, G., Peralta-Yahya, P.P., George, A., Holmes, B.M., Steen, E.J. *et al.* (2011) Synthesis of three advanced biofuels from ionic liquid-pretreated switchgrass using engineered *Escherichia coli. Proceedings of the National Academy of Sciences, USA*, **108**, 19949–19954.

Nate Brown, FAA, 18 May 2012 http://www1.eere.energy.gov/bioenergy/pdfs/10_brown_roundtable.pdf

Frankel, E.N. (1998) *Lipid Oxidation*. The Oily Press, Dundee, Scotland.

Howard, T.P., Middlehaufe, S., Moore, K., Edner, C., Kolak, D.M. *et al.* (2013) Synthesis of customized petroleum-replica fuel molecules by targeted modification of free fatty acid pools in *Escherichia coli. Proceedings of the National Academy of Sciences, USA*, **110**, 7636–7641.

International Energy Agency (2014) *World Energy Outlook 2014*. IEA, Paris.

International Transport Forum/OECD (2012) *Transport Outlook 2012*. ITF/OECD, Paris

Knothe, G. (2005) *Fuel Processing Technology*, **86**, 1059–1070.

Kung, Y., Runguphan, W. and Keasling, J.D. (2012) From fields to fuels: recent advances in the microbial production of biofuels. *ACS Synthetic Biology*, **1**, 498–513.

Lange, J.P., Ayoub, P.M., Petrus, L., Gosselink, H., Price, R. *et al.* (2010) Valeric biofuels: a platform of cellulosic transportation fuels. *Angewandtke Chemie – International Edition*, **43**, 4479–4483.

Mittelbach, M. (2009) Process technologies for biodiesel production. In: *Biofuels* (Soetaert, W. and Vandamme, E.J. eds), Wiley, Chichester, UK, pp. 77–93.

Owen, K., Coley, T. and Weaver, C.S. (2005) *Automotive Fuels Reference Book*, 3rd edition. SAE International.

Steen, E.J., Kang, Y., Bokinsky, G., Hu, Z., Schirmer, A. *et al.* (2010) Microbial production of fatty-acid-derived fuels and chemicals from plant biomass. *Nature*, **463**, 559–562.

Stern, N. (2006) *Stern Review on the Economics of Climate Change*. HM Treasury, London http://www.hydrocarbonprocessing.com/Article/3292494/US-Army-tests-isobutanol-bio-jet-fuel-for-helicopters.html
http://www.biofuelsdigest.com/bdigest/2012/07/03/air-force-tests-gevos-isobutanol-based-biojet
http://www.shell.com/global/future-energy/natural-gas/gtl/products.html

3

Anaerobic Digestion

John Bombardiere[1] and David A. Stafford[2]

[1] West Virginia State University, USA
[2] Enviro-Control Ltd., Devon, UK

Summary

Production of methane (biogas) by anaerobic digestion (AD) is one of the longest established biofuel technologies. AD was used in sewage works in the UK and USA in the late 19th century and the methane was then employed for street lighting. We now know much more about the organisms responsible for this process – the Archaea – and about the biochemical pathways that are involved. AD is now used to treat a wide range of types of biomass, including human sewage, farm and food wastes and biomass from agriculture. Several different digester designs are used, depending on the primary substrate and the scale of operation. One more recent development is to link ethanol production with AD, where the latter is used to treat the biomass left over after the readily accessible sugars have been fermented. Concerns about fuel security and climate change have led to increased investment and therefore wider application of AD in many countries worldwide. In this chapter, we focus mainly on developments in the USA and in the EU, including the UK.

3.1 History and Development of Anaerobic Digestion

3.1.1 Introduction

Micro-organisms possess great versatility in recycling complex molecules contained in organic wastes into useful products. They are able to oxidise such materials using oxygen (O_2) as well as other oxidants such as sulphate (SO_4) and nitrate (NO_3). In the process known as Anaerobic Digestion (AD), some are able to use carbon dioxide (CO_2) as the final oxidant or hydrogen acceptor and can convert the most oxidised form of carbon into the most reduced, methane (CH_4). These organisms are known as Methanogens. Methanogenesis thus provides a biofuel but the process also reduces the amount of contained carbon in waste mixtures and thus increases the relative concentrations of essential nutrient elements such as nitrogen, phosphorus and potassium. The value of microbial methane generation was recognised in late Victorian times (see Chapter 1), when anaerobic digesters were first used to treat sewage sludge and to provide biogas for lighting in Exeter (UK) and Boston (USA). Remarkably, an anaerobic digester was also installed in the late 19th century at a leprosy hospital in India, where the methane produced from the treatment of human sewage was used to light the hospital itself. Further engineering applications grew first in municipal areas during the early to mid-20th century and later in farming and food waste applications. In these more modern developments, the methane is not burned to give light as it was in Victorian times. Instead, the reducing power of methane can be harnessed to drive the generation of electricity using gas engines.

As mentioned above, treatment of sewage and other biological wastes by AD concentrates essential nutrients present in the original substrate. Thus the solid/liquid residue can often be further recycled in the form of organic fertilisers for which there is a growing market, thus adding to the value of the process. We note in passing that this also reduces the consumption of fossil fuels which are used in the 'conventional' synthesis of many fertilisers.

3.1.2   Mixtures of Micro-Organisms

The only organisms capable of generating methane are all members of the Archaea (*i.e.* archaebacteria). However, in the initial stages of metabolism of the substrates, eubacteria and micro-fungi are very much involved. They participate in the breakdown of proteins, nucleic acids, complex carbohydrates, including cellulose and hemi-celluloses (see Chapter 4), lipids and lignin as well as simpler compounds such as sugars, fatty acids and amino acids. Thus the overall process requires a consortium of many and varied organisms. It starts with hydrolysis of the larger more complex molecules, followed by fermentation and then acetogenesis (acetic acid is the prime intermediate molecule), and bio-methanation, forming the end gaseous products of CO_2 and CH_4, as well as water. And of course, the biological function of all such physiological processes is to produce energy[1] for the organisms, so that synthesis of cellular material can lead to growth and reproduction.

Model bio-reactor systems have been used in studies of the dynamics and metabolism of these complex consortia of micro-organisms, showing that reactor design can affect the dynamic relationships and the equilibrium composition of microbial populations. Early biochemical measurements of methane production in anaerobic digesters[2] laid the foundation for the further control of the microbial processes using intelligent engineering[3]. A knowledge of the way in which the engineering of design types can influence control and efficiency of anaerobic digester activities has been of particular benefit in the successful applications of AD to produce bio-fuels from organic wastes. Other benefits include the significant reduction in pathogenic viruses, bacteria and protozoa that may be contained in wastes[4] with the possibility of making significant savings in disease treatment, which will be obtained when utilising AD systems in waste treatment.

----

1 Hughes (1980).
2 Whitmore *et al.* (1986).
3 Espinosa-Solares *et al.* (2006).
4 Rhunke *et al.* (2004).

3.2 Anaerobic Digestion: The Process

3.2.1 General Biochemistry

AD involves the utilisation of microbial metabolism to effect a change from a reduced form of the materials to a mixture of alcohols and fatty acids before conversion to the biogas mixture. The micro-organisms are contained in gas-tight containers with sizes varying from a few cubic metres to up to $8,000 \, M^3$ in volume. They operate at meso-philic (30–36 °C) or thermophilic temperatures (50–60 °C). Details of the biochemistry and microbiology have been shown elsewhere[5] The overall result for the digestion of glucose for example would be as follows:

$$C_6H_{12}O_6 \rightarrow 3CH_4 + 3CO_2$$

3.2.2 Design Types

In order to shed light on the biochemical and biological control of the microbial pro-cesses involved in AD, mathematical modelling has been used to study the complex inter-organism reactions. These studies can enable predictions to be made on the per-formance and efficiency of digesters on specific waste streams. For example, because livestock farms may produce much organic material in the form of slurries and manures, a large potential organic resource is available to cycle the contained carbon through to methane which can be used to generate power. It is possible to consider AD applications for single farm units or combine wastes from different farms to benefit from economy of scale, where a single unit would not be economically justifiable. A single pig farm unit would need to have 10,000 pigs, a cattle farm about 200 milking cows and a poultry farm about 250,000 birds before they become efficient enough to make a reasonable return from selling the power and making fertiliser. In general then, it is important to consider each specific application, including type and volume of waste stream to be treated and to consider the type and capacity of the digester that will be most suitable. We therefore now describe in more detail several different types of reactor.

3.2.3 Complete Mix Design

The digester vessel is usually a round insulated tank (see Figure 3.1), above or partly below ground and made from reinforced concrete, steel (glass-lined, stainless or coated), or fibre glass. Heating coils with circulating hot water can be placed inside the digester or, depending on the consistency of the feedstock, the contents can be circulated through an external heat exchanger to maintain desired temperatures. The latter is the preferred option for ease of maintenance. The contents can be mixed with a motor-driven mixer, a liquid recirculation pump or by using compressed biogas. The authors prefer gas mixing, to provide a more gentle process to preserve microbial integrity, as well as to reduce the power input to the overall process. A gas tight cover (floating or fixed) traps the biogas, which accumulates at the top from where it may be piped away. The complete mixed

---

5 Stafford and Hughes (1981).

Figure 3.1 Complete Mix Anaerobic Digester. Photo by DAS.

digester processes organic materials with a 3% to 10% total solids input range. The Hydraulic Retention Time (HRT – *i.e.* the time taken for the waste to be treated) is usually 8 to 15 days. Complex mix digesters can be operated in the mesophilic and thermophilic ranges, with the latter being more increasingly employed. A pilot complete mix system is shown above being used for the treatment of high strength poultry wastes.

3.2.4   Plug Flow Digesters

A plug-flow digester vessel is a long, narrow (typically a 5:1 ratio of length: width), insulated and heated tank made of reinforced concrete, steel or fiberglass with a gas tight cover to capture the biogas (see Figure 3.2). They can operate at mesophilic or thermophilic temperatures. The plug flow digester often has no internal mixing (though some mixing at the input end is advised for high strength wastes), and can take wastes of 8–15% total solids. These units are excellent designs for a scrape manure management system[6] or poultry waste, with some dilution, or using mixed farm wastes. Retention times vary between 15 and 20 days. The manure in a plug flow digester does not mix longitudinally through the digester but flows through as a plug, moving towards the exit whenever new manure is added. Manure that has been treated discharges over an outlet weir arranged to maintain a gas-tight atmosphere.

---

6 See *e.g.* http://www.manuremanagement.cornell.edu/Pages/General_Docs/Events/4.Andy.Lenkaitis.pdf

Figure 3.2 Pilot-scale plug flow anaerobic digester, showing the easy loading ramp and the submerged digester. Photo by DAS.

The effluent flows out with a much reduced concentration of solids and can be collected for use as organic fertiliser, although in some applications it may need further treatment. The biogas is collected centrally at the apex and routed to a power generator. The system as described is more low-cost and can be used 'on-farm' for treatment of smaller volumes of manure compared with the complete mix system described above. It is the design of choice for small applications because of its simplicity and low cost.

3.2.5 High Dry Solids AD Systems

High dry solids units essentially use a dry digestion process with high input of total solids. High loading rates are claimed with smaller digester volumes because input water is not required. The digesters usually have no phase separation and no crust or sediments are formed. Some units have a digester with a conical outlet with no accumulation of solids within the digester. It is usually a one-phase process with no mixing, which also reduces energy input. The feed often requires continued inoculation of active microbes, but with operation at the high end of the thermophilic range, pathogen kill and weed seed removal are assured. The capital outlay is often high because of the need for more complex feed pumps and controls, but this is balanced by the advantage of treatment of wastes with high concentrations of total solids without the need to add water. Applications include the treatment of municipal or domestic garbage, either on its own or in admixture with high strength manures.

3.2.6   Upflow Anaerobic Sludge Blanket (UASB)

The UASB digester evolved from the anaerobic Clarigester, an early type of upflow digester, and was pioneered in The Netherlands. Other variant types include the expanded granular sludge bed (EGSB) digester. In the latter, micro-organisms accumulate around particles to form individual microcosms. In UASB systems, a bed of particulate microcosms (where the starting granules can be carbon or even sand particles) is elevated to form a blanket of granular sludge which is suspended in the tank. Soluble wastes from brewing or sugar production processes are ideal candidates for the system, where wastewater flows upwards through the blanket while being processed by the anaerobic micro-organisms. The upward flow combined with the settling action of gravity suspends the blanket, sometimes with the addition of flocculants to aid aggregation. The commissioning time to produce a good active bed can be as long as three to six months. Small sludge granules begin to form whose surface area is covered in aggregations of bacteria. The micro-organisms are thus supported on the granules which, when they reach a particular size, slough off to retain the integrity of the microbial aggregates.

The technology needs complex monitoring when put into use to ensure that the sludge blanket is maintained; ammonia concentrations must be carefully controlled as the bed microbial aggregates can be adversely affected and be washed out. The blanketing of the sludge provides for a two-phase solid and hydraulic (liquid) retention time in the digesters. Solids (the active biomass microbes) can remain in the reactors for up to 90 days, whereas the hydraulic retention time can be as low as hours in some applications. UASB reactors are suited to the treatment of dilute waste water streams with low concentrations of suspended solids and can be used in the mesophilic temperature range to reduce input power requirements.

3.2.7   Anaerobic Filters

An anaerobic filter is a fixed-bed biological reactor; waste water flows through the filter, particles are trapped and organic matter is degraded by the biomass that is attached to the filter material. The filter material is usually of commercial plastic packing normally associated with aerobic waste water treatment. The treatment can be used for treating the supernatant from settled sewage at ambient temperatures or can be used mesophilically at higher rates of treatment. They are simple devices for soluble waste waters and can be used instead of UASB units or in conjunction with them as a polishing device to capture the remaining biogas locked in the untreated organic effluent. The units comprise a sedimentation tank followed by one or more filter chambers. Typical filter material sizes range from 12 to 55 mm in diameter and the material will provide up to $300 \, m^2$ of surface area per m^3 of digester. By providing a large surface area for the bacterial mass, there is increased contact between the organic matter and the active biomass that effectively degrades it. The filters can be operated in either up-flow or down-flow usage. The hydraulic retention time is low (0.5 to 1.5 days) and removal of suspended solids and reduction of biological oxygen demand (BOD) can be as high as 85–90% and provides for a smaller footprint than other types of digesters. However, only liquid wastes (albeit containing *dissolved* solids) are suitable for treatment with this type of digester.

3.3 Commercial Applications and Benefits

3.3.1 In the United Kingdom

In the UK, the Department of Food, Rural Affairs and Agriculture (DEFRA) has an action plan to produce a zero-waste economy with AD as one of the options to treat organic materials. The estimated potential for producing heat and electricity is between 3 and 5 TeraWatt[7]-hours by 2020. Debt financing of £10 million has been made available over 4 years to stimulate investment in AD. In helping to develop a sustainable farming sector, the proposals will help to provide a low-carbon-fertiliser agriculture and will also help with the sustainability of fertiliser components such as phosphate.

At the end of 2014, the treatment capacities of the existing 54 AD plants in the UK (not including sewage sludge treatment facilities) were as follows:

- 534,200 tonnes of commercial waste;
- 382,000 tonnes from food and drink manufacture; and
- 136,156 tonnes in farm-based plants.

Output capacity of these plants is 35 MW of electrical power, which is about 1% of the total UK usage.

The treatment capacity of the existing (in late 2014) 146 AD plants that treated sewage sludge was 1,100,000 dry tonnes per year. This provided an output capacity of 110 MW, while the output capacity of 50 further sewage-treatment AD plants currently at the planning stage is 70 MW. So, although AD plants in the UK currently produce only a proportion of the projected 2020 total of electricity, there is movement towards that goal. Furthermore, attitudes are changing with regard to the incentives for using AD. The UK public is now expressing a positive attitude towards renewable energy and towards environmental sustainability. The advantages of AD can be determined fiscally in such a way that the burden of owning and operating digesters can be spread between those who benefit from them and this will contribute to a wider use of AD in renewable energy generation.

3.3.2 In the USA

In the USA, the Chicago Climate Exchange (CCX), has an active, voluntary and legally binding[8] integrated trading system to reduce greenhouse gases (GHGs). Projects set up to produce biogas can sell credits through the CCX after producing proof of eligibility. Details can be seen on the CCX website[9].

The USA has a more extensive agricultural system than European countries, with for example 14 times the expanse of cropland and 8 times as many cattle as in Germany. Thus a network of 88,000 digesters could produce 223 million MWh of electricity, representing over 5% of total electricity production[10]. However, in late 2014, that potential was far from being fulfilled, as reported by the US Departments of Agriculture and of

---

7 TeraWatt (TW) = 10^{12} Watts
8 Entry into the scheme is voluntary but once entered, the requirements are legally binding.
9 https://www.theice.com/ccx
10 Stamakatos (2012).

Energy (USDA, USDoE). Only about 2,300 plants were producing biogas, but there was a potential to immediately develop a further 13,000 sites. The greatest opportunities for AD were identified as being part of livestock manure-based systems, with potential to develop over 8,000 sites producing 7.2 million m^3 of biogas nationwide. Of those currently operational, the majority were sewage works digesters (1,241), followed by landfill gas capture (636), and manure-based digester systems (239). Most of the manure-based digester systems were located on dairy farms (191)[11].

3.3.3 In Germany

In Germany, there is a very strong push from the Government to develop renewable energy production. Furthermore, the decision to phase out nuclear power plants following the accident at Fukushima (see Chapter 1) has added further impetus. Energy from biomass, including biodegradable waste, is planned to play a large part in this move to renewable energy. At present, Germany intends to use biomass for almost 8% of its electricity and heat, (see below), with AD contributing a significant proportion of this. In 2011, digestion of agricultural feed stocks produced 17 million MWh of electricity, representing 3% of total production. Put in more economic terms, 6,800 AD plants supplied nearly 5 million homes with electricity, created €6 billion in revenue and employed 20,000 people. The projected growth of this industry up to 2020 shows 25,000 digesters supplying Germany with 96 million MW-hours, 17% of total electricity production, powering 22 million homes, creating €26 billion in revenue and employing 85,000 people.

3.3.4 Overall Benefits

As we have just seen from just three example countries, AD has the potential to bring a range of benefits, from the environmental to the social. This is because AD can 'turn waste into energy' *via* the production of biogas (methane), very often alongside a bonus of producing nutrient-rich compost or fertiliser pellets. The technology is flexible, is available today and does not need years of subsidy-driven research to make it viable; it has proved to be a reliable process for the last 75 years on a commercial scale, with innovations on design having made the process much more efficient in a range of applications (see above). Thus, many different types of organic wastes, singly and in admixture, are now utilisable sources of feed for the digesters.

   Cost reductions in waste treatment, coupled with the production of biogas that can run generators, thus helping to offset treatment costs, are often the major drivers for the adoption of AD in sewage treatment. AD was first used for sewage treatment at the end of the 19th century (Section 3.1.1; see also Chapter 1) and on a much wider scale in the UK since the early years of the 20th century. It is now very well established such that in 2007, 66% of all sewage sludge was treated by AD. However, there are indications that AD is falling out of favour for several cities, because the digested sludge can no longer be used for agricultural purposes due to metal and potential microbial contamination of crops.

---

11 USDA/USEPA/USDoE (2014).

Although there are currently few AD plants used for treating organic wastes other than sewage, a significant growth is expected in its use for treatment of food wastes and agricultural manures. Improvements in the process are being brought on-line as R&D translates into practice. The future may well reside in improved sludge pretreatment technologies such as thermal and enzyme hydrolysis prior to AD processing. For example, a plant in northern England carries out thermal hydrolysis at 165 °C and 6 bar (600 kPa) pressure for 1 hour to provide a homogeneous, more soluble feed of secondary sludge for a thermophilic anaerobic digester[12]. The process reduces the sludge disposal volume by 50% compared with conventional AD and also improves the biogas quality. The plant also produces 4.7 MW of electrical power.

3.4 Ethanol Production Linked with Anaerobic Digestion

As mentioned in Chapters 1, 8 and 17, ethanol is a widely produced biofuel. It is made by the fermentation of glucose, which in the USA (now the country producing the highest volume of bio-ethanol) is derived from the stored starch and oligosaccharides in the seeds (grains) of corn (maize).

Many ethanol plants have been set up in the corn belt of the USA and, as noted in Chapters 8 and 17, there has been a tension between growth of corn for ethanol production and growth of corn for animal and human nutrition. Indeed, in 2012, for the first time, more corn was harvested for ethanol production than for animal nutrition and it has been claimed (see Chapter 17) that the switch to ethanol production was one of the factors that drove up the price of food so strongly in 2008. However, it is not these issues, important as they are, that concern us in this chapter. Rather, we focus on financial aspects of the technology itself. Despite the enthusiastic uptake of the technology, there are doubts about its ongoing profitability. However, profitability may be restored if the by-products of ethanol production can be used rather than being wasted. The first of these is 'corn stover', the waste plant material that is left after the grain has been harvested. This is of course a by-product of the harvesting of any grain crop and increasingly is being used as a biomass fuel, for example in the production of ethanol from hydrolysed cellulose (see Chapter 5). The second consists of the grains from which the starch and oligosaccharides have been removed to go into the ethanol production process. This material is known as dried distillers grain (DDG) and is usually sold as animal feed.

The second main by-product is the residue left after fermentation and using this as a substrate for AD helps ensure the profit margins of ethanol production from corn. Typically, the total residue after fermentation and distillation is sent to a thermophilic anaerobic digester where biogas and fertiliser are produced. The biogas can provide up to 100% of the fuel needs of the ethanol plant and provides a solid co-product that can be sold as a fertiliser, aquaculture feed, or (with some on-going certification) animal feed. Depending on the quantity of the residue that is processed, there may be additional gas for injection into the natural gas pipeline system. This integrated operation stabilises throughout the life of the ethanol production plant, the cost of the fuel used in

---

12 Neave (2012).

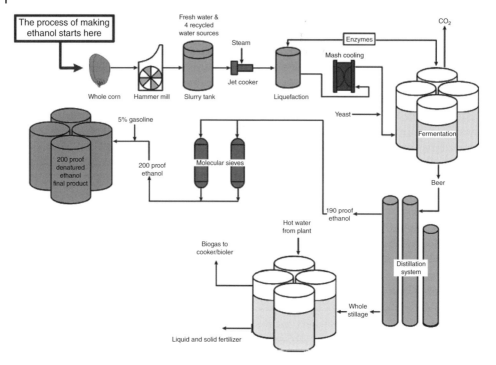

Figure 3.3 Anaerobic digester integration with ethanol production. © EnviroControl, Ltd.

producing ethanol, as well as providing the valuable co-products mentioned above. The integrated system is shown in the Figure 3.3.

3.5 Financial and Economic Aspects

In Western economies, variations in the commodity markets can influence adoption of many technologies. AD is clearly one of those technologies, especially when feedstocks for the process have multiple uses. For example, the 2014 commodity price for corn DDG (see above) from ethanol production in the USA was close to $100 per tonne, down from $200 per tonne in 2013. Coupled with the 2014 spot price for natural gas at $3.78 per MMBtu[13], lower by-product prices made AD more attractive in 2014 than just a year earlier, but less attractive than the period between 2006 and 2007, when natural gas prices fluctuated between $7 and $13 per MMBtu and DDG prices were around $100 to $125 per tonne. The best option currently is to turn the biogas into electrical power.

Fertiliser prices also influence the implementation of the technology. With increased demand for fertilisers in the developing world and increasing use in the USA and EU, farmers have been willing to pay historically high prices for manures. This helps AD to be more profitable.

---

13 One Btu = 1055 Joules or 0.0003 kW-hours

A further example helps to illustrate the influence of economic factors. In the southern US state of Georgia in the Spring of 2014, the price per tonne for poultry litter varied from $10 to $50 per tonne, with most farmers paying between $20 and $35[14]. If the litter price as sold to the fertiliser market is low, AD has an advantage (by producing power and fertiliser), in over-selling the litter just for fertiliser; if high, the farmer may just as well sell the litter rather than spend capital on producing power at a loss. One advantage of AD compared to other 'manure to energy' technologies such as combustion, is preservation and concentration of nutrients through the system. Nutrients such as phosphorus and calcium can be concentrated in the solid fraction of digester effluent at levels per dry tonne basis up to three times that of raw manures[15]. Potential for significant revenues from the effluent may thus offset the price of higher value feed stocks in AD systems.

Each application of AD is now a matter of environmental and commercial necessity and if the economy of scale can be matched to the appropriate AD design, profitability can be assured whilst providing benefits to the wider community. Many AD units are now providing a payback within three to eight years and are thus potentially fundable with 'private' money; furthermore, government grants may in some cases tip the balance in favour of profitability, especially where specific environmental and public health issues are improved.

3.6 UK and US Government Policies and Anaerobic Digestion – An Overview

As noted in Chapter 1 and earlier in this chapter, Anaerobic Digestion (AD) is a technology that produces fuel without making demands on land used for food production. The opportunities that fuel production *via* waste treatment promise include substantial reductions in waste disposal costs as well as providing useful sellable by-products (in addition to the biogas itself). AD technology has thus caught the attention of several governments. In the UK, for example, the Government set up WRAP (Waste Resources Action Programme)[16] to aid the application of a whole range of waste treatment and recycling technologies, including AD, with both financial and interactive support. The following statement embodies their approach:

> WRAP's vision is a world in which resources are used sustainably. Its 'strapline' is 'Working together for a world without waste' and its declared mission statement is to accelerate the move to a sustainable resource-efficient economy through:
>
> - *reinventing how we design, produce and sell products;*
> - *rethinking how we use and consume products; and*
> - *redefining what is possible through recycling and re-use.*

---

14 Dunkley *et al.* (2014).
15 Liedl *et al.* (2006).
16 The UK Government actually set up WRAP as a Charity (No. 1159512), so it is in effect a 'quango' rather than an office of the Government.

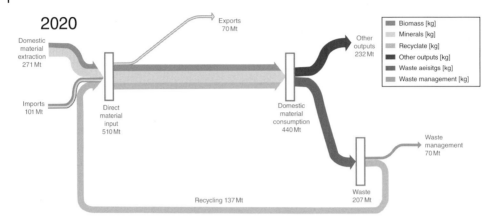

Figure 3.4 The headline '30 Mt less materials going into the economy' is close to WRAP's 2010 research: *Securing the Future: The Role of Resource Efficiency 3.*

AD is clearly one of the technologies that helps to drive government policy and which will encourage a new industry that covers energy, recycling and fertiliser production for use in agriculture. The main overall aim therefore is to provide waste stream treatment *via* a circular economy, with government at the centre of decision-making and implementation, but with private applications at the 'coal-face'. Changes in the future of waste handling and benefits are shown in Figure 3.4.

In the USA, it was considered timely early on to support such biomass recycling technologies, with a new bill in 2003 when the US Department of Agriculture (USDA) announced the availability of $23 million in funds to assist farmers, ranchers and rural businesses to purchase renewable energy systems and make energy efficiency improvements. In the same year, the California Renewables Portfolio Standard (RPS) programme was established. The goal of the RPS programme was for 20% of California's energy to come from renewable energy sources. To reach that goal, the legislation required California's investor-owned utilities (IOU) to increase their procurement of renewable energy by at least 1% per year. Whilst this was not achieved, several AD units have been set up since then on farms to recycle manure and to treat waste; thus the generation of renewable energy in the State was carried forward. The added bonus of reducing air pollution, especially with large dairy farms being close to cities, helped the reception of such technologies by local populations.

3.7 Concluding Comments

As we mentioned earlier, the generation of methane by AD is a technology that dates back over 100 years. The beauty of the technology is that it can be used at a variety of levels of sophistication, from basic systems suitable for rural agrarian communities in less developed countries to complex systems, some of which have been discussed in this chapter. One recent and fascinating example of the former is the use by the Maasai people of Kenya to treat the blood and other waste from slaughtered cattle[17]. The methane

---

17 Mayes (2015).

is used to drive the generation of electricity to run the slaughterhouse and is also sold for use as a cooking fuel. There are clearly benefits all round here, from economic to environmental. The latter benefits arise because trees are no longer chopped down for firewood or to make charcoal for use as cooking fuels. Another recent example comes from an industrialised, developed country, namely the UK. One of the buses operating in the city of Bristol runs on methane generated by the AD of food waste and human sewage[18].

In general, the marriage of well designed and engineered anaerobic digesters and an understanding of the requirements of the microbial consortia that are relied upon to metabolise the organic molecules in waste materials, provides the end-user with a system which will reduce pollution whilst generating energy-rich biogas. The 'add-ons' of fertiliser production with considerable reduction in pathogens, especially in thermophilic AD units, ensures efficiency in external energy usage and public health benefits. The process has become the backbone of energy production within agriculture and will contribute significantly to the alternative energy profile of several fossil fuel replacement programmes. The use of mixtures of different organic waste streams will play an increasing part in obviating waste treatment costs, as well as increasing the range of applications of the technology.

Selected References and Suggestions for Further Reading

Dunkley, C. *et al.* (2014) *University of Georgia Extension Bulletin No 1386.*

Espinosa-Solares, T., Bombardiere, J., Chatfield, M., Domashko, M., Easter, M. *et al*, (2006). Macroscopic mass and energy balance of a pilot plant anaerobic bioreactor operated under thermophilic conditions. *27th Symposium on Biotechnology for Food and Chemicals: Applied Biochemistry and Biotechnology*, **129–132**, 959–968.

Hughes, D.E. (1980) What is anaerobic digestion?: an overview. In: *Anaerobic Digestion*, (Stafford, D.A., Wheatley, B.I. and Hughes, D.E. eds), Applied Science Publishers, London, pp. 1–14.

Korres, N., O'Kiely, P., Benzie, J.A.H., West, J.H., eds (2013) *Bioenergy Production by Anaerobic Digestion using Agricultural Biomass and Organic Wastes*. Routledge, Abingdon, UK/New York.

Liedl, B.E., Bombardiere, J. and Chatfield, J.M. (2006) Fertilizer potential of liquid and solid effluent from thermophilic anaerobic digestion of poultry waste. *Water Science and Technology*, **53**, 69–79.

Mayes, R. (2015) http://www.design trend.com/articles/34417/20150111/maasai-generating-biogas-animal-blood.htm

Neave, G. (2009) Advanced Anaerobic Digestion: More Gas from Sewage Sludge. http://www.renewableenergyworld.com/rea/news/article/2009/04/advanced-anaerobic-digestion-more-gas-from-sewage-sludge

Pullen, T. (2015) *Anaerobic Digestion: Making Biogas, Making Energy*. Routledge, Abingdon, UK/New York.

Rhunke, T. *et al.* (2004) *Proceedings of the 10th World Congress in Anaerobic Digestion, Montreal, Canada*, 1843–1844.

---

18 http://www.firstgroupplc.com/news-and-media/latest-news/2015/18-03-15.aspx

Shang, Y., Johnson, B.R. and Sieger, R. (2005) Application of the IWA Anaerobic Digestion Model (AD1), for simulating full-scale anaerobic sewage sludge digestion. *Water Science and Technology*, **52**, 487–492.

Stafford, D.A. and Hughes, D.E. (1981) Microbial production of fuels. In: *Energy – Present and Future Options* (Merrick, D. and Marshall, R. eds). Wiley, Chichester, UK, pp. 69–92.

Stamatakos, E. (2012) Exploring Biogas, an Untapped Source of Clean Renewable Energy. http://eponline.com/articles/2012/01/12/exploring-biogas-an-untapped-source-of-clean-renewable-energy.aspx?admgarea=Features

US Department of Agriculture, US Environmental Protection Agency, US Department of Energy (2014) Biogas Opportunities Roadmap, Voluntary Actions to Reduce Methane Emissions and Increase Energy Independence. http://www.epa.gov/climatechange/Downloads/Biogas-Roadmap.pdf

Whitmore, T.N. *et al.* (1986) *Biomass*, **9**, 29–35.

4

Plant Cell Wall Polymers*

Stephen C. Fry

Institute of Molecular Plant Sciences, School of Biological Sciences, Edinburgh, UK

Summary

The cell wall is an important factor in the use of plant biomass to generate liquid biofuels. It needs to be broken down in order for its components to be available for fermentation and other biofuel-generating pathways. This chapter provides a comprehensive account of the cell wall in living plants, covering biological functions, chemical composition, evolution, biosynthesis and natural modification. Where appropriate, differences between different plant groups are considered.

4.1 Nature and Biological Roles of Primary and Secondary Cell Walls

Plant biomass has a long history of use as fuel, primarily as fuel to be burned. However, plant biomass can also give rise to fuel by other means, including anaerobic digestion and fermentation to produce methane (Chapter 3) and ethanol (Chapter 5). These process involve breakdown of the plant cell wall into more readily metabolisable components and thus it is important in the quest for biofuels to have an understanding of the cell wall – the outer skin of a plant cell, controlling its shape and size, and providing a defensive barrier.

The cellulose-rich cell wall is a non-living but nevertheless metabolically active cellular 'coat'. It is a thin fabric (often about 0.1 μm thick in young plant tissues), usually flexible, but strong enough to resist the tearing which a turgid protoplast might otherwise inflict on its wall. The inner surface of the wall is in contact with the lipid/protein-rich *plasma membrane*, which is less than 0.01 μm thick. The wall's outer surface is usually coated with a pectin-rich *middle lamella*, through which neighbouring cells adhere.

Wall material deposited by a young, growing cell constitutes the *primary* wall layer. Any additional wall material, deposited between the plasma membrane and the primary wall after cell expansion has ceased, is a *secondary* wall layer. The latter may eventually become very thick (*e.g.* up to 20 μm), but does not increase in surface area.

---

* Manuscript for this chapter received 31 May 2013.

Biofuels and Bioenergy, First Edition. Edited by John Love and John A. Bryant.
© 2017 John Wiley & Sons Ltd. Published 2017 by John Wiley & Sons Ltd.

The primary cell wall contains microfibrils, bundles of cellulose molecules, which lie periclinally (parallel with the plasma membrane), but whose orientation in the other two dimensions governs the direction(s) in which the cell will grow (= irreversibly increase in volume). A cell whose microfibrils are laid down in various (or 'random') orientations is destined to expand into an approximate sphere (*e.g.* a potato tuber parenchyma cell). On the other hand, if most of the microfibrils have one preferred orientation, then that cell will elongate in a direction perpendicular to the microfibrils (*e.g.* a cortical parenchyma cell in a stem). The wall thereby directly dictates a plant cell's shape.

That walls dictate cell shape and cell size can be shown experimentally if the cell wall is digested away: the resulting naked protoplast is always spherical, and has a variable volume governed purely by the osmotic pressure of the medium in which it is bathed.

Different cell-types in plant organs differ drastically in wall thickness and in the visible patterns of any wall thickenings, correlating with biological function, as seen for instance in meristematic and parenchymatous cells (thin walls; rapid growth), collenchyma, sclerenchyma and cork (thick walls; physical protection), xylem vessel elements (rings, helices and scales of thickening; conduction of water under tension), epidermis (thick on the atmospheric side; limiting stem and leaf growth and preventing water loss), and cotyledons (in some species the walls being thickened with specific polysaccharides as food reserves).

As the outermost layer of the cell, the wall is in a position to provide physical defence against mechanical damage and pathogen attack. The cell wall and middle lamella also implement the softening which is characteristic of ripening in some fruits and of abscission of some leaves. Finally, cell wall components serve as locked-up precursors of hormone-like signalling molecules known as oligosaccharins. Plant cell walls are thus biologically interesting structures, and there are numerous reasons for wanting to understand their composition, synthesis and post-synthetic behaviour *in vivo*.

4.2 Polysaccharide Composition of Primary and Secondary Cell Walls

4.2.1 Typical Dicots

The plant cell wall is sometimes misleadingly called the 'cellulose wall', but this is a gross over-simplification. More than half the dry mass of most plant cell walls is contributed by non-cellulosic polymers, collectively called the wall *matrix*, and it should be remembered that cell walls are often about 70% water. Wall polysaccharides are of three major classes: pectins, hemicelluloses and cellulose, whose extraction and physical properties are summarised in Figure 4.1. The proposed biological roles of pectins, hemicelluloses and cellulose, and their overall sugar building-blocks, are summarised in Figure 4.2. The principal interconnections (glycosidic bonds) between the monosaccharide residues are shown in Figure 4.3. Details of the chemical composition of wall polysaccharides are given in Table 4.1. The major polysaccharides will first be described for dicots, then the differences seen in other plant taxa will be outlined.

Pectins, which are largely confined to primary walls, are best defined as polysaccharides rich in α-GalA residues.[1] They usually also contain α-Rha, α-Ara and β-Gal and

---

1 Sugar abbreviations: Api, D-apiose; Ara, L-arabinose; Fuc, L-fucose; Gal, D-galactose; GalA, D-galacturonic acid; Glc, D-glucose; GlcA, D-glucuronic acid; Man, D-mannose; Rha, L-rhamnose; Xyl, D-xylose. All sugars are in the 6-membered pyranose ring-form unless marked '*f*' for furanose.

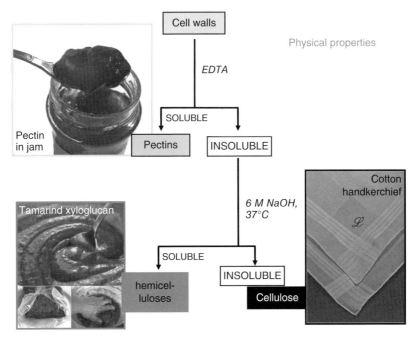

Figure 4.1 Fractionation of the cell wall into three major polysaccharide classes, and their physical properties.

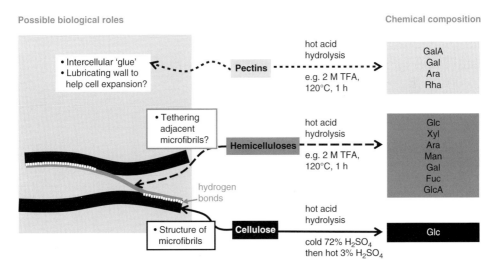

Figure 4.2 Principal biological roles and monosaccharide building blocks of three major classes of cell-wall polysaccharides.

numerous minor sugar residues. Pectins can be partially extracted from the cell wall at room temperature with 'Ca^{2+}-stripping' chelating agents (*e.g.* EDTA). They are more completely extracted by heating, for example in aqueous ammonium oxalate, pH 4–5, which however partially cleaves the polymer backbone. Pectins are built of interconnected 'domains', including some or all of homogalacturonan, rhamnogalacturonan

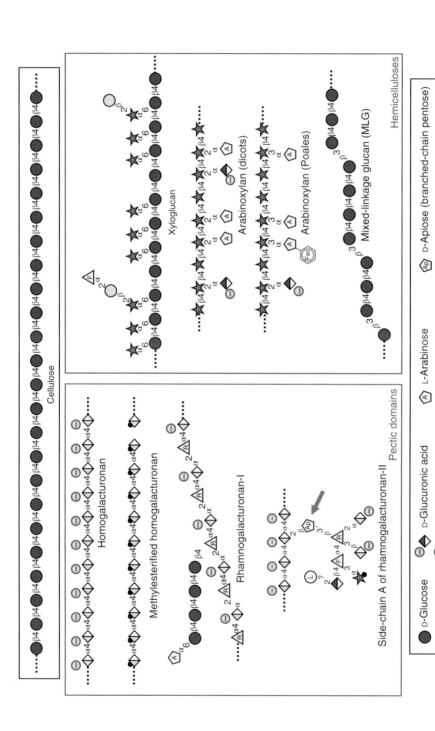

Figure 4.3 Cartoons of representative segments of the major wall polysaccharides. The Greek letters (α or β) indicate the anomeric configuration of the sugar residue; the number represents the position on the adjacent sugar residue to which a glycosidic bond has been made. Fer = ferulic acid residue; – in circle = negative charge; arrow = apiose residue capable of forming a borate bridge; ······ = continuation of polysaccharide backbone. From Fry *et al.* (2011a).

Table 4.1 Major polysaccharides of 'the' land-plant cell wall. For further details, see Fry (2011b).

Polysaccharide	%[*]	Net charge	Principal residue(s) of backbone	Linkage(s) in backbone	Main residues present in side-chains	Main linkages between side-chain and backbone	Other notes
Microfibrils							
cellulose	30	0	βGlc	(1→4)	none	–	Water-insoluble. Hydrogen-bonded within microfibril
Wall matrix							
Pectic domains							
homogalacturonan domain	15	–	αGalA	(1→4)	none	–	Some GalA residues Me-esterified, some 2- and/or 3-O-acetylated, some both
rhamnogalacturonan-I domain	10	–	αGalA, αRha	[-GalA-(1→2)-Rha-(1→4)-]	βGal, αAra, > αFuc, βXyl, βGlcA	Gal-(1→4)-Rha, Ara-(1→4)-Rha	Some GalA residues 2- or 3-O-acetylated, or 2,3-di-O-acetylated. Side-chains rich in Gal-(1→4)-Gal and Ara-(1→5)-Ara linkages
rhamnogalacturonan-II domain	1–4	–	αGalA (no Me-esters)	(1→4)	αAceA, βApi, αAra, αArap, βAra, βDha, αFuc, βGal, L-Gal, αGalA, βGalA, βGlcA, α(?)Kdo, αMeFuc, αMeXyl, αRha, βRha	Api-(1→2)-GalA, Kdo-(1→3)-GalA, Dha-(2→3)-GalA, Ara-(1→3)-GalA	Exceedingly complex structure; five main types of side-chain. MeFuc and AceA are O-acetylated. One Api residue can stably esterify with borate, cross-linking two RG-II molecules. EPG-resistant

(Continued)

Table 4.1 (Continued)

Polysaccharide	%[*]	Net charge	Principal residue(s) of backbone	Linkage(s) in backbone	Main residues present in side-chains	Main linkages between side-chain and backbone	Other notes
Hemicelluloses							
xyloglucans	20	0	βGlc	(1→4)	αXyl, βGal (± Ac ester), αFuc (± αAra, βAra, βXyl, α-L-Gal)	Major: αXyl-(1→6)-Glc, Minor: Ara-(1→2)-Glc, βXyl-(1→2)-Glc	See Table 4.2 for repeat units. Some 6AcGlc, 6AcGal and 5AcAra. The βGlc may be O-acetylated, especially in Poales
xylans, including heteroxylans	8	−	βXyl	(1→4)	αAra, αGlcA, (± βXyl, β-D-Gal, ?-L-Gal, …)	Ara-(1→2)-Xyl, Ara-(1→3)-Xyl, GlcA-(1→2)-Xyl	Some 2AcXyl or 3AcXyl. In Poales, some Fer-Ara and 2AcAra
mannans, including heteromannans	±	0	βMan, ± βGlc	(1→4)	± αGal	Gal-(1→6)-Man	Well known in secondary walls of xylem. Some Man residues 2- or 3-O-acetylated
mixed-linkage glucans (MLGs)	0	0	βGlc	(1→4), (1→3)	none	–	Only in Poales and *Equisetum*
callose	±	0	βGlc	(1→3)	none?	–	Mainly in wounded tissues, wall-regenerating protoplasts, and phloem sieve-tubes

[*]Typical % of dry weight in a dicot primary wall.
Abbreviations: See footnote 1: unless otherwise indicated, the sugar is the enantiomer (D- or L-) and ring-form (-*p* or -*f*) listed in the footnote. Ac = acetyl; Me = methyl. ± = May be absent.

(RG) I, RG-II and xylogalacturonan. The homogalacturonan domain (a linear chain of (1→4)-linked α-GalA residues) is partially methyl- and acetyl-esterified; after de-esterification it can be completely hydrolysed to mono- and oligosaccharides with endo-polygalacturonase; this treatment disconnects and releases the other pectic domains in a convenient form for detailed study.

Hemicelluloses are not solubilised from the wall by cold chelating agents, nor generally by heating at pH 4–5, but they are extracted by aqueous strong alkali (optimally 6 M NaOH at 37°C). The principal hemicelluloses of dicots are xyloglucans, xylans, mannans and glucomannans, with (1→4)-linked backbones of β-Glc, β-Xyl, β-Man and β-Glc + β-Man residues respectively. Xyloglucans and xylans predominate in primary and secondary walls respectively. There are, in addition, side-chains (*e.g.* α-Xyl, β-Gal and α-Fuc in the case of dicot xyloglucans); details are given in Table 4.1. Many hemicelluloses carry a small proportion of acetyl ester groups, which however are lost during extraction with NaOH. Hemicellulose names can be embellished, depending on how precisely one wishes to specify their side-chains, for example 'glucuronoarabino-xylans'. Details of hemicellulose structures are best studied after 'enzymic dissection' to yield repeating oligosaccharides; for example, xyloglucan is hydrolysed by XEG to give an informative range of mainly hepta- to decasaccharides (Figure 4.4), which can be purified chromatographically and structurally characterised by MS and NMR. Xyloglucan oligosaccharide sequences thus discovered are represented by a short-hand (Table 4.2); the major examples in most dicots are XXXG and XXFG. Some specialised cells have large amounts of a specific hemicellulose (*e.g.* xyloglucan in tamarind seed cells), which,

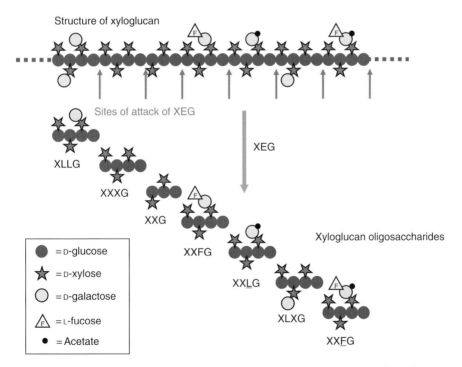

Figure 4.4 Enzymic dissection of xyloglucan to release repeat units that can be readily purified and characterised. XEG = xyloglucan endo-glucanase.

Table 4.2 Repeat units of xyloglucan. Code letters used for reporting xyloglucan sequences. Oligosaccharides are named by listing the code letters in sequence along the (1→4)-β-D-glucan backbone of xyloglucan, from non-reducing to reducing end. Square brackets indicate branching. From Franková and Fry (2012).

Code	Structure
A	α-D-Xyl-(1→6)-[α-L-Ara*f*-(1→2)]-β-D-Glc*p*[*]
B	α-D-Xyl-(1→6)-[β-D-Xyl-(1→2)]-β-D-Glc*p*[*]
C	α-D-Xyl-(1→6)-[α-L-Ara*f*-(1→3)-β-D-Xyl-(1→2)]-β-D-Glc*p*[*]
D	α-L-Ara*p*-(1→2)-α-D-Xyl-(1→6)-β-D-Glc*p*[*]
E	α-L-Fuc-(1→2)-α-L-Ara*p*-(1→2)-α-D-Xyl-(1→6)-β-D-Glc*p*[*]
F	α-L-Fuc-(1→2)-β-D-Gal-(1→2)-α-D-Xyl-(1→6)-β-D-Glc*p*[*]
F̲	α-L-Fuc-(1→2)-β-D-Gal-(1→2)-α-D-Xyl-(1→6)-β-D-Glc*p*[*] with acetate on the Gal
G	β-D-Glc*p*[*] with no side-chain attached
G̲	β-D-Glc*p*[*] with acetate on O-6
Gol	glucitol (the former reducing terminus after reduction, e.g. with NaBH₄)
J	α-L-Gal-(1→2)-β-D-Gal-(1→2)-α-D-Xyl-(1→6)-β-D-Glc*p*[*]
L	β-D-Gal-(1→2)-α-D-Xyl-(1→6)-β-D-Glc*p*[*]
L̲	β-D-Gal-(1→2)-α-D-Xyl-(1→6)-β-D-Glc*p*[*] with acetate on the Gal
M	α-L-Ara*p*-(1→2)-[β-D-Gal-(1→4)]-α-D-Xyl-(1→6)-β-D-Glc*p*[*]
N	α-L-Ara*p*-(1→2)-[β-D-Gal-(1→6)-β-D-Gal-(1→4)]-α-D-Xyl-(1→6)-β-D-Glc*p*[*]
P	β-D-GalA-(1→2)-[β-D-Gal-(1→4)]-α-D-Xyl-(1→6)-β-D-Glc*p*[*]
Q	β-D-Gal-(1→4)-β-D-GalA-(1→2)-[β-D-Gal-(1→4)]-α-D-Xyl-(1→6)-β-D-Glc*p*[*]
S	α-L-Ara*f*-(1→2)-α-D-Xyl-(1→6)-β-D-Glc*p*[*]
T	β-L-Ara*f*-(1→3)-α-L-Ara*f*-(1→2)-α-D-Xyl-(1→6)-β-D-Glc*p*[*]
U	β-D-Xyl-(1→2)-α-D-Xyl-(1→6)-β-D-Glc*p*[*]
V	α-D-Xyl-(1→4)-α-D-Xyl-(1→6)-β-D-Glc*p*[*]
X	α-D-Xyl-(1→6)-β-D-Glc*p*[*] (= isoprimeverose)

[*] The β-D-Glc marked with an asterisk in each structure is part of the (1→4)-β-glucan backbone of the xyloglucan.

not being firmly bound to cellulose, can be extracted by heating at neutral pH. Once extracted from the wall, many hemicelluloses (like pectins) remain soluble when neutralised and dialysed against pure water.

Cellulose is a rigid, unbranched polymer of (1→4)-linked β-Glc residues. Cellulose molecules are bundled into microfibrils, the 'scaffolding' of both primary and secondary walls. It is insoluble, even in 6 M alkali.

Dicot cells with predominantly *secondary* walls have an overall polysaccharide composition (Northcote, 1972) in which the primary wall contribution (thus pectins and xyloglucan) is negligible. For example, dicot xylem (hardwood) typically comprises (dry weight basis) 33–50% cellulose, 20–30% methylglucuronoxylan and 0.5–5% glucomannan. Cork cell-wall polysaccharides are basically similar. Cotton epidermal hair ('fibre') secondary walls are almost pure cellulose.

For documenting the sugar building blocks of polysaccharides, acid hydrolysis followed by chromatography is often used. For pectins and hemicelluloses, 2 M trifluoroacetic acid (TFA; selected because it is volatile and thus easily dried off after hydrolysis) at 120°C for 1 h is usually satisfactory. However, owing to its crystallinity, microfibrillar cellulose must be dispersed in cold concentrated acid (*e.g.* 72% H_2SO_4) before it can be 'Saeman' hydrolysed (in hot 3% H_2SO_4). A useful alternative to acid hydrolysis is digestion with 'Driselase', a fungal enzyme preparation that digests most of the glycosidic bonds in plant cell-wall polysaccharides (including microfibrillar cellulose), but with two useful exceptions: α-Xyl residues in xyloglucan (which thus yields the diagnostic disaccharide, α-Xyl-(1→6)-Glc, known as isoprimeverose) and the whole of RG-II (which is thus recovered essentially intact).

4.2.2 Differences in Certain Dicots

Some orders of dicots exhibit subtle differences in their polysaccharides. For example, in the Solanales (potato, tomato, tobacco, etc.), xyloglucan carries very few α-Fuc but numerous α-Ara residues instead. In the Solanales and some members of the Lamiales, the usual xyloglucan building plan, based on XXXG (*e.g.* XXXG itself, XXFG, XLLG, etc.), is replaced by one based on XXGG (XSGG, LLGG, etc.) and XXGGG (XSGGG, SXGGG, etc.).

In the Caryophyllales (*e.g.* spinach, beet, pearlwort, etc.), pectins carry feruloyl and 4-coumaroyl groups esterified to some of the α-Ara and β-Gal residues, predominantly at the non-reducing termini of the RG-I domains. Diagnostic oligosaccharides (Fer-Gal-Gal and Fer-Ara-Ara) are released from these by digestion with 'Driselase' (which lacks feruloylesterase activity).

4.2.3 Differences in Monocots

Most monocots have cell-wall polysaccharide compositions similar to those of dicots, although aquatic monocots of the order Alismatales (*e.g.* duckweeds) possess a unique pectic domain, apiogalacturonan, containing highly acid-labile Api residues.

However, a monocot order of great economic importance, the Poales (including grasses, cereals, reeds and related plants), have some major differences from the dicots. The primary walls of most poalean cells (but not usually cell suspension-cultures) have a relatively low content of xyloglucan and of the pectic domains, and proportionately more xylans. Furthermore, poalean xyloglucan has a very low (though not zero) α-Fuc content, and poalean xylans are highly substituted with esterified ferulic and *p*-coumaric acid groups. And the Poales possess an additional hemicellulose, mixed-linkage glucan (MLG), which is completely absent from the dicots. MLG has an unbranched backbone of β-Glc residues, most of the glycosidic bonds being (1→4), as in cellulose; however, roughly 30% are (1→3), the latter introducing flexibility into the polysaccharide chain so that it is water-soluble (once extracted from the cell wall with alkali). MLG is particularly abundant in the walls of young, rapidly expanding poalean cells, and it is partially broken down once the cell has reached its full size. As with xyloglucan in tamarind seeds, hot-water-extractable MLG is an abundant reserve polysaccharide in cereal grains. MLG is readily analysed by use of 'lichenase', which specifically recognises and cleaves a (1→4)-bond immediately following a (1→3) bond, thus giving a series of diagnostic oligosaccharides, predominantly the trisaccharide, β-Glc-(1→3)-β-Glc-(1→4)-Glc (abbreviated as G3G4G).

4.2.4 Differences in Gymnosperms

Gymnosperms (Figure 4.5) are non-flowering seed plants, for example conifers. The gymnosperms studied to date have primary wall polysaccharides similar to those of dicots, including xyloglucan with the repeat-units such as XXXG, XXFG and XLFG. Conifers and cycads tend to be richer in mannans, though gnetophytes (possibly the gymnosperms closest related to angiosperms) seem to buck this trend. The secondary walls of conifer xylem (softwood) show some differences from those of dicots (hardwood); for example, conifer (galacto)-glucomannans are acetylated but the xylans are not. Larchwood contains water-soluble (thus not strictly cell-wall) arabinogalactans.

4.2.5 Differences in Non-seed Land-plants

Ferns and their allies (= monilophytes), lycophytes and bryophytes – being of less economic significance – have been less intensively studied than seed plants, though this situation is beginning to be rectified as the evolutionary significance of such knowledge comes to be recognised (Figure 4.5). Like dicots, all non-seed land-plants possess xyloglucan, xylans, mannans, pectins and cellulose, though in different quantities. For example, bryophytes are poor in xylans. There are also qualitative differences: for example, moss xyloglucan is unusual in being rich in acidic GalA residues. *Equisetum* xyloglucan also has unusual sequences. Nevertheless, the hornworts, which like mosses are bryophytes, have remarkably dicot-like xyloglucan, rich in XXFG repeats.

Monilophytes possess the complex pectic domain RG-II, though sometimes with MeRha in place of Rha. Bryophytes (liverworts, mosses and hornworts) possess small but measureable amounts of RG-II. Acid hydrolysis of total walls yields large amounts of MeRha in the case of bryophytes and homosporous lycophytes (*e.g. Lycopodium*), but not heterosporous lycophytes (*Selaginella*), monilophytes or seed-plants. Some of this MeRha is a component of RG-II, though the majority must arise from a more abundant, unidentified polysaccharide. In contrast, MeGal is plentiful in all lycophytes, but not in the bryophytes, monilophytes or seed-plants.

Compared with seed plants, many non-seed land plants are rich in mannose – very rich in *Psilotum* and *Equisetum*, less so in leptosporangiate ferns.

4.2.6 Differences in Charophytes

Charophytic algae (Figure 4.5) and land plants are closely related, together constituting the Streptophyta, well resolved from other green algae, the Chlorophyta. Charophytes share with land plants the possession of cellulose and probably mannans. However, there are few other polysaccharides that can be confidently said to be common to *all* charophytes. Pectic α-GalA residues are present in three charophytic orders (Coleochaetales, Charales and Zygnematales), but probably not in the Klebsormidiales or Chlorokybales. Rha residues also occur, suggesting but not proving the existence of rhamnogalacturonans.

There is immunological evidence for a xyloglucan-like polymer in the Charales, but chemical analyses have so far failed to confirm the existence of true xyloglucan (with isoprimeverose residues) in charophytes. Acid hydrolysis shows that *Coleochaete* walls contain very little xylose at all, indicating that xyloglucan, if any, must be quantitatively minor. There is preliminary evidence for MLG in some charophytes, for example the Zygnematales. However, lichenase digestion failed to yield oligosaccharides from *Chara*

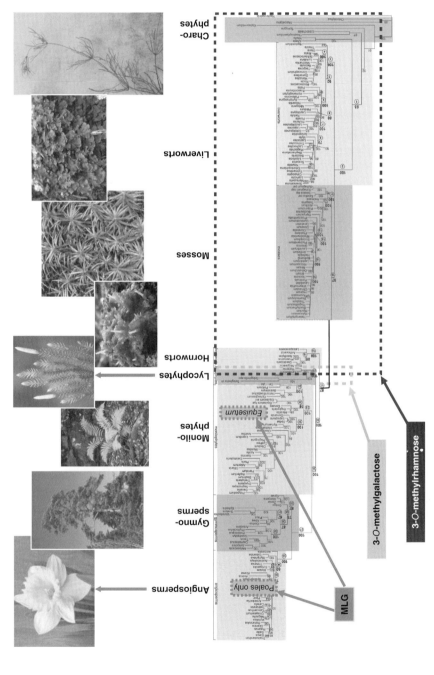

Figure 4.5 Overview phylogenetic tree of the streptophytes (land plants + charophytes). The restricted taxonomic distribution of selected wall components is highlighted: MLG only in the Poales and *Equisetum*; 3-*O*-methylrhamnose in the charophytes, bryophytes and homosporous lycophytes; 3-*O*-methylgalactose in the homosporous and heterosporous lycophytes. Phylogram from Qiu, *et al.* (2007).

or *Coleochaete* walls, indicating undetectable MLG. Much remains to be discovered about charophytic cell walls – information that would promote our understanding of the evolutionary foundations of the land-plant wall.

4.3 Post-synthetic Modification of Cell-wall Polysaccharides

4.3.1 Cross-linking of Cell-wall Polysaccharides

Polysaccharides in a plant cell wall are generally stable, structural molecules; however, they do participate in interesting and important metabolic reactions *in vivo*. Such reactions include those that bring about cross-linking (Figure 4.6). Given that most matrix polysaccharides (pectin and hemicelluloses) remain soluble in pure water once extracted from the wall with chelating agents or alkali, it follows that they must have been cross-linked to each other within the architecture of the intact wall. Information on cross-linking is valuable in our quest to understand how the wall is built, and how it may be modified *in vivo* to permit cell growth and post-harvest for biomass exploitation.

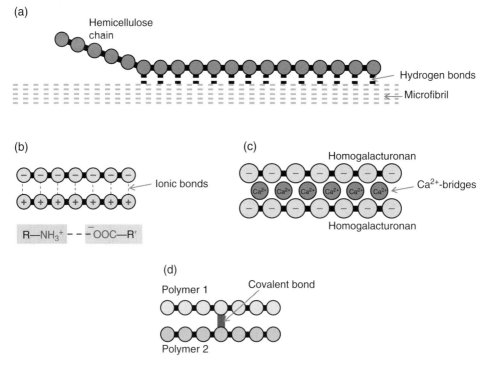

Figure 4.6 Covalent *vs* non-covalent cross-links in the plant cell wall. Three types of non-covalent cross-link are: (a) hydrogen bonds, *e.g.* between a hemicellulose (*e.g.* xyloglucan) and a cellulosic microfibril; (b) ionic bonds between an anionic polymer (*e.g.* pectin) and a cationic polymer (*e.g.* extensin); (c) calcium-bridges between two non-esterified homogalacturonan domains. Non-covalent bonds are individually weak but constitute strong cross-links owing to their large number. Covalent cross-links (d) are strong but few.

The term 'cross-link' in this chapter implies an individual chemical bond (non-covalent, Figure 4.6a–c; or covalent, Figure 4.6d) that joins two otherwise separate polymers; it is not a whole molecular chain that may inter-connect two other structures (*e.g.* a xyloglucan chain tethering two microfibrils; Figure 4.2), which would be the usage implied when the term 'cross-linking glycans' is employed for what are more generally called hemicelluloses.

After being secreted by the protoplast, a new polymer molecule may quickly become non-covalently cross-linked to existing wall material. For example, a newly secreted hemicellulose chain may *hydrogen-bond* to exposed surfaces of cellulosic microfibrils (Figure 4.6a). Such hemicellulose–cellulose bonding occurs without the help of enzymes, as one can demonstrate *in vitro* simply by adding a hemicellulose solution to pure cellulose. Although a single H-bond is weak and transient, the numerous H-bonds that form between two large polysaccharide chains add up to a very firm attachment (Figure 4.2), as in *Velcro*®. In this way, hemicellulose chains, especially xyloglucans, probably tether adjacent microfibrils; this is feasible because a xyloglucan molecule (chain length ~300 nm) is much longer than the spacing (~30 nm) between adjacent microfibrils. This model gives hemicelluloses a key position in wall architecture.

Charged wall polymers can form *ionic* bonds with those of the opposite charge (Figure 4.6b). For example, acidic pectins will ionically cross-link to extensins (cationic glycoproteins; see later). In addition, two negatively charged pectin domains (specifically, non-methylesterified homogalacturonans) can be cross-linked non-enzymically to each other *via* the divalent inorganic cation, Ca^{2+}, in an 'egg-box' arrangement (Figure 4.6c).

Besides these non-covalent cross-links, various *covalent* types have been investigated (Figure 4.6d). In contrast to non-covalent cross-links, which are weak but many, covalent ones are strong but few. Even a single covalent cross-link can securely interconnect two huge polymer chains at least as strongly as any of the glycosidic linkages within the polysaccharides' backbones. For example, the ferulate side-chains present on some arabinoxylans (especially in the Poales, grasses) and rhamnogalacturonans (in the Caryophyllales) can undergo oxidative phenolic coupling to form stable dimers known as diferulates (strictly, dehydrodiferulates) (Figure 4.7). Diferulate bonds may well covalently cross-link wall polysaccharides, although it is difficult to exclude the possibility that they form intra-chain loops. The enzymes catalysing oxidative phenolic cross-linking are often presumed to be peroxidases (with H_2O_2 as electron acceptor), though phenol oxidases ('laccases'; with O_2 as electron acceptor) may also contribute. *In-vivo* radio-labelling experiments in cultured maize cells show that feruloyl coupling is initiated within the protoplast (probably in the Golgi bodies or vesicles), and continues within the cell wall after polysaccharide secretion.

Tyrosine residues of certain cell-wall glycoproteins, notably extensins (see later) undergo similar oxidative coupling to form the dimer, isodityrosine, which may itself then dimerise to the tetramer, di-isodityrosine (Figure 4.8).

Other proposed covalent cross-links that could be formed by reactions occurring in the cell wall itself include ester or amide bonds between the –COOH groups of α-GalA residues in pectins and the –OH or $–NH_2$ groups of other polymers (Figure 4.9). Such *O*- or *N*-galacturonoyl linkages, however, remain to be definitively demonstrated *in vivo*. Theoretically, they could be formed by the wall-localised enzyme, pectin methylesterase (PME), acting in 'transacylase' mode.

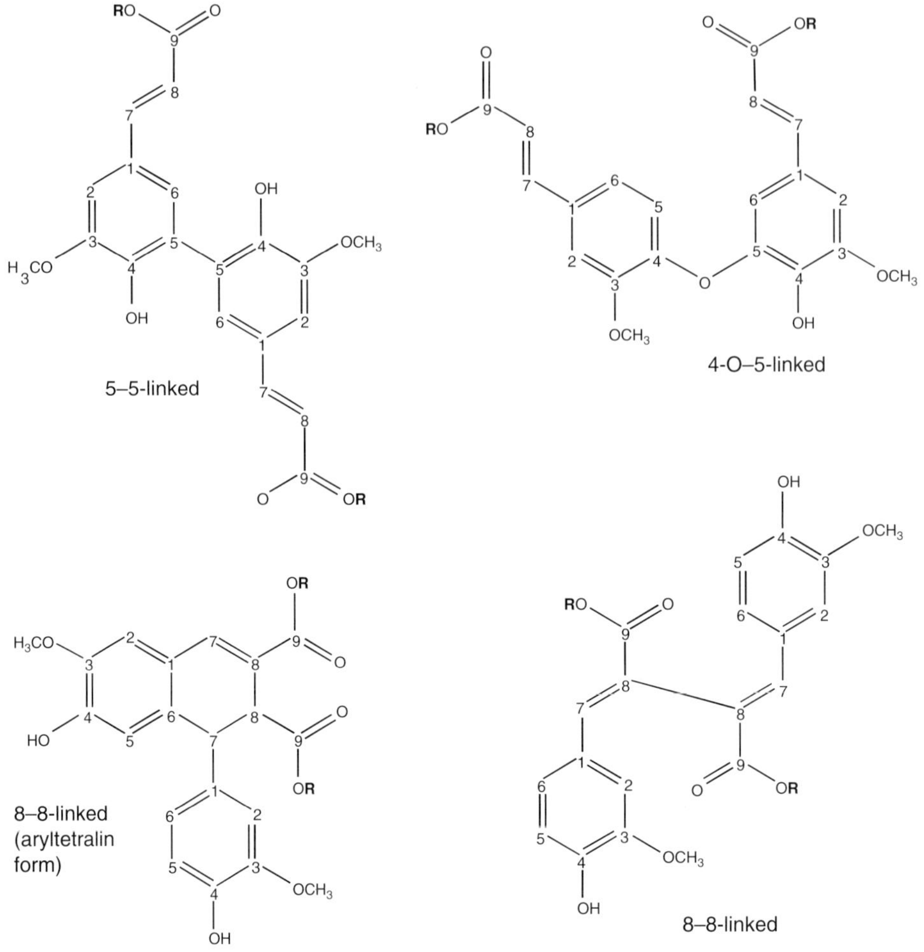

Figure 4.7 Four of the possible diferulates (dehydrodiferulates) found esterified to plant cell-wall polysaccharides. **R** represents a long polysaccharide chain.

4.3.2 Hydrolysis of Cell-wall Polysaccharides

Plant cell walls contain numerous enzyme activities that, theoretically, could cut cell-wall polysaccharide chains by hydrolysis. This statement is particularly well documented in primary cell walls, where polysaccharide cleavage might be expected to usefully 'loosen' the cell wall, enhancing cell expansion or fruit softening. Such enzymes are classified into two broad types: glycanases, which are endo-hydrolases, cleaving the backbone of the polysaccharide somewhere in mid-chain:

●●●●●●●●●●●●●●●●●● + H₂O → ●●●●●●●●●●●● + ●●●●●,

and glycosidases, which are exo-hydrolases, progressively releasing monosaccharides by nibbling from the non-reducing end of the polysaccharide:

■●●●●●●●●●●●●●●●● + H₂O → ■ + ●●●●●●●●●●●●●●●●.

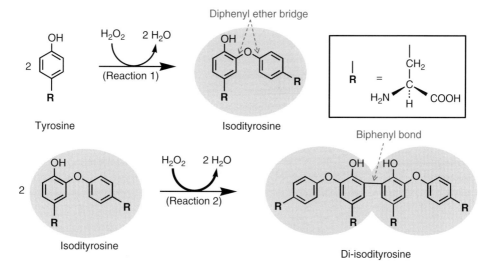

Figure 4.8 Oxidative coupling of tyrosine residues, as demonstrated in the cell-wall glycoprotein, extensin. Reaction 1 may produce an inter- or intra-polypeptide bridge; steric considerations show that Reaction 2 can only produce an *inter*-polypeptide cross-link.

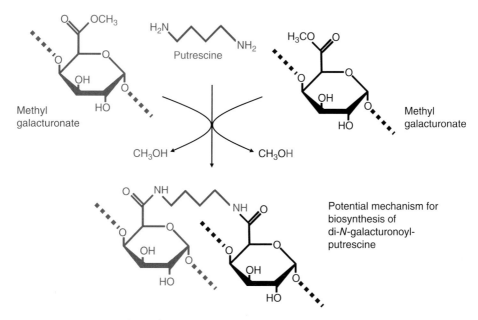

Figure 4.9 The proposed formation of amide (isopeptide) cross-links between pectic homogalacturonan domains via a bifunctional amine such as putrescine. Theoretically this reaction could be catalysed by a PME acting in transacylation mode. The by-product, when PME catalyses either hydrolysis or transacylation, is methanol.

It is easy to see how glycanases could serve a role in cell-wall loosening, since they may cleave an architecturally significant polysaccharide, such as a tethering hemicellulose, by a single cut, thus locally untethering two microfibrils. It is more difficult to rationalise the biological significance of the apparently innocuous exo-attack by glycosidases (but see next section discussing transglycosylation).

Table 4.3 Major cell wall-localised glycanase activities found in plants.

Enzyme activity	Mid-chain bond cleaved (underlined)	Polysaccharide whose backbone could potentially be hydrolysed
(1→4)-α-Galacturonanase [pectinase, endo-polygalacturonase]	... (1→4)-α-GalA-(1→4)-α-GalA-(1→4)...	pectic homogalacturonan (non-methyl-esterified regions)
(1→4)-β-Glucanase [cellulase]	... (1→4)-β-Glc-(1→4)-β-Glc-(1→4)...	cellulose, MLG; sometimes xyloglucan
Xyloglucan endo-hydrolase (XEH)	... (1→4)-β-Glc-(1→4)-β-Glc-(1→4)... [with α-Xyl on O-6 of 2nd Glc]	xyloglucans
(1→3)-β-Glucanase [laminarinase]	... (1→3)-β-Glc-(1→3)-β-Glc-(1→3)...	callose (laminarin), glucans of invading fungi
(1→4)-β-Mannanase	... (1→4)-β-Man-(1→4)-β-Man-(1→4)...	mannans (including galactoglucomannans, etc.)
(1→4)-β-Xylanase	... (1→4)-β-Xyl-(1→4)-β-Xyl-(1→4)...	xylans (including glucuronoarabinoxylans, etc.)
Chitinase	... (1→4)-β-GlcNAc-(1→4)-β-GlcNAc-(1→4)...	chitin of invading fungi

Cell-wall-localised glycanases, and the polysaccharides whose backbones they may cleave, are listed in Table 4.3. The list covers most of the important structural polysaccharides of the plant cell wall, possibly the only significant exceptions being the rhamnogalacturonans. This is not to say that the plant can totally lyse its cell walls; indeed, there is relatively little experimental evidence that the wall enzymes' activities routinely exert any real action *in vivo*. Nevertheless, the existence of such enzyme activities strongly suggests that they evolved because of a biological function that they can fulfil. Among the clearest evidence for glycanase action *in vivo* is the partial disappearance of MLG from cereal tissues that have recently completed their period of cell expansion. In contrast to this, *Equisetum* MLG does not undergo turnover, but persists and even increases in quantity during ageing, suggesting that MLG plays fundamentally different roles in cereal and horsetail cell walls.

Cell-wall-localised glycosidases, and the polysaccharides whose termini they may 'trim', are listed in Table 4.4. Again, the list covers many of the known wall polysaccharides. Note that some glycosidases can only possibly act after another glycosidase has acted first: for example, β-glucosidase can act on a xyloglucan chain only after α-xylosidase has removed a terminal xylose residue, thereby exposing a glucose residue as the new terminus.

Other hydrolytic enzyme activities in the plant cell wall are esterases, the best studied of which is PME. *Arabidopsis* has at least 79 genes encoding putative PMEs, although

Table 4.4 Major cell wall-localised glycosidase activities found in plants.

Enzyme activity	Terminal bond cleaved	Polysaccharides potentially 'pruned' by the enzyme
α-Arabinosidase	α-Araf-(1→...	RG-I domains of pectin, arabinogalactans, (glucurono)arabinoxylans
α-Fucosidase	α-Fuc-(1→...	xyloglucan (near the polysaccharide's non-reducing chain end)
α-Galactosidase	α-Gal-(1→...	galacto(gluco)mannans
β-Galactosidase	β-Gal-(1→...	RG-I domains of pectin, arabinogalactans
β-Galactosidase	β-Gal-(1→...	xyloglucan
α-Galacturonidase [exo-polygalacturonase]	α-GalA-(1→...	homogalacturonan domains of pectin
β-Glucosidase	β-Glc-(1→...	cellulose, MLG, xyloglucan, callose
β-Mannosidase	β-Man(1→...	(galacto)(gluco)mannans
α-Xylosidase	α-Xyl-(1→...	xyloglucan (at the polysaccharide's non-reducing chain end)
β-Xylosidase	β-Xyl-(1→...	(glucurono)(arabino)xylans

few of these have so far been tested for actual enzyme activity. PME attacks the methyl-esterified regions of pectic homogalacturonan domains, usually acting progressively so that a lengthy sequence of (neutral) GalA methyl ester residues is converted into a run of (anionic) GalA residues. This may have two significant consequences:

1) enabling homogalacturonan domains to become cross-linked to each other *via* Ca^{2+} bridges; and
2) rendering a zone of the homogalacturonan chain susceptible to endo-hydrolysis by the relevant glycanase (endopolygalacturonase).

Thus, PME's biological role is enigmatic: does it initiate wall tightening by enabling Ca^{2+} bridging, or does it initiate loosening by facilitating chain cleavage?

Other wall esterases also exist, whose roles may be to remove acetyl or feruloyl ester groups. Wall proteases have also been detected, but their action on wall (glyco)proteins appears to be uncertain.

4.3.3 'Cutting and Pasting' (Transglycosylation) of Cell-wall Polysaccharide Chains

Plant cell-wall polysaccharides are attacked not only by hydrolases but also by enzyme activities that 'cut and paste' polysaccharide chains. The enzyme cleaves one polysaccharide molecule (the donor substrate), and then transfers a section thereof on to a second polysaccharide molecule (the acceptor substrate). By analogy with glycanases and glycosidases, such enzymes can conveniently be classed as transglycanases and transglycosidases – cleaving the donor substrate in mid-chain or at a non-reducing terminus respectively, and thus transferring a poly- or monosaccharide group as the case may be:

transglycanase

$$●●●●●●●●●●●● + ■■■■■■■ → ●●●●●●●●●■■■■■■■ + ●●●●$$

transglycosidase

$$●●●●●●●●●●●●● + ■■■■■■■ → ●■■■■■■■ + ●●●●●●●●●●●●●$$

Best characterised are the transglycanase (endotransglycosylase) activities, especially one specific example, xyloglucan endotransglucosylase (XET). XET is one activity of a class of proteins collectively known as XTHs (xyloglucan endotransglucosylase/hydrolases). *Arabidopsis thaliana* has genes for 33 XTHs, 31 of which possess essentially only XET activity, and two of which (XTH31 and XTH32) appear to be very predominantly XEH-active enzymes (Table 4.3). The expression of XTHs often peaks during episodes of rapid cell expansion, suggesting that they act to loosen the cell wall and/or help to construct the architecture of new wall material. Clear evidence exists from *in-vivo* labelling experiments using the simultaneous application of 'heavy' and 'hot' isotopes (*e.g.* ^{13}C and ^{3}H, respectively) to show that xyloglucan does indeed undergo transglycosylation in the walls of living plant cells.

More recently, it has been established that some plants possess trans-β-mannanase activity ('mannan endotransglycosylase'; cutting and pasting (galacto)(gluco)mannans, *e.g.* in tomato fruit), and trans-β-xylanase activity ('xylan endotransglycosylase'; acting on (glucurono)(arabino)xylans, in many plant tissues, including non-seed land-plants). These activities, probably attributable to mannanases and xylanases respectively, may also reversibly remodel the plant cell wall as proposed for XET activity.

The transglycanases mentioned so far are all 'homo-transglycanases', in which the donor and acceptor substrates are qualitatively similar to each other. In addition, however, 'hetero-transglycanase' activities exist, especially MXE (= MLG:xyloglucan endotransglucosylase), which among land plants appears to be essentially confined to one genus, *Equisetum*. Its presumed role is to graft segments of MLG to segments of xyloglucan, and since its activity peaks in ageing *Equisetum* stems, its significance may be to strengthen wall architecture in mature plants. This would contrast with XET activity which in *Equisetum*, as in most plants, peaks during phases of rapid cell expansion.

Corresponding to the 'nibbling hydrolases' (glycosidases), there are trans-glycosidase activities, including trans-α-xylosidase and trans-β-xylosidase, which, by acting on xyloglucans and xylans respectively, transfer xylose residues, one at a time, from a donor to an acceptor substrate. These may be activities of α-xylosidase and β-xylosidase proteins with an ability to catalyse transglycosylation as well as hydrolysis. However, the transglycosylation:hydrolysis ratio varies with the enzyme source, assayed under identical conditions, indicating that some are better than others at 'cutting and pasting' their substrates rather than just cutting them.

It has been proposed, but not yet definitively established, that plant cell walls possess trans-acylase as well as transglycanase and transglycosidase activities. Potential targets for trans-acylases include pectins and cutin. Acting on methylesterified homogalacturonan domains of pectin, for example, some PMEs might potentially transfer a GalA residue from methanol (*i.e.* the methyl ester) either onto an –OH group of some other polysaccharide (forming an *O*-galacturonoyl ester cross-link) or onto an –NH$_2$ group of a protein or polyamine (forming an *N*-galacturonoyl amide (isopeptide) cross-link) (Figure 4.9).

The possible 'cutting and pasting' of cutin is discussed later.

4.4 Polysaccharide Biosynthesis

4.4.1 General Features

Plant cell wall polysaccharides are synthesised from 'high-energy' donor substrates (nucleoside diphosphate sugars; NDP-sugars) in nucleophilic substitution (S$_N$2) reactions. The basic synthetic reaction is:

- *glycosyltransferase* or *polysaccharide synthase*
- NDP-sugar + (sugar)$_n$ → NDP + (sugar)$_{n+1}$

The NDP-sugar has an electrophilic centre (carbon atom no. 1 of the sugar), which is transferred to a nucleophilic –OH group in a nascent polysaccharide chain; the NDP moiety is a good leaving group. The enzymes catalysing the reactions are generally called polysaccharide synthases if they act 'processively' (*i.e.* manufacturing the polysaccharide's lengthy backbone by repeatedly adding identical sugar residues one-by-one without the enzyme detaching from the growing chain), and glycosyltransferases if they non-processively add side-chain sugar residues. Important NDP-sugars include UDP-α-D-Glc, UDP-α-D-Gal, UDP-α-D-GlcA, UDP-α-D-GalA, UDP-α-D-Xyl, UDP-β-L-Ara*f*, UDP-β-L-Rha, UDP-α-D-Api*f*, GDP-α-D-Man, GDP-β-L-Fuc and GDP-β-L-Gal. The enzymes add new sugar residues to the growing polysaccharide chains in the form of new non-reducing termini – described as 'tailward growth'. Some of the enzymes are 'retaining' transferases: for example, the xylosyltransferases involved in xyloglucan biosynthesis employ UDP-α-D-Xyl as donor and add a new α-D-Xyl residue as a side-chain to the nascent polysaccharide. Others are 'inverting': xylan synthase likewise takes UDP-α-D-Xyl as donor but adds a new β-D-Xyl residue to the growing polysaccharide xylan backbone. All the enzymes of wall polysaccharide biosynthesis are membrane-associated.

4.4.2 At the Plasma Membrane

Only two cell-wall polysaccharides are well established to be produced by synthases at the plasma membrane: cellulose and callose. MLG was also recently suggested to be produced at the plasma membrane, but the balance of evidence suggests that this is incorrect. Cellulose is manufactured by terminal complexes ('rosettes', visible by electron microscopy in the plasma membrane), each rosette comprising 36 CESA protein molecules in the form of six hexamers. CESAs are integral membrane proteins which take cytosolic UDP-Glc and generate extraprotoplasmic cellulose. *Arabidopsis* has ten *CESA* genes, of which *CESA1, 3* and *6* are mainly expressed during primary wall formation, and *CESA4, 7* and *8* during secondary. The 36 protein molecules in a rosette simultaneously spin out one microfibril, probably usually composed of 18 parallel cellulose chains. *In vivo*, the approximately 18 new chains almost instantly crystallise to form a microfibril, visible by EM as a strand emanating from the terminal complex. *cesa1* mutants are compromised in this crystallisation step. Since microfibrils, once formed, are firmly embedded in the wall, the rosette is pushed around within the fluid mosaic plasma membrane by the elongating nascent microfibril. The direction of the rosettes' movement, and hence the orientation in which new microfibrils come to lie on the inner face of the wall, is channelled by microtubules closely underlying the membrane.

4.4.3 In the Golgi System

All pectins and hemicelluloses, except callose, are synthesised in the Golgi body. Some of the polysaccharide synthases that manufacture their backbones are processive and inverting, thus resembling cellulose synthases, and are called CSL (cellulose synthase-like) proteins. Like the CESAs, they belong to the GT2 (glycosyltransferase 2; Cantarel *et al.*, 2009) family; they fall into eight sub-families (CSLA to H). Like CESAs, the CSLs have multiple transmembrane domains and are inverting transferases (*e.g.* donor = GDP-α-D-Man; product = $(β$-D-Man$)_n$). Some of them (*e.g.* CSLAs) have the NDP-sugar binding site and the catalytic centre on the cytosolic side of the Golgi membrane, others (*e.g.* CSLBs) on the Golgi lumen side; both deliver their polysaccharide product into the Golgi lumen. Work is in progress to match the eight CSL families to the various polysaccharide backbones that they synthesise. Evidence available in May 2013 suggested the following roles, in which the enzymes principally synthesise $β$-$(1{\rightarrow}4)$-bonds:

1) CSLA → mannans
2) CSLB → ? (absent in Poales)
3) CSLC → backbone of xyloglucan?
4) CSLD → ?
5) CSLE → ?
6) CSLF → MLG? (only in Poales?)
7) CSLG → ? (absent in Poales)
8) CSLH → MLG? (only in Poales?)

Other polysaccharide backbones are produced by polysaccharide synthases that are *not* CSLs. For example, the $β$-$(1{\rightarrow}3)$-Glc backbone of callose appears to be made by GT48 enzymes, which resemble GT2 enzymes but evolved independently. Homogalacturonan synthases (members of family GT8) differ from the CSLs and GT48 in having only a single transmembrane domain and in being retaining (donor = UDP-α-D-GalA; product = $(α$-D-GalA$)_n$).

On the backbones of most pectic domains and hemicelluloses, additional sugar residues become attached as side-chains. The side-chains are added by non-processive glycosyltransferases. For example, during xyloglucan synthesis, α-Xyl side-chains are added to $β$-Glc residues of the backbone by GT34 enzymes. Only those $β$-Glc residues that have themselves very recently been incorporated into the nascent polymer's backbone can be xylosylated; equally, further elongation of the xyloglucan backbone by addition of new Glc residues appears to depend on successful α-xylosylation of existing Glc residues at or near the non-reducing end. Similar conclusions have been reached in studies of galactomannan biosynthesis, in which α-Gal residues are added only near the growing non-reducing terminus of the $β$-mannan backbone, also by GT34 enzymes. In contrast, α-Fuc side-chains appear to be added along the length of a pre-formed xyloglucan core, not only at the non-reducing terminus, by GT37 enzymes.

The non-sugar groups of wall polysaccharides (methyl, acetyl and feruloyl esters, and methyl ethers) are also added to polysaccharide chains within Golgi cisternae. The donor substrates for methyl, acetyl and feruloyl residues are *S*-adenosyl-methionine, acetyl-coenzyme-A and probably feruloyl coenzyme A, respectively.

The oxidative dimerisation of feruloyl side-chains to form diferuloyl groups (potentially inter-polymeric bridges; Figure 4.7) occurs partly in the cell wall after the polysaccharides' secretion. However, the process begins within the plant protoplast, on feruloyl-polysaccharides that are still located within the endomembrane system. Thus, some feruloyl-polysaccharides are secreted ready cross-linked, in the form of a large 'coagulum' – a Golgi vesicle's worth of covalently cross-linked feruloyl arabinoxylans, which would differ fundamentally from free feruloyl arabinoxylans secreted without prior cross-linking. Free feruloyl arabinoxylan molecules would be capable of inveigling into the existing wall material and then 'clamping' in place by oxidative phenolic coupling. Thus, the distinction between intra- and extra-protoplasmic coupling is likely to have large functional significance.

It is possible that other types of polymer–polymer bonding also occur intra-protoplasmically, for example the borate cross-linking of RG-II domains.

Glycosidic bonds *cannot* form true 'cross-links' as defined in this chapter. A glycosidic bond simply attaches the (single) reducing *end* of one polysaccharide chain to some point on another chain; the product is then effectively a single, though larger, polysaccharide molecule. Such 'end-linking' (as distinct from cross-linking) is probably responsible for joining together the discrete domains of pectin, *e.g.*:

$$\text{RG-I} \rightarrow \text{HGA} \rightarrow \text{RG-I} \rightarrow \text{HGA} \rightarrow \text{RG-II} \rightarrow \text{HGA}$$

(where → is a glycosidic bond and HGA = homogalacturonan). In contrast, a true cross-link between two pectic strands, as in the fictitious structure:

$$\ldots\text{HGA} \rightarrow \text{RG-I} \rightarrow \underline{\text{HGA}} \rightarrow \text{RG-II} \rightarrow \text{HGA}$$

$$\downarrow$$

$$\ldots\text{HGA} \rightarrow \text{RG-I} \rightarrow \text{HGA} \rightarrow \text{RG-II} \rightarrow \text{HGA},$$

is impossible through glycosidic bonding, since the HGA shown underlined would need to have had *two* reducing termini – a chemically unknown scenario. It is assumed, but with little experimental evidence, that pectic domain end-linking (as in HGA→RG-I) occurs within the protoplast before the pectic polysaccharide is secreted into the wall.

In dicot cell-cultures, about 50% of all the (neutral) xyloglucan is bonded to acidic pectic domains, probably RG-I, most likely via glycosidic 'xyloglucan→RG-I' end-links. *In-vivo* radiolabelling experiments show that such bonding occurs before secretion into the wall, thus presumably within the Golgi system, such that the xyloglucan arrives at the cell wall ready-bonded to the RG-I.

4.4.4 Delivering the Precursors – Sugar Nucleotides

Looking further back into the life history of wall polysaccharides, we can explore the source of NDP-sugars – the high-energy donor substrates for polysaccharide biosynthesis. In the cases of cellulose synthase and callose synthase (in the plasma membrane) and certain Golgi-localised enzymes (matrix polysaccharide synthases and glycosyl-transferases) whose catalytic centres are exposed to the cytosol, cytosolic NDP-sugars are required. In the cases of enzymes whose catalytic centres face the Golgi lumen, the NDP-sugars have to be located within the Golgi lumen. Sugar nucleotides are initially formed in the cytosol and also extensively interconverted there; many NDP-sugars are then transported via specific carrier proteins through the Golgi membrane into the lumen; and some may also be interconverted within the lumen.

Known pathways of NDP-sugar production and interconversion are summarised in Figure 4.10. There are examples on this metabolic map where two or more alternative routes potentially lead to the same NDP-sugar end-product (*e.g.* UDP-GlcA and GDP-Man; Figure 4.10). Given that genes and enzymes may exist that would enable both pathways to occur, distinguishing which pathway actually does predominate required *in-vivo* radioisotope tracer experiments – most informatively if *two* isotopes (^{3}H and ^{14}C) were simultaneously infiltrated into the pathways via different entry points (Figure 4.11). The ^{3}H:^{14}C ratio in the sugar residues of the polysaccharide end-products can then indicate which parts of the reversible NDP-sugar interconversion pathways supplied the majority of the precursors actually used in wall biosynthesis. For example, the Man and Fuc residues of wall polysaccharides acquired a ^{3}H:^{14}C ratio that indicates that the major source was fructose 6-phosphate (thus via the phosphomannose isomerase pathway), and not glucose 1-phosphate (via a postulated GDP-Glc 2-epimerase pathway). Likewise, the isotope ratio seen in wall-bound GalA, Xyl and Ara residues indicated that UDP-GlcA is predominantly formed directly from UDP-Glc (via the UDP-Glc dehydrogenase pathway), and not from inositol (via the alternative *myo*-inositol oxygenase pathway) (Figure 4.11).

4.5 Non-polysaccharide Components of the Plant Cell Wall

4.5.1 Extensins and Other (Glyco)Proteins

Plant cell walls contain structural proteins as well as polysaccharides. They include extensins (basic, hydroxyproline-rich glycoproteins), proline-rich proteins and glycine-rich proteins. Extensins have a protein backbone with a most unusual sequence; for example, the first extensin to be studied in this respect (from tobacco) was predicted from its gene sequence to contain 19 repetitions of SPPPPKK (or SPPPP or SPPP), 15 of TPVYK, and 7 of PYYPPH. Overall, the extensin (excluding the transit peptide) has an amino acid composition highly skewed from that of a typical protein: *viz.* P 127, Y 46, K 38, S 28, V 20, T 16, H 15, A 3, E 1, L 1, N 1, Q 1, C 0, D 0, F 0, G 0, I 0, M 0, R 0, W 0 [2]. In fact, more than 90% of the 'P' (proline) residues predicted by the gene sequence are post-translationally modified to hydroxyproline, most of which also carry a tetrasaccharide of arabinose. As with feruloyl polysaccharides, extensin molecules become covalently cross-linked to each other by oxidative coupling — in this case, via some of their tyrosine residues, forming a dimer (isodityrosine), trimer (pulcherosine) and tetramer (di-isodityrosine) (Figure 4.8). Some isodityrosines form tight intra-polypeptide loops as distinct from inter-polypeptide cross-links; however, steric considerations show that the trimer and tetramer can only constitute true

---

2 See Appendix for single-letter abbreviations of amino acids.

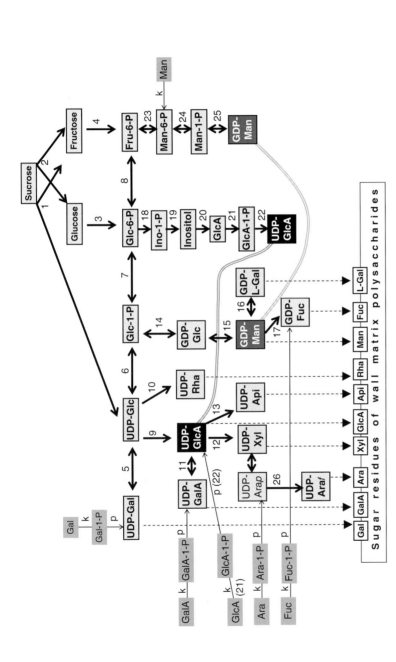

Figure 4.10 Intermediary metabolism leading to the 'activated' sugars, nucleoside diphosphate sugars, the precursors of cell-wall polysaccharides. Note that UDP-GlcA and GDP-Man could each potentially arise via two competing routes. Grey boxes (at left) indicate scavenger pathways, capable of bringing free monosaccharides (e.g. released during polysaccharide turnover, or experimentally added in radioactive form) back into central metabolism and thus *de-novo* polysaccharide synthesis. Comparable scavenger pathways do not exist for xylose, rhamnose or apiose. Numbered enzymes in the primary pathways (bold arrows) are: 1. sucrose synthase; 2. invertase; 3. hexokinase; 4. fructokinase; 5. UDP-Glc 4-epimerase; 6. UDP-Glc pyrophosphorylase; 7. phosphoglucomutase; 8. phosphoglucose isomerase; 9. UDP-Glc dehydrogenase; 10. UDP-Rha synthase; 11. UDP-GlcA 4-epimerase; 12. UDP-GlcA decarboxylase; 13. UDP-Api synthase; 14. GDP-Glc pyrophosphorylase; 15. GDP-Glc 2-epimerase; 16. GDP-L-Gal synthase (GDP-Man 3,5-epimerase); 17. GDP-Fuc synthase (two enzymes); 18. *myo*-inositol phosphate synthase; 19. *myo*-inositol phosphatase; 20. *myo*-inositol oxygenase; 21. glucuronokinase; 22. UDP-GlcA pyrophosphorylase; 23. phosphomannose isomerase; 24. phosphomannomutase; 25. GDP-Man pyrophosphorylase; 26. UDP-Ara mutase. Enzymes in the scavenger pathways are ATP-dependent kinases (k) and UTP- or GTP-dependent pyrophosphorylases (p).

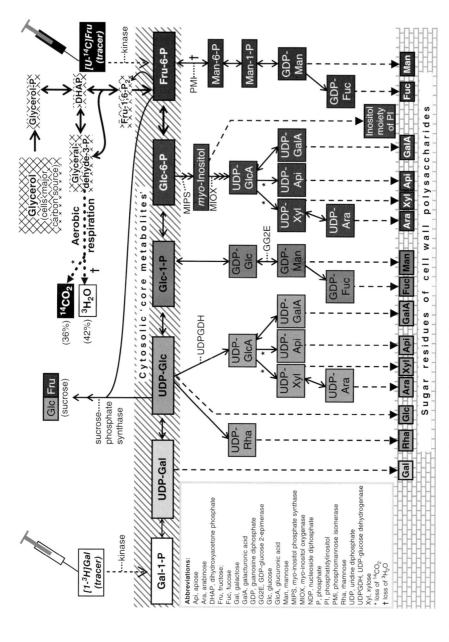

Figure 4.11 An experimental strategy for quantitatively estimating which of two competing pathways actually contribute(s) to wall polysaccharide biosynthesis in cases where competing pathways potentially exist (cf. Figure 4.10). Living plant cells, grown with glycerol as major carbon source, are simultaneously fed [³H]galactose (white rectangle) and [U-¹⁴C]fructose (black rectangle). As the two isotopes mingle in the intermediates, different ³H:¹⁴C ratios are established (represented by various shades of grey between the 'white' and 'black' tracers), which are preserved in the polysaccharide end-products. The isotope ratio in the end-product indicates from which intermediate it principally arose. Diagram based on Sharples and Fry (2007).

(inter-chain) cross-links. Cross-linked extensin probably contributes to the wall's inextensibility and indigestibility.

As noted in earlier sections, plant cell walls also contain diverse enzymes, which are non-structural (glyco)proteins, although some of them become very firmly linked to other structural wall components, probably polysaccharides.

4.5.2 Polyesters

The shoot epidermis in all land plants has a multifunctional extracellular cuticle containing cutin – a net-like, hydrophobic, aliphatic polyester composed of hydroxy-fatty acids. Cutin synthesis often peaks before or during periods of rapid growth, so it does not of itself prevent cell expansion.

The cuticle is significant in limiting the ability of agrochemicals to enter foliage and fruits. Natural roles of cutin include protection against physical abrasion, defence against microbial ingress and herbivory, prevention of desiccation (in conjunction with waxes), protection against UV damage (in conjunction with phenolic components), control of gas exchange, mechanical constraint of seed germination and pollination, and the delineation of organ boundaries in apical bud primordia.

Cutin may also sometimes restrain, and thus control, plant growth; this idea is supported in submerged rice coleoptiles by cutinase-treatment experiments.

The major building-block of cutin is often 10,16-dihydroxyhexadecanoic acid, whose −OH groups then have other fatty acyl residues esterified to them, potentially producing a branched polyester. Glyceryl and *p*-coumaroyl esters are also present. Cutin has often been postulated to be bonded to polysaccharides (especially pectins) and polyphenols, contributing to epidermal wall architecture; however, information on such cross-links is sparse.

As cutin is water-insoluble, it must be made *at* its destination in the apoplast, but until recently it was unclear what energy source might drive this biosynthesis. Cutin synthase (encoded by tomato gene *CD1*) is a recently discovered apoplastic acyltransferase, which makes the ester bonds of cutin in the absence of ATP or CoA. It catalyses a transesterification in which the acyl donor is a water-soluble 2-monoacylglycerol (MAG) such as 2-(10,16-dihydroxyhexadecanoyl)glycerol. The C_{16} acyl group is transferred from the glycerol moiety onto an −OH group in an acceptor substrate (R−OH):

transacylase

$$\text{acyl–glycerol } + \text{ R–OH} \leftrightarrow \text{acyl–R } + \text{ glycerol,}$$

which would normally be the nascent cutin polymer (or, for the first step, another MAG molecule). Repetition of this type of reaction ultimately generates high-M_r cutin. The energy for biosynthesis is provided intracellularly by the acyl-CoA-dependent synthesis of MAGs. Cutin synthase of tomato belongs to a large family of serine 'esterase/lipase/transacylases', typically having a Gly-Asp-Ser-Leu (GDSL) active site near the *N*-terminus.

Once cutin has been deposited, cutin synthase or a similar enzyme may catalyse cutin-to-cutin 'cutting and pasting' – a transacylation reaction in the cuticle analogous in many ways to the XTH-catalysed transglycosylation reaction in the primary cell-wall matrix.

Suberin is another polyester, but typically found in cork and endodermal cell walls. Its construction is broadly comparable with that of cutin, but it tends to have a higher content of C_{18}, rather than C_{16}, monomers.

4.5.3 Lignin

Lignin typically occurs in sclerenchyma and xylem cell walls, often making up about a third of the dry weight of mature wood. It is first deposited in the middle lamella, later in the primary walls, and finally, in the largest amounts, in secondary walls.

Lignin is a polymer formed by the oxidative coupling of aromatic monolignols: coniferyl (C), sinapyl (S) and *p*-coumaryl (= *p*-hydroxycinnamyl;(H) alcohols, inter-linked via diverse bonds, especially C–O–C (ethers) and C–C (*e.g.* biphenyls), comparable to the bonds between the ferulate residues shown in Figure 4.7. Grass lignins contain C, S and H, dicot lignins mainly C and S, and conifer lignins mainly C units.

By partially replacing the H_2O in the cell-wall matrix, lignin waterproofs the xylem side-walls. It is highly resistant to digestion by the enzymes of microbes and animals. Mechanically, it renders the xylem wall resistant to buckling (*e.g.* under the weight of a tree) and tearing (*e.g.* when a branch sways in the wind).

Lignin is synthesised by homolytic reactions, not nucleophilic substitutions. The monolignols (*e.g.* coniferyl alcohol; $C_{10}H_{12}O_3$) are not 'activated' as in the case of glucose in UDP-Glc. Instead, the monolignols are oxidised by H_2O_2:

peroxidase

$$2C_{10}H_{12}O_3 + H_2O_2 \rightarrow 2(C_{10}H_{11}O_3)^{\cdot} + 2H_2O$$

or by O_2

oxidase (laccase)

$$4C_{10}H_{12}O_3 + O_2 \rightarrow 4(C_{10}H_{11}O_3)^{\cdot} + 2H_2O$$

to form free radicals (indicated by the "'"), which then non-enzymically polymerise. The source of H_2O_2 for lignin synthesis remains uncertain: it may arise from O_2 by the action of wall-bound oxidases.

4.5.4 Silica

Although itself inorganic, the silica phase, which is abundant in certain cell walls (especially in the Poales and *Equisetum*), is intimately associated with the organic matrix. Plants take up soluble silicic acid ($Si(OH)_4$; a very weak acid of pK_a 9.8) from the soil and transport it in aqueous solution through the xylem into the shoots, where silica deposition is to some extent enforced by the increase in concentration due to evaporation of the H_2O. When the Si concentration in a silicic acid solution at pH 4–8 exceeds approximately 2 mM, condensation reactions form insoluble silica non-enzymically. However, heavy silicification can also occur in roots and at stem-nodes, which are not major evaporative sites; thus the plant can control when and where silicification occurs, presumably by means of organic constituents, ultimately under genetic control.

Silicas from different organisms vary greatly in density, hardness, solubility, viscosity and composition, further indicating a steering influence of the organic components. In plants, cationic proteins (possibly extensins) may promote silica nucleation. Few attempts have been made to characterise Si–polysaccharide interactions in plants, but it is known that Si will bind remarkably stably to citrus pectin (~1 Si atom per 60 GalA residues). The Si–pectin complex remains stable inside a dialysis sac, even after moderately severe acid and alkali treatments (0.2 M HCl or 1 M NaOH at room temperature), until the pectin is digested by pectinase. Ether- or ester-like derivatives of silicic acid ('silanolates') may play a structural role within the cell wall.

4.6 Conclusions

In conclusion, a good deal of detailed knowledge currently exists concerning the composition, assembly and *in-vivo* modification of the plant cell wall. Nevertheless, significant gaps in our understanding have been highlighted, in particular concerning wall evolution and the molecular architecture of the cell wall matrix – the nature and origin of cross-links that hold the individual wall polymers together as a coherent fabric of great strength, yet also flexibility and extensibility. A reliable model of the cell wall, and of its formation and evolution, would be a valuable weapon in our current endeavours to harness the energy and chemical resources locked up in vast quantities in cheap and under-utilised biomass, as discussed in Chapter 5.

Acknowledgements

The author thanks the BBSRC and the Leverhulme Foundation for research support in these research areas.

Appendix

Single-letter Abbreviations for Amino Acids

A	alanine	M	methionine
C	cysteine	N	asparagine
D	aspartic acid	P	proline
E	glutamic acid	Q	glutamine
F	phenylalanine	R	arginine
G	glycine	S	serine
H	histidine	T	threonine
I	isoleucine	V	valine
K	lysine	W	tryptophan
L	leucine	Y	tyrosine

Selected References and Suggestions for Further Reading

Literature review conducted in May 2013.

Albersheim, P., Darvill, A., Roberts, K., Sederoff, R. and Staehelin, A. (2011) *Plant Cell Walls: from Chemistry to Biology*. Garland Science, New York.

Burton, R.A., Gidley, M.J. and Fincher, G.B. (2010) Heterogeneity in the chemistry, structure and function of plant cell walls. *Nature Chemical Biology*, **6**, 724–732.

Cantarel, B.L., Coutinho, P.M., Rancurel, C., Bernard, T., Lombard, V., and Henrissat, B. (2009) The Carbohydrate-Active EnZymes database (CAZy): an expert resource for glycogenomics. *Nucleic Acids Research*, **37**, D233–D238.

Carpita, N.C. (1996) Structure and biogenesis of the cell walls of grasses. *Annual Review of Plant Physiology and Plant Molecular Biology*, **47**, 445–476.

Carpita, N.C. (2012) Progress in the biological synthesis of the plant cell wall: new ideas for improving biomass for bioenergy. *Current Opinion in Biotechnology*, **23**, 330–337.

Darvill, A., Augur, C., Bergmann, C., Carlson, R.W., Cheong, J. *et al.* (1992) Oligosaccharins – oligosaccharides that regulate growth, development and defence responses in plants. *Glycobiology*, **2**, 181–198.

Eklöf, J.M. and Brumer. H. (2010) The XTH gene family: an update on enzyme structure, function, and phylogeny in xyloglucan remodelling. *Plant Physiology*. **153**, 456–466.

Franková, L. and Fry, S.C. (2011) Phylogenetic variation in glycosidases and glycanases acting on plant cell wall polysaccharides, and the detection of transglycosidase and trans-β-xylanase activities. *Plant Journal*, **67**, 662–681.

Franková, L. and Fry, S.C. (2012) Trans-*alpha*-xylosidase, a widespread enzyme activity in plants, introduces (1->4)-*alpha*-D-xylobiose side-chains into xyloglucan structures. *Phytochemistry*, **78**, 29–43.

Franková, L. and Fry, S.C. (2013) Darwin Review: Biochemistry and physiological roles of enzymes that 'cut and paste' plant cell-wall polysaccharides. *Journal of Experimental Botany*, **64**, 3519–3550.

Fry, S.C. (1986) Cross-linking of matrix polymers in the growing cell walls of angiosperms. *Annual Review of Plant Physiology*, **37**, 165–186.

Fry, S.C. (1995) Polysaccharide-modifying enzymes in the plant cell wall. *Annual Review of Plant Physiology and Plant Molecular Biology*, **46**, 497–520.

Fry, S.C. (2004) Tansley Review: Primary cell wall metabolism: tracking the careers of wall polymers in living plant cells. *New Phytologist*, **161**, 641–675.

Fry, S.C., Franková, L., and Chormova, D. (2011a) Setting the boundaries: Primary cell wall synthesis and expansion. *The Biochemist*, **33**, 14–19.

Fry SC (2011b) Chapter 1: Cell wall polysaccharide composition and covalent crosslinking. Annual Plant Reviews Vol. 41, *Plant Polysaccharides, Biosynthesis and Bioengineering*, pp 1–42, edited by Peter Ulvskov, Blackwell.

Hongo, S., Sato, K., Yokoyama, R., and Nishitani, K. (2012) Demethylesterification of the primary wall by pectin methylesterases provides mechanical support to the Arabidopsis stem. *Plant Cell*, **24**, 2624–2634.

McNeil, M., Darvill, A.G., Fry, S.C. and Albersheim, P. (1984) Structure and function of the primary cell walls of higher plants. *Annual Review of Biochemistry*, **53**, 625–663.

Northcote, D.H. (1972) Chemistry of plant cell wall. *Annual Review of Plant Physiology*, **23**, 113–132.

O'Neill, M.A., Ishii, T., Albersheim, P. and Darvill, A.G. (2004) Rhamnogalacturonan II: structure and function of a borate cross-linked cell wall pectic polysaccharide. *Annual Review of Plant Biology*, **55**, 109–139.

Park, Y.B. and Cosgrove, D.J. (2012) A revised architecture of primary cell walls based on biomechanical changes induced by substrate-specific endoglucanases. *Plant Physiology*, **158**, 1933–1943.

Popper, Z.A., editor (2011) *The Plant Cell Wall: Methods and Protocols*. Springer, New York.

Popper, Z.A. and Fry, S.C. (2003) Primary cell wall composition of bryophytes and charophytes. *Annals of Botany*, **91**, 1–12.

Qiu, Y., Li, L., Wang, B., Chen, Z., Dombrovska, O. *et al.* (2007) A non-flowering land plant phylogeny inferred from nucleotide sequences of seven chloroplast, mitochondrial, and nuclear genes. *International Journal of Plant Sciences*, **168**, 691–708.

Rose, J.K.C., Braam, J., Fry, S.C. and Nishitani, K. (2002) The XTH family of enzymes involved in xyloglucan endotransglucosylation and endohydrolysis: current perspectives and a new unifying nomenclature. *Plant and Cell Physiology*, **43**, 1421–1435.

Rose, J.K.C. ed. (2003) *The Plant Cell Wall*. CRC Press/Blackwell, Oxford.

Scheller, H.V. and Ulvskov, P. (2010) Hemicelluloses. *Annual Review of Plant Biology*, **61**, 263–289.

Sharples, S.C. and Fry, S.C. (2007) Radio-isotope ratios discriminate between competing pathways of cell wall polysaccharide and RNA biosynthesis in living plant cells. *Plant Journal*, **52**, 252–262.

Szymanski, D.B. and Cosgrove, D.J. (2009). Dynamic coordination of cytoskeletal and cell wall systems during plant cell morphogenesis. *Current Biology*, **19**, R800–R811

Ulvskov, P. ed. (2011) *Plant Polysaccharides, Biosynthesis and Bioengineering*. Blackwell, Oxford

Varner, J.E. and Lin, L-S. (1989) Plant cell wall architecture. *Cell*, **56**, 231–239.

Verbelen, J.-P. and Vissenberg, K. eds (2007) *The Expanding Cell*. Springer, Berlin.

Yeats, T.H., Martin, L.B., Viart, H.M., Isaacson, T., He, Y. et al (2012) The identification of cutin synthase: formation of the plant polyester cutin. *Nature Chemical Biology*, **8**, 609–611.

5

Ethanol Production from Renewable Lignocellulosic Biomass

Leah M. Brown, Gary M. Hawkins and Joy Doran-Peterson

University of Georgia, Athens, USA

Summary

Ethanol was first used as a motor fuel over 100 years ago but fell out of favour as the oil industry expanded. However, interest in ethanol as a fuel has increased greatly in the past 30 years for economic and political reasons and because of the desirability of decreasing the use of fossil fuels in order to reduce the extent of climate change. To date, most fuel ethanol comes from fermentation of sucrose from sugar cane and other sugar crops and from the carbohydrates of cereal grains, especially corn (maize). However, the latter raises concerns that crops that can be used for animal or human nutrition are being used to make fuel. Other sources of fuel ethanol are therefore being sought, including waste plant biomass left over after harvest and biomass grown specifically as a biofuel. However, the sugars in waste biomass are locked up in complex molecules. The plant material needs biological and physio-chemical pre-treatment to release the sugars. Furthermore, several of these sugars are not amenable to fermentation by conventional yeasts. The use of other microbial species, in some cases genetically modified, is required to ferment these sugars. In spite of these problems, there is very active interest in production of ethanol from plant biomass, and biomass refineries have been opened in the USA, Brazil and several countries in the EU.

5.1   Brief History of Fuel-Ethanol Production

Ethanol is a high-octane motor fuel produced from the sugars, starches and cellulose found in plant materials. This biodegradable fuel has been in use for over a century, since the Model T Ford was launched in 1906 (Figure 5.1a). Henry Ford recognised some of the desirable qualities of ethanol, particularly that it came from renewable bio-logical materials and was relatively easy to produce. The Model T was designed to run on either gasoline or ethanol, depending upon what was available, so this was really the first 'flex fuel' car! In 1919, prohibition caused challenges for using ethanol as a fuel because it now had to be altered so as to be undrinkable, adding costs to its use. Gasoline

Biofuels and Bioenergy, First Edition. Edited by John Love and John A. Bryant.
© 2017 John Wiley & Sons Ltd. Published 2017 by John Wiley & Sons Ltd.

Figure 5.1 (a) Henry Ford's Model T. Source: National Library on The Commons. (b) Environmental awareness of industrial processes. Source: 1970s USPS postage stamps. (c) Closed pumps along Interstate 5 in Oregon in the United States, October 1973. Photo by David Falconer as part of the EPA's DOCUMERICA series in The National Archives.

(petrol in UK usage) soon emerged as the dominant transportation fuel, at least in part due to discovery of many large oil fields in the early 1900s. Gasoline was cheap. In 1933, prohibition ended and 'gasohol', a blend of gasoline and 10% ethanol became available, although cheap gasoline dominated the fuel market.

We now 'fast forward' to the 1960s and 1970s, when there was widespread evaluation of the impact of human activity on the environment (Figure 5.1b; see also Chapter 1). The consequences of industrialism with respect to air and water quality were beginning to be examined in earnest. This shift in public awareness sparked thinking about alternatives to the *status quo* use of petroleum-derived products, especially the large-scale use of gasoline as the primary liquid transportation fuel. Questions began to arise regarding emissions, air quality and quality of life.

In the 1970s, two so-called energy crises occurred (1973 and 1979), the first caused by an oil embargo[1] and a severe price increase in oil by the Organization of the Petroleum Exporting Countries (OPEC) (Figure 5.1c). This resulted in severe gasoline shortages, long queues at the pump, rationing, higher inflation and a heightened sense of national security issues with respect to energy. The 1979 crisis was a result of a reduction in oil production following the Iranian revolution. This reduction was actually quite small, about 4%, but it resulted in panic buying, especially in the USA. Gasohol, the 10% ethanol blend first used in the 1930s, was reintroduced in the USA in 1979 in order to help ameliorate the impact of the 'energy crisis'. People began to consider 'alternative fuels' such as ethanol on a larger scale. Research into ethanol production and other alternative fuels began to emerge in the 1970s. Once the energy crises with petroleum began to resolve, however, widespread interest in ethanol began to fade, although some research continued through the 1980s and 1990s. However, widespread interest in ethanol has been renewed recently, in part because many countries are net importers of crude oil and are therefore extremely vulnerable to price fluctuations that can have a crippling effect on a region's liquid transportation fuel supply and subsequently the entire economy of the country.

---

1 The Arab members of OPEC imposed an oil embargo on the US, on some EU countries including the UK, Canada, The Netherlands and Japan, because of their (perceived) support for Israel in the '6-day war'. The embargo lasted from October 1973 to March 1974. The UK avoided rationing of fuel because oil from the North Sea field had come on stream.

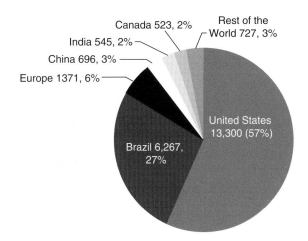

Figure 5.2 Global ethanol production in 2013 worldwide in millions of gallons by country, millions of gallons, and percentage of global production (data source USDA -FAS). Note: 1US gallon = 3.785 litres.

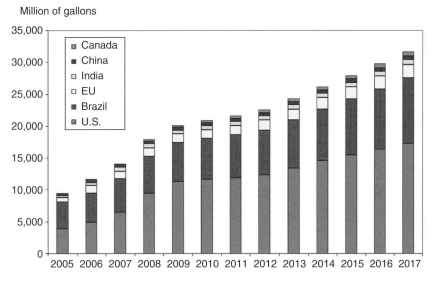

Figure 5.3 World ethanol production, millions of US gallons, from 2005–2008 with estimates up to 2017. Source: FAPRE 2008 and World Agricultural Outlook. For conversion to litres, see Figure 5.2.

Although many of the world's economies have experienced challenges with financial health, the global ethanol industry continues to be a bright spot. World fuel ethanol production in 2013 was over 85 billion litres[2], with the USA and Brazil the largest producers (Figure 5.2). Africa is expected to see the greatest increase in production in the next few years, although the overall production capacity in Africa is still relatively low. Europe currently produces close to 5 billion litres of ethanol. Ethanol production worldwide is expected to continue to increase (Figure 5.3).

---

2 1 litre = 0.264 US gallon or 0.22 UK gallon.

5.2   Ethanol Production from Sugar Cane and Corn

Most ethanol produced in the USA comes from fermentation of corn starch and in Brazil from fermentation of sugar cane sucrose. While the basic steps have remained the same for generations, recent refinements make today's processes very efficient. For example, in the USA, the amount of thermal energy required to make a given volume of ethanol has fallen to 36% since 1995, electricity use is down 38%, and water use has been cut in half (now at about 2.7 litres of water per litre of ethanol produced). Compare this with the controversial oil production from Canadian tar sands, where the production of 1 litre of oil requires 8 to 10 litres of water. Furthermore, the US Department of Agriculture's latest research determined that 1 unit of energy invested in the corn ethanol production process results in production of 2.3 units of usable energy in the form of ethanol.

In Brazil, sugar cane is milled with water to extract the juice containing sugar. The fibrous portion remaining, called bagasse, is burned to provide energy required for the overall process[3]. The yeast, *Saccharomyces cerevisiae*, is used in both sugar cane sucrose and corn starch fermentations to produce ethanol. Ethanol produced from fermentation of sugar cane juice or from corn starch is of course the same molecule.

Corn ethanol production uses two basic processes: dry milling and wet milling. Dry milling (Figure 5.4) uses the whole corn kernel ground into flour and processed without

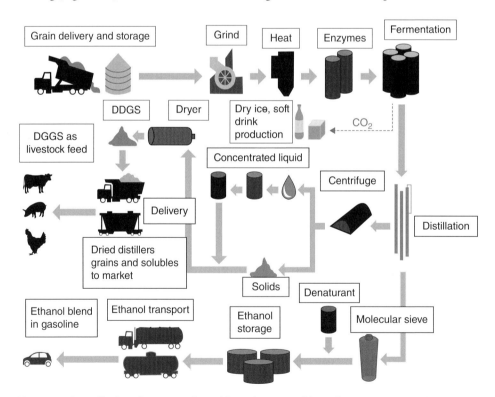

Figure 5.4 Dry mill ethanol process. Adapted from the Renewable Fuels Association (2011).

---

3 Although Brazil has recently opened a refinery for producing ethanol from bagasse.

separating the different parts of the grain. This flour is mixed with water and enzymes are added to convert the starch polymer to simple sugar (dextrose) for the yeast to consume. Ammonia is used to keep the pH constant and as a nutrient for the yeast. This material is processed at high temperatures to kill off bacterial contaminants. After cooling and transfer to fermenters, yeast is added and fermentation of the sugars to ethanol and carbon dioxide continues for 40–50 hours. Carbon dioxide is captured and sold for dry ice manufacture or for use in production of carbonated soft drinks. Temperature and pH are controlled to benefit the yeast. Ethanol is produced as a product of the yeast's metabolism and the liquid from the fermenter is transferred to distillation columns where the ethanol is evaporated off (distilled). Ethanol is concentrated using distillation and further dehydrated using molecular sieves to create anhydrous ethanol. A denaturant such as gasoline is added to make the ethanol undrinkable and then it is shipped to gasoline retailers. The remaining materials are dried to produce dried distillers grains with solubles (DDGS) and sold as a nutritious animal feed.

Corn wet milling processes involve soaking the grain in water and dilute acid to separate the grain into the germ, fibre, wet gluten and starch. Corn oil is extracted from the germ and the soaking water or 'steeping liquor' is concentrated, dried with the fibre and sold as corn gluten feed. The protein or gluten component is dried to produce corn gluten meal, a highly desirable feed ingredient in poultry farms. The starch can then be fermented to ethanol by yeasts, in a process very similar to that described for corn dry milling, or processed into corn syrup, or dried and sold as modified corn starch.

5.3 Lignocellulosic Biomass as Feedstocks for Ethanol Production

5.3.1 The Organisms

5.3.1.1 Saccharomyces cerevisiae, *as the Fermenting Organism*

Saccharomyces cerevisiae, the yeast used in most ethanol production facilities, metabolises sugars present in corn starch and sugar cane juice in the absence of oxygen to produce ethanol and CO_2. This process is outlined in Figure 5.5. In order to calculate the potential maximum amount of ethanol that could be produced from glucose, use the following formula: 180 g of glucose is converted to 92 g of ethanol and 88 g of carbon dioxide. Dividing the formula weight of ethanol by the formula weight of glucose equals 0.511. Therefore the maximum yield of ethanol from glucose is 51.5 wt%, or for every 100 g of glucose, the most ethanol that could be produced is 51.5 g. Some calculations assume 10% of the carbohydrate will be converted to microbial cell mass. This value is used to estimate the amount of ethanol that could be produced from a certain amount of biomass or starting carbohydrate. Theoretically, any plant biomass could be used as a substrate for ethanol production, because all plant biomass contains potentially fermentable sugars.

5.3.1.2 Fermenting 5-Carbon and 6-Carbon Sugars

Many types of lignocellulosic biomass including grasses, sugar cane bagasse and hardwoods, contain high concentrations of 5C (pentose) sugars (see Chapter 4). Several approaches have been taken to address the fermentation of sugars, especially 5C sugars, unavailable to traditional *Saccharomyces cerevisiae* strains.

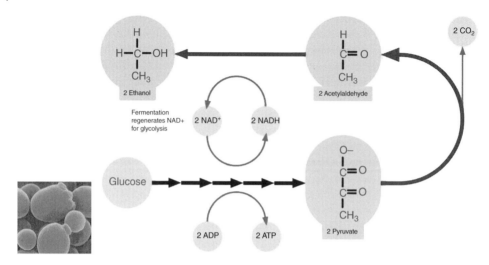

Figure 5.5 *S. cerevisiae* ethanol production with glucose as the carbon source or substrate.

One way to overcome the limitations of *S. cerevisiae* is to use metabolic engineering to add the needed genes for fermentation of other types of sugars. During the 1980s and 1990s, Dr Nancy Ho at Purdue University added genes for xylose (a 5C sugar) metabolism to *S. cerervisiae*. In 1993, her group produced a genetically engineered yeast that could effectively ferment both the 6C sugar glucose (the normal substrate) and the 5C sugar xylose. She has continued to refine and improve this organism for commercial production[4] and in 2004, the Ho-Purdue yeast was licensed by Iogen in Ottawa, Canada for their cellulosic ethanol demonstration facility and it is still in use. Using Iogen's process, approximately 313 litres of ethanol are produced per dry tonne of straw. About one-third of the straw cannot be fermented and is burned to generate power to run the facility.

Another approach to fermenting all of the sugars, including the 5C sugars which occur in plant biomass, was undertaken by Dr Min Zhang of the National Renewable Energy Laboratory. She and her colleagues used a bacterium, *Zymomonas mobilis*, for conversion of sugars to ethanol. *Z. mobilis* brings about a direct conversion of pyruvate to ethanol, with only one intermediate, acetaldehyde. Conversion of pyruvate to acetaldehyde is catalysed by the enzyme pyruvate decarboxylase (PDC), requiring thiamine pyrophosphate (TPP) as a cofactor and releasing carbon dioxide as a by-product. Next, acetaldehyde is reduced to ethanol *via* an alcohol dehydrogenase (ADH), which requires reduced nicotinamide adenine dinucleotide (NADH) as a cofactor. *Z. mobilis* is an obligate anaerobic fermenter, able to flourish only where oxygen is limited. *Z. mobilis* is ethanol tolerant like *S. cerevisiae* and relatively high concentrations of ethanol may be obtained in fermentation broths, facilitating recovery of the ethanol.

However, *Z. mobilis* also 'suffers' from limited sugar usage, like *S. cerevisiae*. Dr Zhang and her team were able to engineer *Z. mobilis* to enable it to use 5C sugars, thus broadening the substrate range and utility of this organism in commercial processes[5].

---

4 *E.g.* Moniruzzaman *et al.* (1997); Ho *et al.* (1998).
5 *E.g.* Mohagheghi *et al.* (1998).

Similarly, a 5C-fermenting *Z. mobilis* strain for ethanol production is in use in the Dupont Cellulosic Ethanol demonstration plant in Tennessee. To generate pentose fermentation in *Z. mobilis*, two primary pathways are incorporated. First, xylose isomerase, an enzyme that metabolizes xylose into xylulose is added; xylulose is subsequently phosphorylated to xylulose-5-phosphate, an intermediate of the pentose-phosphate pathway. Sugars from the pentose-phosphate pathway can be cycled into the glycolysis pathway[6] and thus generate pyruvate for fermentation. As with many enzymes in these glycolytic and fermentative reactions, this enzyme also requires the oxidative-reductive capacity of NADH or NADPH to successfully perform this conversion. A xylose-fermenting strain of *Z. mobilis* has been engineered by transforming it with two *E. coli*-derived operons encoding xylose assimilation-encoding enzymes, as well as pentose phosphate pathway enzymes (transaldolase and transketolase), thus allowing the organism to ferment a previously unutilized, primary 5C sugar component in hemicellulose (see Chapter 4). Second, three enzymes are utilised consecutively to metabolise arabinose, another five-carbon component of hemicellulose. The first step involves arabinose isomerase, which catalyses the isomerisation of arabinose to ribulose. The second step involves ribulokinase, which phosphorylates ribulose to form ribulose 5-phosphate. From here, ribulose 5-phosphate can directly enter the pentose phosphate pathway, or it can be converted into another pentose phosphate pathway intermediate by the enzyme ribulose 5-phosphate 4-epimerase. This enzyme catalyses the epimerisation of ribulose 5-phosphate into xylulose 5-phosphate, which can enter the pentose phosphate pathway at several steps. An arabinose-fermenting strain of *Z. mobilis* has also been engineered by incorporating the genes encoding these enzymes along with those encoding transaldolase and transketolase from *E. coli*.

In 1989, Dr Lonnie Ingram, at the University of Florida, developed an alternative approach that in 1991 was awarded US patent number 5 million[7]. Ingram's approach used a common bacterium, *E. coli*, because this micro-organism can already use 6C and 5C sugars. The strain of *E. coli* generally used can produce some ethanol, though not nearly as much as *S. cerevisiae* or *Z. mobilis*. Ingram and his colleagues engineered *E. coli* by adding two genes encoding the ethanol pathway from *Z. mobilis*. In this fashion they developed a micro-organism with a broad substrate range, able to use different types of sugars found in lignocellulosic biomass that could also produce ethanol in concentrations high enough for cost-effective distillation. The genes for these two enzymes, pyruvate decarboxylase (PDC) and alcohol dehydrogenase (ADH), have been successfully transformed into *E. coli* from *Zymomonas mobilis* under a single promoter, and were subsequently referred to as the PET (production of ethanol) operon. *E. coli* strain KO11 contains the PET operon on its main chromosome (not on a plasmid), allowing it to generate ethanol in a more efficient manner by essentially bypassing its own mixed acid fermentation pathway and exhibiting greatly increased ethanol production.

Mixed acid fermentation employed by several bacterial genera contributes significantly to reduction in ethanol yields. In this process, degraded lignocellulose is fermented into several acidic products, the most common of which are acetate, lactate and formate. Knocking out the genes encoding the enzymes in these acid-producing pathways would thus be beneficial to ethanol production and has been explored by several researchers.

---

6 See, *e.g.* Bryce and Hill (1997).
7 Ingram *et al.* (1991).

Acetate production involves the enzymes phosphate acetyltransferase (PTA), which converts acetyl coenzyme-A into acetyl phosphate and acetate kinase (ACK), which in turn converts acetyl phosphate into acetate. Knocking out the genes encoding phospohate acetyltransferase and acetate kinase has been successfully performed in *Thermoanaerobacterium saccharolyticum*, a thermophilic anaerobic bacillus, as well as in *E. coli*.

In addition, the genes encoding L-lactate dehydrogenase (LDH) have been knocked out to prevent the production of lactate. L-lactate dehydrogenase catalyses the reduction of pyruvate to lactate (while oxidising NADH to NAD⁺). Removing these genes caused the engineered *T. saccharolyticum* to utilise a homo-ethanol pathway (*i.e.* a pathway to ethanol using the organisms own enzymes rather than adding transgenes), significantly increasing ethanol yields while also reducing cellulase loading during hydrolytic saccharification steps. This deletion has also been observed in *E. coli*, resulting in reduced growth of the micro-organism and thus increased availability of fermentable carbohydrates to improve final ethanol concentrations.

Formate production is catalysed by the enzyme pyruvate-formate lyase (PFL). Formate can also be further oxidised to form CO_2 (and H_2) by various enteric bacteria. The pyruvate-formate lyase pathway does not produce exclusively formate: coenzyme A is also utilised to form acetyl-CoA, which can yield ethanol, acetate and various other fermentation products, depending on the organism. In addition, the utilisation of the pyruvate-formate lyase pathway generates a redox imbalance in NADH consumption and production (4 NADH consumed and 2 NADH produced), thus negatively affecting the organism's ethanol yield by depleting its necessary cofactor. Deletion of genes encoding pyruvate-formate lyase (among those encoding enzymes for other metabolic pathways) in *E. coli* led to reduced cellular growth rate and significantly increased ethanol yields.

All of the researchers mentioned above have continued to improve upon their micro-organism of choice and a wealth of other possible biocatalysts are being investigated. For example, *Clostridium thermocellum* and *C. phytofermentans* are obligate anaerobes possessing enzymes required for biomass degradation. Academic laboratories and companies are endeavouring to improve the production of ethanol from these bacteria.

5.3.2 Lignocellulosic Biomass

Renewable lignocellulosic feedstocks include crop and forestry residues, dedicated energy crops (both woody biomass and grasses) and industrial wastes. Lignocellulosic biomass is composed of:

a) lignin, a polyphenolic non-carbohydrate polymer that is often used to generate energy to drive the overall process;
b) cellulose, a glucose polymer;
c) hemicellulose, variable in composition depending on the type of plant biomass; and
d) pectin, variable in structure and amount dependent upon the type of plant biomass (see also Chapter 4).

Cellulose, hemicellulose and pectin may all be fermented to produce ethanol using organisms capable of metabolising all of the carbohydrates present in these components. This is an important consideration, as *Saccharomyces cerevisiae* used in corn ethanol fermentations is usually limited to fermenting only six-carbon sugars such as glucose. If the biomass in question has large quantities of other sugars present, then the actual yield of ethanol produced could be much lower than the calculated maximum theoretical possible yield (see Table 5.1). In the example presented below, *S. cerevisiae* is only able to ferment the 6C component of pine and birch. Roughly 288 litres of ethanol per dry tonne of each type of biomass may be obtained by fermentation of just the cellulose component. In pine, the hemicellulose is composed of mostly polymers of 6C sugars, so *S. cerevisiae* can ferment those sugars too. In birch, most of the hemicellulose consists of polymers of 5C sugars that cannot be fermented by traditional *S. cerevisiae*. The theoretical yield of ethanol production from birch is about 500 litres per dry tonne of biomass. However, the actual ethanol yield when using *S. cerevisiae* to ferment the biomass is only about 313 litres per dry tonne. In order to gain an additional 180 litres of ethanol per dry tonne of birch, it is necessary to use an organism capable of fermenting both 6C and 5C sugars.

Some examples of theoretical yield estimates for selected feedstocks are presented in Table 5.2. The actual amount of ethanol produced may approach the theoretical maximum, or it may fall short. Theoretical yields are just that, estimates that are theoretically possible under ideal conditions using a micro-organism that can metabolise all sugars present in the biomass.

Whatever the type of biomass, it is important to consider the methods of growing and harvesting the material to be delivered to the processing plant. Sustainability is key to a commercially successful renewable lignocellulosic biomass conversion process. Researchers are involved in trying to understand and address the potential environmental impacts of biofuels production activities in order to optimise processes, encourage the benefits and mitigate any concerns (see Chapters 16 and 17 for further discussion). Production of any form of energy, including obtaining and transporting petroleum, results in impacts to the environment that must be considered. Ideally, with a

Table 5.1 Actual *vs* calculated theoretical ethanol yield using traditional *Saccharomyces cerevisiae* as the fermenting organism. Component content presented on a % weight basis.

Component	Wood		Ethanol (litres per dry tonne)	
	Pine (softwood)	Birch (hardwood)	Pine (383–504 max. theoretical yield)	Birch (416–500 max. theoretical yield)
Cellulose (6C)	38–44	40–41	288	288
Glucomannan (6C)	11–20	2–5	112	25
Actual ethanol yield using *Saccharomyces*			400	313
Xylan 5C	7–10	20–25	62	166
Other carbohydrates	0–5	0–4	18	15
TOTAL Ethanol from 6C and 5C carbohydrates using an organism able to ferment all sugars			480	493

Table 5.2 Theoretical yield estimates for selected feedstocks.

Biomass	Theoretical yield of ethanol (litres/dry tonne)
Pine (softwood)	383–504
Birch (hardwood)	416–500
Forest thinnings	78–85
Forest residue	Highly variable
Hybrid poplar	325–458
Corn grain	458–520
Corn stover (stalk)	416–470
Rice straw	396–458
Sugar beet pulp	96–529

Source: US Department of Energy Technologies Office, Theoretical Ethanol Yield Calculator and Biomass Feedstock Composition and Property Database.

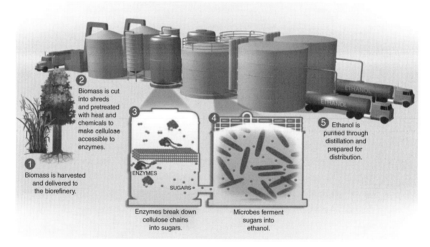

Figure 5.6 Overall schematic for ethanol production from lignocellulosic biomass. Source: US Department of Energy.

lignocellulosic biomass process, the feedstock will be produced in an environmentally aware manner and the overall industrial process will be designed to minimise water use and maximise recycling of water, chemicals and other components of the process. For example, today's corn ethanol reduces greenhouse gas emissions by an average of 34%, with some estimates as high as 48%, compared to gasoline, even when the corn ethanol process is 'penalised' for speculative land-use change emissions.

A schematic of lignocellulosic ethanol production is presented in Figure 5.6. First, the biomass is harvested and delivered to the processing plant or 'biorefinery'. Most lignocellulose requires some sort of chopping or shredding to reduce particle size, followed by a chemical and physical pretreatment to help open up the fibre structure for attack by enzymes. Microbes such as *Saccharomyces cerevisiae* and others can then ferment these sugars into ethanol and the ethanol is distilled, dehydrated and prepared for distribution.

5.3.3 Pretreatment of Lignocellulosic Biomass

Pretreatment of lignocellulose is required to open the fibrous structure of the plant biomass (Figure 5.7). The 'lignin' part of lignocellulose is a heterogeneous phenolic polymer covalently bound to hemicellulose sugars. Lignin prevents plants from collapsing under their own weight. In addition, because lignin is composed of hydrophobic subunits, water can be transported efficiently throughout the vasculature of the entire plant. Intact lignin also protects the plant biomass against microbial attack, but in so doing it reduces the accessibility of cellulose to the hydrolytic enzymes during an ethanol production process such as the one described in Figure 5.6.

Acids and bases may be used to chemically disrupt bonds. There are four main types of bonds in lignocellulosic biomass: ether bonds, ester bonds, carbon-to-carbon bonds and hydrogen bonds. The bonds provide linkages within the individual components of lignocellulose (intra-polymer linkages) and connect the different components to form the lignocellulose complex (inter-polymer linkages). Dilute acid hydrolysis uses low concentrations of various acids (sulphuric acid, phosphoric acid, nitric acid or hydrochloric acid) and relatively high temperatures to break these bonds in the plant biomass. These acids (or others) may be used in combination with steam 'explosion', using high heat and pressure to disrupt bonds. Some processes use concentrated acids requiring acid-resistant equipment and recirculation of the acids. Ammonia fibre expansion (AFEX) uses concentrated ammonia under high pressure to break hemicellulose bonds and facilitate enzymatic digestion of biomass. 'Organosolv' pulping pretreatment extracts lignin from ligocellulose using organic solvents or their aqueous solutions. Perhaps the most gentle pretreatment uses liquid hot water under pressure (not steam), where the acids present in the plant biomass are liberated *via* the hot water and act to some degree as a dilute acid pretreatment.

All forms of pretreating lignocellulose may produce potentially inhibitory compounds. These compounds are generated from the various polymers present in biomass as they are degraded *via* the harsh physical and chemical actions during pretreatment

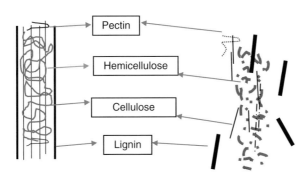

Figure 5.7 Disruption of pectin, hemicellulose, cellulose and lignin using physical and chemical pretreatment of biomass.

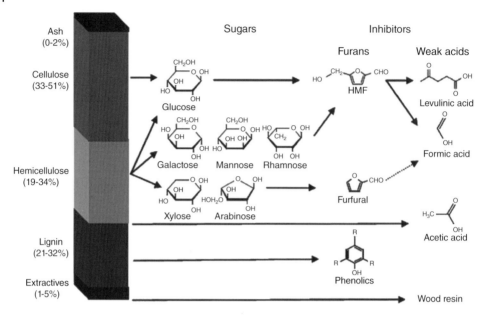

Figure 5.8 The production of inhibitory compounds from the various polymers of biomass during pretreatment. Source: Sun and Cheng (2002).

(Figure 5.8). These compounds are typically categorised by their chemical structure as aromatics, aliphatic acids or furans. Aliphatic acids and furans are sugar degradation products that are generated from the cellulose and hemicellulose. Acetic acid, which is typically found in greater concentration than the other acid inhibitors, is generated as a sugar degradation product or it can be released from the hemicellulose during pretreatment, as this is acetylated to varying degrees depending on the type of biomass being pretreated. Aromatic inhibitors are released from lignin during degradation and separation from the cellulose. Because lignin is a large, complex molecule with many different structural motifs, the variety of aromatics released is much greater than that seen in the other two classes of compounds. Further confounding the fermentation of biomass is the fact that the various inhibitors produced during pretreatment can have synergistic effects; the combination of multiple inhibitors in the media can cause greater inhibition than any single compound (see next section). Compound concentrations can be reduced and fermentation performance improved using a variety of inhibitor abatement or 'clean-up' techniques. While inhibitor removal does increase ethanol yields, there is an additional cost associated with this clean-up step. This may be prohibitively high and hurt the economic performance of the biomass fermentation process.

5.3.4 Effect of Inhibitory Compounds on Fermenting Microorganisms

Various inhibitory compounds can have negative effects on fermenting micro-organisms, with the effects of furan compounds hydroxymethylfurfural (HMF) and furfural on *Saccharomyces* the best understood of the biomass-derived inhibitors. Furfural reduces the growth rate of cells, particularly under aerobic conditions, and will inhibit the activity of cellular enzymes, particularly alcohol, aldehyde and

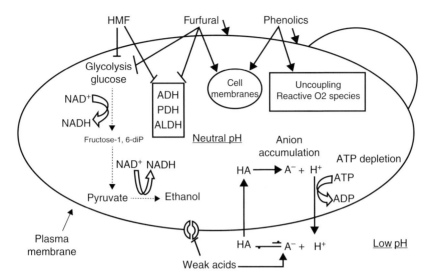

Figure 5.9 Effects of certain inhibitory compounds on *Saccharomyces cerevisiae*. 'Blocks' indicate the inhibition of processes or enzymes, while arrows indicate damage or inhibition of cellular components without a specific known target. Adapted from Sun and Cheng (2002).

pyruvate dehydrogenases. These enzymes are involved in the metabolism of glucose and the production of ethanol. When furfural and HMF are added into culture media together, synergistic effects can be observed: the inhibition of cellular activity observed is greater than in media containing an equal concentration of only one of the inhibitory compounds. *Saccharomyces* can reduce these furans to their alcohol analogues, thus removing the aldehyde inhibitors from the medium and leaving their alcohol forms, which are less toxic. Exposure of *Saccharomyces* to the inhibitory compounds found in pretreated biomass also causes a range of other effects including reactive oxygen damage, mitochondrial disruption, DNA damage and membrane instability (Figure 5.9).

The aliphatic acids can inhibit cellular activity by acting as uncouplers. In this process, the undissociated form of the acid will diffuse into the cell; the acid then will dissociate, losing a proton. This dissipates the proton motive force the cell uses to generate ATP and hampers cellular growth and metabolic activity while also reducing the pH of the cytosol. The cell must then correct for this pH imbalance by ATP-dependent pumping of protons out of the cell, a process that depletes the cell's reserves of energy. Acetic acid also has inhibitory effects on certain glycolytic enzymes, particularly enolase (phosphopyruvate hydratase; and catalyses the conversion of 2-phosphoglycerate to phosphenolpyruvate, the penultimate step in glycolysis. The aliphatic acids have also been shown to inhibit the activity of transport proteins associated with aromatic amino acid import.

The effects of aromatic inhibitors on *Saccharomyces* are not as well understood as the other two classes of compounds, due at least in part to the much greater variety of compounds present in this class. They are known to destabilise cellular membranes. They have also been shown to act as uncouplers, dissipating the proton gradient the cell uses to generate ATP in a fashion similar to the acid compounds. Certain aromatic compounds target cellular membranes and disrupt their ability to act as selectively permeable barriers. The complex structure of lignin results in the release of a wide variety of aromatic compounds with many different functional groups during pretreatment. In general, aldehyde and ketone compounds show greater inhibition of cellular activity than acid compounds, with alcohol substituted aromatics showing the least inhibition of *Saccharomyces*.

Ethanol-producing *E. coli* also convert toxic furfural to the less toxic alcohol form. Acetate concentrations as low as 0.5 g/L can reduce the growth of *E. coli* by 50%. Modification of the cAMP receptor protein (CRP) increased its survival time in acetate and in higher concentrations of ethanol. Several micro-organisms, including *E. coli*, have developed a coupled amino acid carboxylase/antiporter system that exports protons to return intracellular pH back to optimum levels.

Lignin-derived phenolic by-products including ferulic acid, vanillin, guaiacol and 4-hydroxybenzaldehyde compromise the fluidity, permeability and selectivity of cellular membranes. Many can also release reactive oxygen species which can denature enzymes, damage structural proteins, cause DNA mutagenesis and induce programmed cell death. Even at relatively low concentrations, these compounds often constitute the most potent of inhibitors.

Solutions to the detrimental effects of inhibitory compounds on fermenting micro-organisms include physical, chemical or biological ameliorations, mostly designed to remove the compounds or convert them to a less toxic form. Adaptive evolution, exposing the micro-organisms to ever-increasing concentrations of inhibitory compounds, has been successfully used to force the organism to adapt to the harsh environment. Both *Saccharomyces cerevisiae* and *E. coli* have been subjected to adaptive evolution to generate superior performing ethanol producers able to perform in a toxic soup of inhibitory compounds.

5.4 Summary

First-generation production methods for ethanol, from sugar or starch produced by food or animal feed crops (*e.g.* corn, sugar cane and sugar beet sucrose, etc.) have remained the same for generations, but recent refinements make today's processes more efficient. The amount of thermal energy required to make a litre of ethanol has fallen, electricity use is down, water use has been cut, and co-products make up a valuable revenue stream as animal feed. Second-generation lignocellulosic ethanol is chemically identical to first-generation ethanol (CH_3CH_2OH); however. the composition of the lignocellulose itself is much more complex and resistant to degradation to fermentable sugars; therefore the overall process is also more complex. Lignocellulosic biomass is abundant, can be sustainably produced and harvested; however, pretreatment is necessary to open up the fibrous structure for enzymatic digestion to produce monomeric sugars for fermentation. Lignocellulosic ethanol production has been the focus of ongoing research efforts since the 1970s and investments are continuing in the USA, Brazil and Europe to propel this industry forward. While research is ongoing, cellulosic ethanol is now being produced on a commercial scale worldwide.

5.5   Examples of Commercial Scale Cellulosic Ethanol Plants

5.5.1   Beta Renewables/Biochemtex Commercial Cellulosic Ethanol Plants in Italy, Brazil, USA and Slovak Republic

Beta Renewables commercial scale cellulosic ethanol plant at Crescentino, Italy, was officially opened in October 2013, and is currently the world's largest advanced biofuels refinery, with an annual production capacity of 75 million litres of cellulosic ethanol. Shareholders of Beta Renewables are Biochemtex, a company of the Mossi & Ghisolfi Group (M&G), TPG Capital, and Novozymes. The plant is based on the patented Proesa™ process, and uses Novozymes enzyme technology to convert local wheat straw, rice straw and *Arundo donax* (Giant reed) to ethanol. Lignin, extracted during the production process, is used at an attached power plant, which generates enough power to meet the facility's energy needs, with any excess green electricity sold to the local grid.

5.5.2   Poet-DSM 'Project Liberty' – First Commercial Cellulosic Ethanol Plant in the USA

POET-DSM Advanced Biofuels, LLC, is a 50/50 joint venture between Royal DSM, The Netherlands, and POET, LLC, based in Sioux Falls, South Dakota. The company was developed as a cooperative effort to commercialise innovative technology to convert corn crop residue into cellulosic bioethanol. In September 2014, production commenced at the $275 m 'Project Liberty' cellulosic ethanol facility. The plant will produce about 75 million litres per year of cellulosic ethanol from corn stover and cob, and shares infrastructure with the adjacent 190 million litres per year corn starch ethanol plant.

5.5.3   Abengoa Hugoton, Kansas Commercial Plant and MSW to Ethanol Demonstration Plant, Salamanca

In October 2014, Abengoa Bioenergy Biomass of Kansas opened its commercial plant producing 100 million litres of cellulosic ethanol and 21 MW of renewable energy from biomass (mixture of agricultural waste, non-feed energy crops and wood waste). In June 2013, Abengoa inaugurated its municipal solid waste (MSW) to cellulosic ethanol plant in Salamanca, Spain. The plant has a capacity to treat 27,500 tonnes of MSW from which it will obtain up to 1.5 million litres of bioethanol for use as fuel. The demonstration plant uses technology developed by Abengoa to produce second-generation biofuels from MSW using a fermentation and enzymatic hydrolysis treatment. During the transformation process, the organic matter is treated in various ways to produce organic fiber that is rich in cellulose and hemicellulose, which is subsequently converted into bioethanol.

Previously, Abengoa provided its proprietary process technology and the process engineering design for a BCyL Biomass Plant in Salamanca. Managed by Abengoa, the biomass plant was completed in December 2008 and has been fully operational since September 2009. It was the world's first plant to utilize this technology on such a large scale. Abengoa also heads the LED (Lignocellulosic Ethanol Demonstration) project funded by the European Commission and developed by a consortium of five companies from four different countries, with a plant in Arance, France. In January 2012, Abengoa announced that its cellulosic ethanol technology would be used to produce ethanol from sugar cane straw and bagasse in Brazil, as part of the Industrial Innovation Program for the Sugar Energy Sector.

Selected References, Suggestions for Further Reading and Useful Websites

Advanced Ethanol Council (2013) Biofuels Digest – Cellulosic Biofuels Industry Progress Report 2012/2013. http://ethanolrfa.3cdn.net/d9d44cd750f32071c6_h2m6vaik3.pdf

Bloomberg New Energy Finance (2012) Moving towards a next-generation ethanol economy.http://about.bnef.com/white-papers/moving-towards-a-next-generation-ethanol-economy-report/

Bryce, J.H. and Hill, S.A. (1997) In: *Plant Biochemistry and Molecular Biology*, 2nd edition (Lea, P.J. and Leegood, R.C. eds), Wiley, Chichester, UK, pp. 1–28.

Dien, B.S., Cotta, M.A. and Jeffries, T.W. (2003) Bacteria engineered for fuel ethanol production: current status. *Applied Microbiology and Biotechnology*, **63**, 258–266.

Energy Research Centre of the Netherlands/Wageningen University and Research Centre/ Abengoa Bioenergia Nuevas Tecnologías (2010) Literature Review of Physical and Chemical Pretreatment Processes for Lignocellulosic Biomass. http://www.ecn.nl/docs/library/report/2010/e10013.pdf

He, M.X., Wu, B., Qin, H., Ruan, Z.Y., Tan, F.R. *et al.* (2014) *Zymomonas mobilis*: a novel platform for future biorefineries. *Biotechnology for Biofuels*, 7, 101.

Ho, N.W.Y. *et al.* (1998) *Applied and Environmental Microbiology*, **64**, 1852–1859.

Ingram, L.O. *et al.* (1991) *Ethanol production by* Escherichia coli *strains co-expressing* Zymomonas *PDC and ADH genes*. US Patent 5,000,000.

Matsushika, A., Inoue, H., Kodaki, T. and Sawayama, S. (2009) Ethanol production from xylose in engineered *Saccharomyces cerevisiae* strains: current state and perspectives. *Applied Microbiology and Biotechnology*, **84**, 37–53.

Mohagheghi, A., Evans, K., Finkelstein, M. and Zhang, M. (1998) Cofermentation of glucose, xylose and arabinose by mixed cultures of two genetically engineered *Zymomonas mobilis* strains. *Applied Biochemistry and Biotechnology*, **70**, 285–299.

Moniruzzaman, M. *et al.* (1997) *World Journal of Microbiology and Biotechnology.* **13**, 341–346.

Peterson, J.D. and Ingram, L.O. (2008) Anaerobic respiration in engineered *Escherichia coli* with an internal electron acceptor to produce fuel ethanol. *Annals of the New York Academy of Sciences*, **1125**, 363–372.

Sun, Y. and Cheng, J.Y. (2002) *Bioresource Technology*, **83**, 1–11.

US Department of Energy (2011) *US Billion-ton Update: Biomass Supply for a Bioenergy and Bioproducts Industry*. http://www1.eere.energy.gov/bioenergy/pdfs/billion_ton_update.pdf

Energy Information Administration http://www.eia.gov/forecasts/aeo/index.cfm

European Biofuels Technology Platform. http://www.biofuelstp.eu/cellulosic-ethanol.html

Renewable Fuels Association (2011): http://ethanolrfa.org

6

Fatty Acids, Triacylglycerols and Biodiesel

John A. Bryant

College of Life and Environmental Sciences, University of Exeter, Exeter, UK

Summary

Although 'biodiesel' has been used to a limited extent for more than a century, interest in it has increased markedly over the past 20 to 25 years. This has resulted in more widespread use, with several countries including the EU enacting regulations about the contribution that biodiesel should make to the total amount of diesel fuel that is used. Biodiesel is derived from the triacylglycerols that many plants store in their seeds and/or fruit. The triacylglycerols are converted to fuel by trans-esterification with methanol. Suitability of these fuels for particular climatic conditions depends largely on the mix of fatty acids in the triacylglycerols. The biosynthetic pathway for triacylg-lycerols contains several features that are suitable targets for genetic modification and some progress has been made in that direction. Concerns have been expressed about the sustainability of production of some biodiesels, especially those derived from crops that are also used for human and/or animal nutrition and those that are derived from crops planted in formerly wild habitats. There are also concerns about productivity in relation to the world's annual consumption of oil-based fuels. Both these factors have led to a search for novel sources of triacyl-glycerols, a search that is beginning to yield positive results.

6.1   Introduction

As we mentioned in Chapter 1, over the past 40 years there has been an increase in the use of liquid fuels from biological sources. Initially this was mainly ethanol produced by fermentation of sugar but over the past 20 years, biodiesel has made an increasing contribution. In a sense, this increased use of biodiesel was foreseen in the 19th century. Rudolf Diesel's first engine ran on peanut oil and there was extensive and active interest in the UK and in mainland Europe throughout the first half of the 20th century on the use of biodiesel as a liquid fuel. However, it is the search for renewable fuels that has come from our understanding of climate change (and for

Table 6.1 Lipid content as percentage of dry weight of a range of lipid-rich seeds.

Species	Lipid content
Peanut, *Arachis hypogea*	48%
Soybean, *Glycine max*	18–22%
Oil-seed rape/Canola, *Brassica napus*	30–48%
Flax/Linseed, *Linum usitatissimum*	35%
Sunflower, *Helianthus annuus*	47%
Oil palm, *Elaeis guineensis*, kernel	35%
Oil palm, *Elaeis guineensis*, mesocarp	70%

some, their understanding of 'peak oil') which has driven the growth in the use of biodiesel over the past 20 years[1].

Biodiesel is derived from triacylglycerols (TAGs; also known as tri-glycerides), esters of glycerol and three fatty acids (FAs). These occur widely in plants as storage compounds in seeds and fruit. In olives, for example, oil makes up 21% of the fresh weight of a drupe, prior to the water loss phase of ripening. At the same stage, TAGs in the mesocarp of the oil-palm drupe make up 56% of the wet weight, approximating to 70% of the dry weight. In seeds, lipid percentage is generally given on a dry weight basis. As shown in Table 6.1, typical contents range between 18% and 48% for some of the widely grown oil-seeds. For a 'newer' oil crop, *Jatropha*, oil yields amounting to 63% dry weight have been claimed, although most authorities give figures between 35% and 50%, depending on variety and growth conditions.

In order to convert these TAGs to biodiesel, they are trans-esterified with methanol to produce fatty acid methyl esters, often referred to simply as FAME. This releases free glycerol which may be used as feedstock for other organic compounds. In Chapter 2, Lionel Clarke sets out the properties of FAME (and hence of the original TAGs) that are needed for use as fuel. The three main features are the cetane value, which relates to combustibility, ignition delay, and ease of starting. In many countries, including all those in the EU, the minimum and maximum cetane values are strictly specified in order to ensure fuel quality. FA chain length may be a factor here, with long-chain FAs being less volatile and therefore less combustible than shorter molecules. The second criterion is performance at low temperatures, defined in various ways but essentially ensuring that the fuel does not start to solidify in very cold weather. This relates to the FA composition of the TAGs, both chain length, degree of saturation and branching. Thus different FA compositions are appropriate for different climates. The third is resistance to oxidation and there is a tension between this requirement and low-temperature performance. Poly-unsaturated FAs tend to confer fluidity on the TAGs (and hence on FAME), but they are very prone to oxidation. A compromise between these two requirements suggests that the ideal FAME has a high proportion of 18:1

---

1 Despite the recent reduction in targets that we note in Chapter 17.

Table 6.2 Percentage fatty acid composition of triacylglycerols in storage lipids.

Species	16:0	18:0	18:1	18:2	18:3	Others
Peanut	11	2	48	32		7
Soybean	11	4	24	54	7	
Oil-seed rape/Canola	4	2	60	22	10	2
Flax/Linseed	3	7	21	16	53	
Sunflower	7	5	19	68	1	
Oil palm kernel	8	3	15	2		72[*]
Oil palm mesocarp	45	4	40	10		1
Jatropha	14	7	45	33		1

[*] Palm kernel oil is rich in short-chain saturated fatty acids: 4% 10:0, 48% 12:0, 16% 14:0

(oleic acid)[2], which is the most abundant FA in the TAGs of oil-seed rape (canola), peanut, *Jatropha* and some algae (see below). *Jatropha* also has high amounts of 18:2 (linoleic acid), which is by far the most abundant FA in soybean and in sunflower TAGs. In palm oil (from the mesocarp), 16:0 (palmitic acid) contributes 45% with 18:1 at 40% of the total FA content of TAGs (see Table 6.2). Furthermore, there is evidence that the FA profile of TAGs in an individual species can vary in different environments.

In the wider picture, two other considerations come into play. The first of these is productivity. How much land does it take to produce a given amount of biodiesel? The second is sustainability, as discussed in more detail in Chapters 16 and 17. Can biofuel crops be grown without reducing the amount of land used for food production? Can they be grown without grabbing land from native farmers? Does their manufacture actually reduce the amount of carbon dioxide released to the atmosphere? These factors come strongly into play, alongside the fuel quality criteria, in driving research on new sources of biodiesel and on modification of currently used sources. I discuss this research in Sections 6.3 to 6.5. In the meantime, I present a brief account of the synthesis of tri-glycerides in plants.

6.2 Synthesis of Triacylglycerol

6.2.1 The Metabolic Pathway

Understanding the pathway for synthesis of TAG helps to identify reactions that may be manipulated in order, for example, to change the chain length or the degree of saturation of FAs to meet specific needs. In addition, understanding the main regulatory mechanisms may enable us to increase yields, although, as highlighted in a recent study[3] by the Crop Genetics group at the John Innes Research Centre (Norwich, UK), this may not be straightforward.

---

2 Note that, as described by Lionel Clarke in Chapter 2, the requirements for jet fuel are different, with a need for shorter FA chains.
3 O'Neill *et al.* (2012).

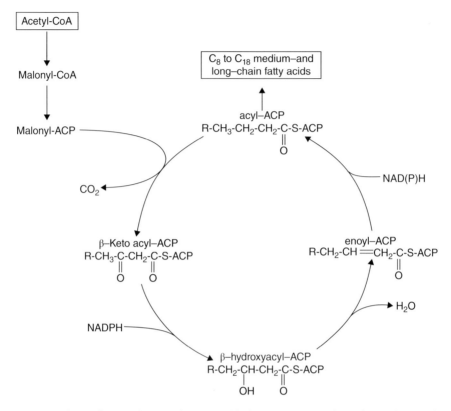

Figure 6.1 Synthesis of FAs in plants. Carbons are added two at a time, each pair being donated form malonyl-CoA in the form of acetyl-CoA. Reproduced, with permission, from Lea P and Leegood R (1999) *Plant Biochemistry and Molecular Biology*, Wiley-Blackwell, Chichester, p128.

Plastids are the sites of FA synthesis, a process that involves two enzymes, acetyl-CoA carboxylase and FA synthetase. Acetyl-CoA carboxylase, with its four catalytic sites[4], is very tightly controlled and appears to be a key regulator of the rate of synthesis. FA synthetase consists of six polypeptides, each of which has a specific role in building up the FA chain. This happens in a cyclic process, adding two carbons at a time, with acetyl-CoA as the original 'feedstock' (Figure 6.1). The first steps in pushing two carbons into the cycle are catalysed by acetyl-CoA carboxylase. Biotin[5] is carboxylated in an ATP-dependent reaction; the source of the CO_2 is the bi-carbonate ion. In the meantime, the acetyl moiety from another acetyl-CoA molecule has been transferred to *acyl carrier protein* (ACP) and this acts as the primer for FA synthesis. The malonyl moiety from malonyl-CoA (synthesised by acetyl-CoA carboxylase, as just described) is then transferred to another molecule of ACP. The original acetate from this malonyl-ACP is then donated to the primer acetyl-ACP, releasing one ACP molecule and the CO_2 that was originally 'fixed' in the carboxylase reaction. This leaves the four-carbon unit acetoacetyl-ACP at the site of synthesis.

---

4 In dicots, the enzyme consists of four polypeptides, each with one catalytic site; in monocots, it is one large (240 kDa) polypeptide with four catalytic sites.
5 Vitamin B$_7$

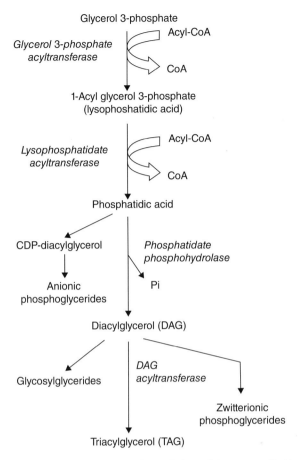

Figure 6.2 The Kennedy pathway for synthesis of triacyl glycerol. Figure supplied by Professor J. L. Harwood, University of Cardiff.

So, while still attached to ACP, this four-carbon acid is successively reduced (with NADPH as the hydrogen donor), loses a water molecule and is again reduced by NADPH. The cycle then starts again with the donation of two more carbons from malonyl-ACP. In most plants, the main end-product of this cyclic process synthesis cycle is 18:0-ACP, with some 16:0-ACP, although, as noted, 16:0-ACP is a major end-product in palm fruit. The plastids themselves contain a desaturase enzyme, which can insert one double bond to give 16:1-ACP and 18:1-ACP.

However, further modification of the FAs can only take place in the endoplasmic reticulum (ER), so the 16:0, 16:1, 18:0 and 18:1 FAs are exported from the plastid after their transfer from ACP to CoA. In the ER, a combination of desaturase and elongase enzymes produces a range of unsaturated FAs, including 16:3, 18:2, 18:3, 20:1 and 22:1. However, our main focus here is on the synthesis of triacyl-glycerol (tri-glycerides). This occurs *via* the 'Kennedy pathway' (Figure 6.2), in which acyl transferase transfers successively two FA molecules from CoA to glycerol-3-phosphate. The phosphate is then removed and the third FA is joined to the glycerol molecule. Acyl transferases exhibit different levels of activity and specificity with different FAs, and thus have a role in determining the FA composition of the triglycerides. For storage in the seed,

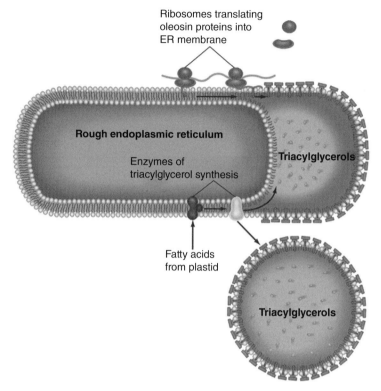

Figure 6.3 Formation of oleosomes. Reproduced, with permission, from Hodson M. J. & Bryant J. A. (2012) *Functional Biology of Plants*, Wiley-Blackwell, Chichester.

the triglycerides accumulate in the space between the outer and inner membranes of the ER. Ribosomes on the ER surface synthesise specialised proteins called oleosins, which are inserted into the outer ER membrane and then the outer membrane containing the tri-glycerides breaks off and rounds up to form oil bodies or oleosomes (Figure 6.3).

6.2.2 Potential for Manipulation

It has been known for over 20 years that plant FA biosynthesis is amenable to modification by GM techniques. Down-regulation and/or over-expression of endogenous genes plus expression of transgenes in a variety of lipid-storing plants showed that specific targeted changes can be made to the FA composition of TAGs. One of the possible difficulties is that while a plant may tolerate extensive changes to the FA composition of storage lipids, the same may not be true of membrane lipids. However, it has long been known that plants which accumulate unusual FAs in their storage TAGs do not direct these to membrane lipids; they are only incorporated into the TAG fraction of the plant's storage lipids and it appears that the same is true of novel FAs produced as a result of genetic manipulation. Indeed, it is becoming apparent that regulation of the synthesis of TAGs may involve more than the basic Kennedy pathway. Furthermore, in recent work on modification of the TAGs of *Camelina sativa*, in order to produce TAGs with a high proportion of omega-3 FAs, it was clear that this species certainly has:

> ... the capacity ... to accumulate high levels of novel fatty acids in TAG and suggests that there may be specific lipid pools, separate to those for membrane synthesis, that channel [FAs] into TAG. How this may be achieved remains unclear: it may involve a subset of specific enzymes or a spatial separation in the ER membrane ...; however, it is clear that manipulation of plant seed oil composition is heavily dependent on the metabolic pathways of the host.[6]

The quotation at the end of the previous paragraph echoes the comment made by Dyer *et al.*, back in 2008[7]:

> A major challenge ... is the spatial organization of the lipid biosynthetic machinery within the various subcellular compartments. Until a better picture of the metabolic channelling occurring within the cell is obtained, the full potential of genetic engineering of plant oil quality cannot be realized.

This is still to a large extent true. Nevertheless, as the work on omega-3-containing TAGs in *C. sativa* has shown, this lack of detailed knowledge has not prevented specific and focused application of GM techniques to plant FA production[8]. However, this has not been extended to any significant extent to the production of biodiesel. At present it is more cost-effective to blend different biodiesels to suit particular climates rather than to expend resources on changing the FA make-up of TAGs in individual species.

6.3 Productivity

As mentioned in other chapters, the demand for liquid fuels is huge. At the time of writing, use of oils in the UK alone is over 93 billion litres per year, contributing to a global yearly total of over 5 trillion litres. Set against that, we have the typical productivity values for the main oil crops (Table 6.3). As discussed in more detail in Chapter 17, using all the UK's arable land for oil-seed rape (canola) would supply less than 10% of the country's oil requirements. Or as Oliver Chadwick, from the UK's Department for Transport, puts it: 'To provide biofuel for one lane of cars requires a strip of land the length of that lane, and 8 km wide'[9]. On a global scale, it is clear that there is a very large gap between current oil usage and the amount that can be obtained from oil crops. Adding ethanol into the equation (Table 6.4) closes the gap slightly, even accounting for lower 'energy yield' (Table 6.5), but the overall shortfall is still enormous. Indeed, reducing our dependence on fossil oil is an objective that involves engineering, economics and social policy in addition to biology. One small example illustrates this: running trains on electricity rather than on diesel oil first allows the trains to be lighter (no need

---

6 Ruiz-Lopez *et al.* (2014).
7 Dyer *et al.* (2008).
8 The first harvests from field trials of omega-3-producing *Camelina* were taken in August 2014.
9 Quoted by Tom Ireland (2013).

Table 6.3 Productivity of selected oil crops, litres/hectare.

Species	Yield of oil[*]
Oil palm	4750
Coconut	2151
Oil-seed rape/Canola	954
Peanut	842
Sunflower	767
Soybean	524

[*] Under optimal conditions

Table 6.4 Productivity of selected ethanol crops, litres/hectare.

Species	Yield of ethanol
Sugar beet (France)	6,676
Sugar cane (Brazil)	6,190
Cassava (Nigeria)	3,834
Sweet Sorghum (India)	3,497
Corn[*] (USA)	3,310
Wheat (France)	2,590

[*] From corn starch, not cellulosic biomass (see Chapter 5); see also Figure 8.2, Chapter 8

Table 6.5 'Energy content' of fuels, KJ/kg.

Petrol	48,000
Diesel	44,800
Ethanol	30,000
Plant oils	39,000–48,000

to carry fuel) and second, allows the use of non-fossil and renewable sources in the generation of the electricity.

Set against this background, increasing biofuel crop productivity can only make a small contribution to closing the gap, but even a small contribution is considered worthwhile. Essentially there are three approaches to this. First, there is agronomic practice. This lies outside the scope of this chapter, except to comment briefly that it reminds us again that growing biofuel crops involves the utilisation of agricultural resources (see Chapters 7, 8 and 17). Second, there is classical breeding. The desired traits are very clear: yield and FA composition of TAGs. A recent study by Ian Bancroft's group at the John Innes Centre, Norwich, UK, focused on six in-bred lines of *Arabidopsis thaliana* (a near relative of oil-seed rape, *Brassica napus*) derived from parental stocks with differing geographical backgrounds and with differences in the yield and composition of TAGs[10]. They identified 219 quantitative trait loci (QTLs), which contributed to TAG accumulation and composition (including the amounts and ratios of the different FAs), of which 81 were significant at the $p < 0.001$ level. In other words, many genes are involved in determining the TAG content of *A. thaliana* seeds and it is important now to identify which are worth focusing on for plant breeding purposes.

The third approach is genetic modification. We have already seen how GM techniques may be used to modify the FA content of TAGs (as in the omega-3-producing *Camelina*), but here we are focusing on yield. In particular, the objective of applying GM techniques to yield is the increase the flow of carbon through acetyl-CoA carboxylase (ACC) into FAs and thence into triglycerides. As mentioned in Section 6.2.1, ACC is a highly regulated enzyme that appears to control the overall rate of the pathway. It is regulated by the ratio of ATP:ADP + AMP (in a similar way to mammalian phospho-fructo-kinase) and by the supply of carbon. In the latter, ACC is inhibited when an inhibitory protein blocks the enzyme's binding to its essential co-factor biotin. The inhibition is reversed in the presence of 2-oxoglutarate and/or pyruvate and/or oxaloacetate. Finally, ACC is subject to feedback inhibition by 18:1-ACP, meaning that when 18:1 (one of the main products of FA synthetase) is not incorporated into TAGs or other lipids, entry into the start of the pathway is inhibited.

Because of its key role in regulation of FA synthesis, ACC has long been regarded as a target for GM in order to increase throughput through the pathway. There has been some success with this approach in bacteria and in lower eukaryotes (see Section 6.4), but so far the results with higher plants have not been spectacular. Over-expression of the sub-units of the plastid ACC (readers will recall that FAs are made in the plastids) did not result in increased yields. However, when the 'eukaryotic-type' cytosolic ACC was targeted to the plastids in potato, a five-fold increase in TAG (albeit starting from a very low baseline) was observed[11]. Similar experiments with *Brassica napus*, oil-seed rape, although resulting in 10- to 20-fold increase in ACC activity, only led to a 5% increase in seed TAG content[12]. In other lipid-storing plants, a more successful approach has been over-expression of the relevant transcription factors[13], which led to an increase in seed lipid content of between 11 and 22% in *Arabidopsis thaliana*.

Two more points need to be made. First, as mentioned above, there are many QTLs that contribute to the synthesis and accumulation of TAGs in seeds. Some of these have already been identified as genes encoding enzymes and carrier proteins involved at specific points in the pathway. This process of matching QTLs to specific genes is an ongoing process and will lead to those genes on which it is worth focusing for plant breeding, whether by 'traditional' or GM methods. Second, there is the question of partitioning of FAs between the synthesis of TAGs and the synthesis of other lipids that have specific roles in plant cells, including phospho- and glyco-lipids, noting that these often have a very different FA composition from the TAGs. The regulation of this partitioning and of the different FA ratios is not clearly understood, except that there is a clear developmental aspect. TAG synthesis, although not confined to seeds, is very

---

10 O'Neill *et al.* (2012).
11 Klaus *et al.* (2004).
12 Roesler *et al.* (1997).
13 Song *et al.* (2013); Liu *et al.* (2014).

much more active in developing seeds than anywhere else. Greater knowledge of this regulation will again help to identify suitable targets for traditional and molecular breeding.

6.4 Sustainability[14]

As mentioned in the previous section and as further discussed in Chapters 7, 8 and 17, there is an obvious tension between growing biofuel crops and growing food crops. The world's human population is growing towards 7.5 billion. Furthermore, it is very probable that we will not be able to increase crop yields sufficiently by 2050 to feed the world's population[15]. These facts present us with an urgent and, many would say, overriding need to retain good arable land for food production. In some circumstances, the clash between food and fuel production can be avoided, as discussed in Chapters 7 and 8. Furthermore, use of 'waste' material after harvest of food crops, either as biomass or as a source of cellulose-derived ethanol (Chapter 5), does not result in a decrease in food production.

Nevertheless, the problem still remains and one further solution has been to use non-agricultural land for biodiesel crops. However, this can also have very negative consequences, as is evident from the growth of oil-palm, especially in Asia. Large areas of rain forest have been destroyed with consequent loss of biodiversity and of carbon sink, leading eventually to degradation of the land (see Chapter 17 for a fuller discussion). If non-agricultural land is to be used, it must be done in areas where the ecological impact is low. This in turn may mean using plants able to grow and produce yield on more marginal land (see next section), as well as algae and bacteria that obviously do not require land. This leads to consideration of novel sources of TAGs for biodiesel production.

6.5 More Recently Exploited and Novel Sources of Lipids for Biofuels

6.5.1 Higher Plants

In respect of biodiesel, the 'big six' higher plant oil producers are oil palm (*Elaeis guineensis*), soybean (*Glycine max*), oil-seed rape (canola, *Brassica napus*), peanut (*Arachis hypogea*), sunflower (*Helianthus annuus*) and *Jatropha curcas*, with other species such as mustard, coconut, flax (linseed), safflower and cotton making smaller contributions. Lipids from all of these species had a range of uses before their utilisation in biofuel manufacture[16]. Many of them are used widely in the food chain, while others have applications in cosmetic manufacture and in industry (*e.g.* as lubricants). Some applications are very specialised. For example, one of the uses of linseed oil (from flax) is oiling cricket bats[17], something of which the author of this chapter has direct knowledge.

---

14 See also Chapter 16, Sustainability of biofuels.
15 Ray *et al.* (2013).
16 Although, as noted earlier, Rudolf Diesel's first engines ran on peanut oil.
17 Some modern cricketers prefer 'poly-coated' bats that do not need to be treated with oil.

As mentioned in Section 6.1 and immediately above, a plant of sub-tropical latitudes, *Jatropha curcas* is now regarded as a useful oil crop. Its yield is 540–680 litres per hectare, depending on cultivar and growth conditions, placing it somewhere between soybean and sunflower. In terms of FA composition, 18:1 predominates, suggesting that it is ideal for use as a source of biodiesel, although the 18.2 content is a little high (see Section 6.1 and Chapter 2). *Jatropha* is a perennial, capable of withstanding extremes of heat and can grow on very poor land. However, if grown on more fertile land, it can be combined with other crops in an inter-cropping system.

Another perennial oil plant that can be grown in difficult environments, including deserts and semi-deserts, is jojoba (*Simmondsia chinensis*), a native of the south-western corner of the USA (despite its somewhat misleading Latin name). Indeed, plantations of jojoba have been established in that region of the USA and also in desert and semi-desert areas in Argentina, Australia, India[18], Israel, Mexico, Peru and several middle-eastern countries. Furthermore, the oil is inedible so any non-food use does not detract from growth of food. Jojoba oil does not consist of TAGs but is actually, technically speaking, a liquid wax, comprised of long-chain FAs (mainly C20 and C22) esterified with long-chain alcohols. Nevertheless, it is easily trans-esterified to form biodiesel, albeit with a different composition from 'conventional' biodiesel. It is resistant to oxidation, has a higher calorific value than conventional biodiesel and performs better at lower temperatures than soybean-derived biodiesel. However, it has as yet achieved little market penetration.

In Australia, *Jatropha* is regarded as a noxious weed and its growth as a biofuel crop is not supported. However, another perennial lipid-storing plant, the leguminous tree, *Milletia pinnata*, is regarded in Australia and India as a useful addition to the range of plants used for sustainable biodiesel production. It is a native of arid regions of Australia and of several other countries in the sub-tropical to tropical parts of Asia. Its seeds contain several toxic compounds and so the tree has never been used as a food crop, although it has been grown as a wind break and to stabilise sandy soils. Depending on genetic lineage and growth conditions, the oil content of the seeds varies between 25 and 40% dry weight. The FA make-up of the TAGs is strongly weighted to mono-unsaturated acids, with 18:1 contributing 49% of the total. This makes *M. pinnata* oil an ideal candidate for use in the manufacture of biodiesel.

6.5.2 Algae

As hinted at earlier, the quest for sustainable biodiesel production has extended to algae and bacteria. In Chapter 2, Lionel Clarke mentions algae as sources of lipids for biofuel production and it is clear that several different species of both fresh-water and salt-water micro-algae have this potential, albeit that this has only been realised during the past few years. However, focusing for the moment on one species, it was not for oil that *Dunaliella salina*, a potential biofuel source, was first cultured commercially. It is the richest natural source of β-carotene and is now cultured on a commercial scale in Australia, China, Israel and the USA to supply this compound. This means that there is already extensive experience in culturing *Dunaliella* which is helpful now that the alga will also be cultured for biofuel production. Also helpful is the completion of the

---

18 In India, it has been planted to stabilise deserts, preventing further desertification.

sequencing of the *Dunaliella* genome, which will be an aid in breeding and genetic manipulation.

In respect of lipid productivity and FA composition of TAGs, there is considerable variation between both different species of micro-algae and under different growth conditions within species. In *D. salina* for example, conditions that induce accumulation of carotene, also induce accumulation of oleic acid (18:1) in the oil globules. Thus the FA profile shown for 'algae' in Figure 2.4 in Chapter 2 is not representative of all micro-algae. There has as yet been very little selection of strains for improvement of productivity, but this situation is changing with the development of selection strategies discussed by Leyla Hathwaik and John Cushman in Chapter 11. However, even at this stage of development, it is clear that some micro-algae can convert 60% of their biomass to oils. Indeed, it is claimed that on a surface area basis, it is already possible for micro-algae to produce 200 times as much yield as a higher plant biofuel crop. Despite such productivity and the research focused on increasing it, it is estimated that by 2020, only 0.2% of 'road fuel' will be produced from algae. This is at least partly related to the capital expenditure relating to large-scale algal culture for biofuel production in relation to the return on investment.

Some micro-algae, typified by *Botryococcus braunii*, do not produce TAGs, even though they accumulate high concentrations of hydrocarbons. This means that they cannot be used as sources of lipids for conventional biodiesel production, but nevertheless they may be valuable feedstock for other types of biofuel, including petroleum-type fuels, although a good deal of research and development will be needed to realise this, as discussed in Chapter 10.

I note in passing that macro-algae are also regarded as sources of biofuels. However, rather than biodiesel, the most likely use of macro-algae is in production of ethanol, as described by Jessica Adams in Chapter 13.

6.5.3 Prokaryotic Organisms

Many prokaryotic organisms produce compounds that may be used as feedstock for different types of biofuel. However, the exact use depends on the types of compound accumulated. For example, cyanobacteria do not accumulate high concentrations of lipids, as discussed by David Lea-Smith and Christopher Howe in Chapter 9. Nevertheless, they are very amenable to genetic manipulation and this has been used to generate strains that synthesise and secrete ethanol and strains that synthesis alkanes, which can be used in production of both diesel and petroleum type fuels.

There are several bacterial species that do accumulate TAGs (albeit that the FA synthesis pathway is slightly different from that in plants) and further are readily amenable to genetic modification, as discussed by Thomas Howard in Chapter 15. This has facilitated the generation of strains of *Escherichia coli* that produce TAGs with a suitable FA content for manufacture of biodiesel. However, perhaps even more exciting is the recent work of John Love and his colleagues at Exeter University, in which GM and synthetic biology approaches were used to modify *E. coli* lipid metabolism to produce alkanes and alkenes suitable as a petroleum substitute[19] (see also Chapter 2).

---

19 Howard *et al.* (2013).

6.6   Concluding Remarks

This chapter has focused mainly on the synthesis of FAs and TAGs in relation to biodiesel production. Established and more novel sources of these compounds have been discussed, along with a brief mention of petroleum-replica biofuels. However, behind all this discussion lies the big problem. We are very dependent on oil and even the most optimistic projections do not see the production of biodiesel and other liquid biofuels as coming anywhere near meeting that dependence. At best, with current and likely future technologies, biodiesel, petroleum-replicas, ethanol and other liquid biofuels are likely to make only relatively small inroads into the world's annual use of oil. However, that does not in any way detract from the value of sustainably produced biofuels. Even a small reduction in the use of fossil fuels is worth pursuing.

Selected References and Suggestions for Further Reading

Dyer, J.M., Stymne, S., Green, A.S. and Carlsson, A.S. (2008) High-value oils from plants. *Plant Journal*, **54**, 640–655.

Harwood, J.L., Ramli, U.S., Tang, M., Quant, P.A., Weselake, R.J. *et al.* (2013) Regulation and enhancement of lipid accumulation in oil crops: The use of metabolic control analysis for informed genetic manipulation. *European Journal of Lipid Science and Technology*, **115**, 1239–1246.

Hodson, M.J. and Bryant, J.A. (2012) *Functional Biology of Plants*. Wiley-Blackwell, Chichester and Oxford, UK.

Howard, T.P., Middelhaufe, S., Moore, K.M., Edner, C., Kolak, D.M. *et al.* (2013) Synthesis of customized petroleum-replica fuel molecules by targeted modification of free fatty acid pools in *Escherichia coli. Proceedings of the National Academy of Sciences, USA*, **110**, 7636–7641.

Hu, Q., Sommerfeld, M., Jarvis, E., Ghirardi, M., Posewitz, M., Seibert, M. and Darzins, A. (2008) Microalgal triacylglycerols as feedstocks for biofuel production: perspectives and advances. *Plant Journal*, **54**, 621–639.

Ireland, T. (2013) Algal biofuels. *The Biologist*, **61**, 20–24.

Klaus, D., Ohlrogge, J.B., Neuhaus, H.E. and Doermann, P. (2004) Increased fatty acid production in potato by engineering of acetyl-CoA carboxylase. *Planta*, **219**, 389–396.

Liu, Y.-F., Li, Q.-T., Lu, X., Song, Q.-X., Lam, S.-M. *et al.* (2014) Soybean GmMYB73 promotes lipid accumulation in transgenic plants. *BMC Plant Biology*, **14**(73).

O'Neill, C.M., Morgan, C., Hattori, C., Brennan, M., Tschoep, H. *et al.* (2012) Towards the genetic architecture of seed lipid biosynthesis and accumulation in *Arabidopsis thaliana. Heredity*, **108**, 115–123.

Pinzi, S., Mata-Granados, J.M., Lopez-Giminez, F.J., Luque de Castro, M.D. and Dorado, M.P. (2011) Influence of vegetable oils fatty-acid composition on biodiesel optimization. *Bioresources Technology*, **102**, 1059–1065.

Podkowrinski, J. and Tworak, A. (2011) Acetyl-coenzyme A carboxylase – an attractive enzyme for biotechnology. *BioTechnologia*, **92**, 321–335.

Ray, D.K., Mueller, N.D., West, P.C. and Foley, J.A. (2013) Yield trends are insufficient to double global crop production by 2050. *PLoS ONE*, **8**(6), e66428.

Roesler, K., Shintani, D., Savage, L., Boddupalli, S. and Ohlrogge, J. (1997) Targeting of the Arabidopsis homomeric acetyl-coenzyme a carboxylase to plastids of rapeseeds. *Plant Physiology*, **113**, 75–81.

Ruiz-Lopez, N., Haslam, R.P. and Napier, J.A. *et al.* (2014) Successful high-level accumulation of fish oil omega-3 long-chain polyunsaturated fatty acids in a transgenic oilseed crop. *Plant Journal*, 77, 198–208.

Shah, S.N., Sharma, B.K., Moser, B.R. and Erhan, S.Z. (2010) Preparation and evaluation of jojoba oil methyl esters as biodiesel and as a blend component in ultra-low sulfur diesel fuel. *BioEnergy Research*, **3**, 214–223.

Sharma, K.K., Schumann, H. and Schenk, P.M. (2012) High lipid induction in microalgae for biodiesel production. *Energies*, **5**, 1532–1553.

Song, Q.-X., Li, Q.T., Liu, Y.F., Zhang, F.X., Ma, B. *et al.* (2013) Soybean GmbZIP123 gene enhances lipid content in the seeds of transgenic Arabidopsis plants. *Journal of Experimental Botany*, **64**, 4329–4341.

7

Development of *Miscanthus* as a Bioenergy Crop

John Clifton-Brown[1], Jon McCalmont[1] and Astley Hastings[2]

[1] Environmental and Rural Sciences (IBERS), Aberystwyth University, Wales, UK
[2] Institute of Biological and Environmental Science, University of Aberdeen, Scotland, UK

Summary

In many ways, giant grasses in the genus *Miscanthus* are ideal energy crops. They are perennial grasses which carry out C4 photosynthesis and thus have the potential to produce good yields of biomass each year. Furthermore, they require little input and thus may be grown on land that is marginal in respect of growth of food crops. Growth of *Miscanthus* as a biofuel crop therefore need not exacerbate 'food *vs* fuel' conflicts. In respect of current use and future potential, the hybrid *Miscanthus x giganteus* is regarded as having the best prospects. Indeed, the number of farmers growing it has been steadily increasing since the early 1990s, although in the UK the area devoted to it is still far short of the 400 KHa projected for this crop. Harvested *Miscanthus* is currently used as a biomass fuel for direct combustion at scales from domestic to industrial, but it can also be used in the production of 'cellulosic ethanol' and as a feedstock for anaerobic digesters. Current research is directed at improvement of the crop itself and at better understanding of crop physiology in relation to agronomic conditions. Both of these will lead to the development of strains for particular situations and for particular applications, increasing the attractiveness of the crop for farmers.

7.1 Introduction

The ideal bioenergy crop will be a rapid-growing, highly productive perennial that is easily harvested and processed. Plants that fix carbon by the C_4 mechanism have some of the highest growth rates and productivities of all terrestrial plants and many of them, such as the 'giant' perennial grasses, also fit other criteria for use as bioenergy crops. Unfortunately, the majority of C_4 species are of tropical or subtropical origin and are poorly adapted to growing in cool and temperate climates such as those found in much of Northern Europe. However, the genus *Miscanthus*, with about 20 species that are spread across parts of Eastern Asia, provides some exceptions to this.

Biofuels and Bioenergy, First Edition. Edited by John Love and John A. Bryant.
© 2017 John Wiley & Sons Ltd. Published 2017 by John Wiley & Sons Ltd.

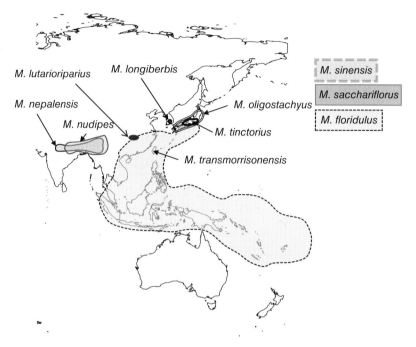

Figure 7.1 Geographical distribution of the major *Miscanthus* species. The distribution of *M. x giganteus* is not fully known, but can potentially be found in regions where *M. sinensis* and *M. sacchariflorus* overlap. Taken from: Clifton-Brown *et al.* (2011, reproduced with permission of the Royal Society of Chemistry).

The approximate geographic distributions of *Miscanthus* species in Eastern Asia are shown in Figure 7.1. The species with the most wide-ranging distribution is *M. sinensis*, stretching from Hebei province, just south of Beijing, to Hong Kong, and including Korea, Taiwan and northern Japan to Ryukus. In contrast, *M. sacchariflorus* has a more restricted distribution than *M. sinensis*, although it embraces a very wide range of climates from Russia, North China, Korea and Japan (Honshu). Both *M. sinensis* and *M. sacchiflorus* are grown as ornamentals. The distribution of *M. floridulus* is confined to latitudes below 30° N, since it is relatively frost sensitive. For *M. lutarioriparius*, often classified as a sub-species of *M. sacchariflorus*, the distribution is more restricted to the vicinity of the Yangtze river system. *M. transmorrisonensis*, which is morphologically rather similar to *M. sinensis*, is found at altitudes above 2,500 m in Taiwan where snow is frequent. *M. longiberbis* is a South Korean endemic species, and it has been proposed that it forms a link between other members of section *Kariyasua* and *M. sacchariflorus* (Ibaragi and Oshashi, 2004). *M. tinctorius* is similar to *M. longiberbis*, but has no awns on its flowers; it is used as a source of traditional yellow colour dye in Japan. *M. nudipes* and *M. nepalensis* have only two anthers (others have three) and are sometimes treated as an independent genus, *Diandranthus*[1]. They are mainly distributed around the Himalayan region discontinuously from other *Miscanthus* members. The *Miscanthus* genus is closely related to *Saccharum* (the genus containing sugar cane, *Saccharum officinarum*) and

---

1 Ibagari (2003).

Figure 7.2 Flowers of *Miscanthus sinensis*. Photo by Miya M, Japan; reproduced from Wikimedia Commons under the terms of the GNU Free Documentation License.

indeed, some *Miscanthus* species have been successfully crossed with *Saccharum* producing intermediate hybrids.

Hybrids also occur naturally within the genus *Miscanthus* itself[2]. The basic chromosome number is 19 but some species are diploid and some are tetraploid. This means that natural hybrids may be sterile because they are triploid, as is the case with the first discovered and best known hybrid, *Miscantheus x giganteus*. It is a hybrid between the diploid *M. sinensis* and the tetraploid *M. sacchiflorus*, which occurs in Japan where the two species sometimes grow in the same area[3].

The use by humans of *Miscanthus* species dates back thousands of years, especially in Japan, where these grasses have been used for forage and thatching. Huge areas were managed by grazing and burning in a manner that has been compared to the management of the prairies by Native Americans (Stewart *et al.*, 2009). In Europe, *Miscanthus* was initially valued for its ornamental appearance. Its tall straight stems and striking silver flowers (Figure 7.2) made it ideal for planting in country estates and large gardens.

---

2 Scally *et al.* (2001).
3 Greef and Deuter (1993).

Figure 7.3 *Miscanthus* growing at Aberystwyth, UK with Dr Elaine Jensen and Dr John Clifton-Brown.

Miscanthus x giganteus, collected from Japan in 1935 by Danish horticulturalists, increased further *Miscanthus*' popularity as an ornamental 'architectural' plant. We note in passing that one of the parent species of giant *Miscanthus x giganteus*, *M. sacchariflorus*, has been used as an ornamental plant in the USA, but is now becoming an invasive nuisance in parts of that country.

However, today, *Miscanthus x giganteus* is better known as a bioenergy crop. This use was first proposed after the oil crisis of the 1970s (see Chapters 1), when the search for alternative fuels began to gather pace. Here was a highly productive, cold-hardy C_4 grass with fast growth rates (Figure 7.3), capable, in some conditions, of growing to over 4 m in a single season, producing considerably more biomass than any of the individual *Miscanthus* species. Further advantages were its ability to grow on marginal land and, under good management, the 15- to 20-year life of individual stands. It had not at that time been subject to any breeding programme in relation to its use as a bioenergy crop (remembering that it is a sterile hybrid), but nevertheless, evaluations to determine the biomass yields of different giant *Miscanthus* clones began across Europe.

7.2 Developing Commercial Interest

In Chapter 1 it was noted that the development of bioenergy crops gained further momentum in the 1990s, because of a widening perception that for both political reasons and as a response to climate change, we needed to be less dependent on fossil fuels. This increased momentum was especially obvious in Europe, where economic policies

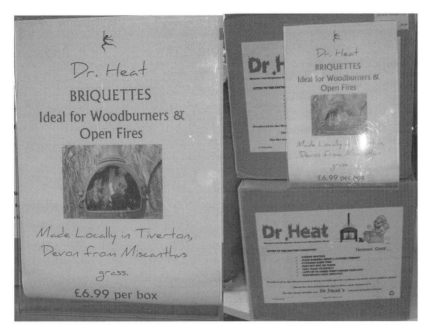

Figure 7.4 Dried and compressed *Miscanthus* on sale for domestic heating at a farm shop near Exeter, UK (Photograph by John Bryant).

to promote the development of energy crops led to increased private sector interest. In Germany for example, TINPLANT Biotechnik und Pflanzenvermehrung GmbH started a breeding programme aimed at bringing alternative hybrids, superior to *M. x giganteus*, to the market. Dr Ralph Pude in Bonn started the *Miscanthus* society (Internationale Vereinigung für *Miscanthus* und mehrjährige Energiegräser (MEG) e.V[4] as a vehicle to translate research into commercial reality.

In the UK, commercial-scale agronomy and the harvest chains required to deliver biomass crops suitable for a variety of end users were pioneered by several start-up companies, the best known being Bical, based in Taunton, UK. Bical developed the agronomy for rhizome planting of *M. x giganteus* and offered growers contracts with power station end-users. Several other organisations and networks have since been set up[5] and growers also sell *Miscanthus* for use in smaller-scale combined heat and power units and for use as a domestic fuel (Figure 7.4). In the latter context, 1 tonne of *Miscanthus* can produce as much heat as ca 485 litres of heating oil.

Furthermore, over the past 20 years or so, field trials and eco-physiological experiments in Europe have led to a better understanding of the exceptional performance of this C_4 plant in temperate climates. Field trials in Ireland and Europe planted with *M. x giganteus* from 1990 onwards were used to parameterise the growth model

---

4 http://www.Miscanthus.de
5 See, for example http://www.terravesta.com/Growing, http://recrops.com/, http://www.miscanthus.org.uk/index.html

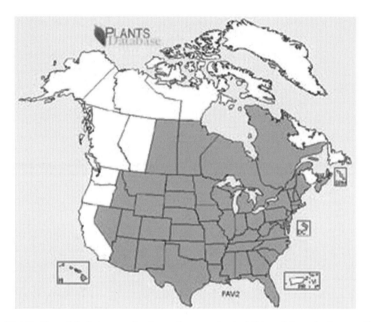

Figure 7.5 Distribution of Switchgrass, *Panicum virgatum*, in the USA and Canada. Source: USDA-NRCS PLANTS database.

MISCANMOD[6]. This model was developed later into a powerful and flexible Fortran version (MiscanFor)[7]. This is being used to predict *Miscanthus* crop performance in different soils under current and future climate conditions using multi-location trials to establish genotype × environment interactions in wild and recently bred novel hybrids, thus adding to the usefulness of this plant as a bioenergy crop[8].

We note in passing that in 1999, in Germany, Frank Möller pioneered the higher value use of *Miscanthus* in light natural sandwich (LNS) construction materials in a private-public funded project. This material was 50% less dense and on a mass basis was stronger than solid wood or plywood. The product idea was acknowledged by an award from lower Saxony in Germany, but it failed to attract sufficient investment to make it a commercial prospect.

During the 1990s, while Europe was focusing on *Miscanthus* as a bioenergy crop, much of the attention in the USA had been on Switchgrass (*Panicum virgatum*), the main giant grass of the American prairies. It is tolerant of a wide range of climatic conditions and thus can be grown in most of the USA (Figure 7.5); like *Miscanthus*, it will grow on more marginal land under low-input agriculture. However, the situation changed in 2000, when Professor Steve Long moved from the University of Essex in the UK to the University of Illinois at Urbana-Champaign. Professor Long took with him his European experience of *Miscanthus* and, together with his research student Emily Heaton, set up field experiments with *M. x giganteus* cloned from plants obtained from the Chicago Botanic Garden. The plants in Illinois established rapidly under the much

---

6 Clifton-Brown *et al.* (2004).
7 Hastings *et al.* (2009b).
8 Hastings *et al.* (2009a).

warmer summer conditions than in Europe and Ms Heaton's father, an Illinois farmer, was so impressed by the demonstration trials on his farm, that in 2002 he provided some financial support to invite Professor Mike Jones and Dr John Clifton-Brown (one of the authors of this chapter) from Dublin, Ireland to engage with Illinois academics, growers and business developers.

In Illinois, plot yields in excess of 50 tonne Dry Matter ha^{-1} were reported from small quadrats[9]. Although the sampling procedure was scientifically acceptable for inter-comparison of different species (the trial compared *Miscanthus* with Switchgrass), yield upscaling errors probably exaggerated the harvestable yield. Another issue in reporting yield was the December harvest time, which is too early for attaining a well senesced, dry crop, low in nutrients, suitable for combustion purposes. Even so, with the probability of spring harvestable yields in excess of 30 t DM ha^{-1} y^{-1}, there was huge potential to produce lignocellulosic biomass from *Miscanthus* in the Mid-West[10].

Yield estimates from the 2004 version of MISCANMOD (see above) were quoted widely and used by commercial developers in Europe. MISCANMOD was also used to project yield in the USA. For Illinois, MISCANMOD projected autumn peak yields[11] from 27 to 44 t DM ha^{-1}. These projected yields were 10–30% lower than actual plot yields reported by Heaton *et al.* in their 2008 paper[10]. However, pre-harvest yield losses are strongly influenced by the delay in harvest time to allow for the crop to ripen. Allowing for full ripening, the MISCANMOD yield estimates were probably a more realistic commercial guide to production than the very high yield estimates reported from the quadrats (see above).

Professor Long and seven others involved in alternative energy projects were invited to the White House to brief US President George W. Bush and advisors about their results. Shortly afterwards, in the 2007 State of the Union Address, the President announced that 'everything from wood chips to grasses to agricultural wastes' should be used to create renewable biofuels. British Petroleum also took a strong interest, and announced an international competition for an Energy Biosciences Institute (EBI). This $350 M Institute was won by a Berkeley-Illinois coalition, headed by Professors Chris Somerville and Steve Long. The Illinois *Miscanthus* programme was a centre-piece of this effort and the support facilitated an explosion of knowledge in the crop in the USA, from genomics (including sequencing of genomes) and molecular understanding of chilling tolerance to ecosystem services, pests and diseases. A core activity of the Institute is in developing the technologies for converting lignocellulose from crops such as *Miscanthus* into next-generation transport biofuels (see Chapter 5). *Miscanthus* thus took its place alongside Switchgrass as a major bioenergy crop in the USA.

Since 2006, a team at Aberystwyth, UK have led Asian germplasm collection missions and have assembled one of the largest *ex-situ Miscanthus* germplasm collections outside Asia. Selection and breeding work has been undertaken with partners in the USA, UK and mainland Europe in a number of interlinked projects. Since 2011, the project GIANT-LINK, a £6.4 m collaboration between public (UK BBSRC and Department of Food, Rural Affairs and Agriculture (DEFRA)) and private partners (CERES, Blankney

---

9 Heaton *et al.* (2004b); Heaton *et al.* (2008).
10 Heaton *et al.* (2008); Dohleman *et al.* (2009).
11 Heaton *et al.* (2004a).

Estates, E.ON and NFU) has pushed *Miscanthus* breeding forward rapidly. Genome and transcriptome sequencing of parental species, along with assembly of high density marker maps, have contributed to this breeding programme[12]. In 2013, seed production of unique parental combinations, discovered through earlier test crosses in Aberystwyth, was started at the field scale by CERES breeder Charlie Rodgers in the USA. These new seed-based hybrids are being trialled in the UK and Europe, at a wide range of sites with differing climates and soils, with institutional and commercial partners. In the UK, a strong market for *Miscanthus* has been established by Terravesta Ltd[13], who are experts in fuel chain logistics for both large-scale power generation and smaller-scale local heating schemes.

Text Box 1 Steps in a Miscanthus breeding programme.

The first step is the *collection and characterisation* of diverse germplasm with traits that could confer advantages in novel hybrids. A breeding programme requires a broad range of germplasm collected across the range of latitudes and geographies, where the target species occurs, to maximise the opportunity to capture a full range of trait diversity. Trait characterisation in *Miscanthus* of different geographical origin is discussed in subsequent sections.

The second step is *hybridisation*. The *Miscanthus* genus is predominantly outbreeding, due to genetic self-incompatibility mechanisms where very low seed numbers are produced *via* self-pollination. For small quantities of seed, paired crosses are made by bagging together panicles from selected parents. For larger quantities of seed, crosses are performed in isolation chambers or in field plots. In either case, a paired cross often results in a seed set on both parents. The quantities of seed produced from a cross depend on many factors, including sexual compatibility, flowering synchronicity, humidity, temperature and plant health. Synthetic varieties[*] are used as the main approach to produce varieties that preserve heterozygosity and minimise inbreeding.

The third step is *ex-situ phenotypic characterisation* of wild germplasm and new hybrids in a range of climates, which is important in understanding genotype x environment interactions. Field evaluations of diverse germplasm in both spaced plots and multi-location trials are used to characterise novel accessions for yield potential and chemical composition. All *Miscanthus* species are perennial, so selections of outstanding crosses can only start to be made reliably after the second growing season. Phenotyping depends on the co-ordination of researchers at different sites and the implementation of standard protocols to ensure inter-site comparisons.

In the fourth phase, *large-scale demonstration trials* are used to develop the agronomic practices which are needed to successfully establish, manage and harvest the crop. Since *Miscanthus* biomass at harvest is low density, pelletisation and high density baling are being developed to improve storage and transport before the crop is used. Traits such as stem diameter are expected to define the most economic method for biomass densification and will consequently feed back into the selection of parents for hybridisation.

[*] Synthetic variety: 'a variety that is maintained from open-pollinated seed following its synthesis by hybridisation in all combinations among a number of selected genotypes.'

---

12 Ma *et al.* (2012); Kim *et al.* (2014).
13 http://www.terravesta.com/Growing

7.3   Greenhouse Gas Mitigation Potential

The main motivation for the work on developing *Miscanthus* as a crop was to use it to produce energy, displacing fossil carbon fuels and reducing greenhouse gas (GHG) emissions. Professor Mike Jones and his co-workers initiated detailed lifecycle analyses (LCA; see also Chapter 16) for various uses of *Miscanthus*, from a biomass fuel to a feedstock for producing biogas, biodiesel, methanol and ethanol, to assess if it did indeed reduce emissions[14]. This work included the impacts on soil carbon stocks of land use change from arable to *Miscanthus* [15] and from grassland to *Miscanthus*[16].

In comparison to arable land used for cereals, where soil organic carbon (SOC) levels will fall every year, SOC under *Miscanthus* plantations on arable lands will generally rise in the 0–30 cm layer to a level that is similar to grasslands. *Miscanthus* biomass harvestable yield levels are directly correlated with the carbon mitigation benefits per hectare as soil organic carbon equilibrium is a balance between the organic material input (leaf fall, root turnover, manure and compost amendments) and the decomposition rate of the organic material in the soil, rather analogous to tipping water into a leaking bucket where the level is a function of the volume tipped and the size of the hole. Higher yields result in higher carbon mitigation[18]. Yields respond more strongly to water input (from the balance of rainfall and evaporation) and are less sensitive to temperature variation in the normal UK climate range of 0 to 30 °C[17]. Furthermore, we are using models to quantify the factors that limit growth potential of different genotypes (*e.g.* to water deficit) in Europe-wide trials such as OPTIMISC, OPTIMA and WATBIO[18].

Miscanthus is a perennial crop and recycles nitrogen (N) from the plant to the rhizome during senescence to use in re-growth the next season, so little N is left in the dead plant material harvested. Consequently, no N fertiliser is normally required. For this reason, N_2O emission from *Miscanthus* is not a significant problem[19], as long as N fertilisation levels are kept low, that is less than 60 kg N ha^{-1}. N use efficiencies in *Miscanthus* stands are high with low off-takes of typically 30 to 70 kg N ha^{-1}, depending on yield and harvest time, although, under some growing conditions, N off-takes and consequently inputs are somewhat higher[20]. There is increasing evidence for N fixation in the soil under *Miscanthus* in some sites[21].

Depending on regional climate and inter-annual variation in climate, annual harvestable yields of *M. x giganteus* range from 10 to 15 tonnes per hectare in the UK. Due to the low energy and chemical input required to manage this perennial plant, our calculations show that the median carbon intensity of the crop is less than 1 g CO_2 equivalent C per MJ of energy, much less than coal (33), oil (22), shale gas (20), Liquefied Natural Gas (LNG) (21) and conventional gas (16)[22]. This results in 5.7 to 9 tonnes C equivalent

---

14 Styles and Jones (2008); Hastings *et al.* (2013).
15 Dondini *et al.* (2009).
16 Clifton-Brown *et al.* (2007); Zatta *et al.* (2014).
17 Richter *et al.* (2008).
18 https://optimisc.uni-hohenheim.de/; http://www.optimafp7.eu/; http://www.watbio.eu/
19 Drewer *et al.* (2012); Roth *et al.* (2013).
20 Shield *et al.* (2014).
21 Dohleman *et al.* (2012).
22 Bond *et al.* (2015).

per hectare per year, saving about 32 gC equivalent per MJ in pellet format[23] if replacing coal as a fuel. Energy output:input ratios for the planting, growing, cutting and baling range from 30 to 50, depending on location and agronomic options[24].

7.4 Perspectives for 'now' and for the Future

The agricultural land areas that can be devoted to energy crops such as *Miscanthus* depend on a plethora of factors. For the last century, increasing global food demands are being met by improving yields on the better grade lands through intensification, which combines modern molecular breeding and agronomy. The challenge is to make modern agricultural intensification sustainable. This has been the subject of the UK Royal Society's 2009 Report 'Reaping the Benefits', which concludes that intensification should continue sustainably through increased investment in crop science. This investment should ensure that there is sufficient food production from better grade land well into the future.

At a very practical level for the UK, Lovett *et al.* (2014) attempted with a 'constraint mapping' approach implemented in GIS (Geographic Information System) to calculate the potential land resource for energy crops[25]. The seven constraints include:

1) roads, rivers, lakes and urban areas;
2) slopes with gradients greater than 15%;
3) areas of cultural heritage;
4) designated areas (*e.g.* national parks);
5) woodlands;
6) peat soils; and
7) natural habitats.

These were combined to produce 'prohibition' areas at a spatial resolution of 1 ha. In the UK, land is graded according to its versatility in agriculture usage. Grades 1 and 2 are the highest quality lands, and these should not be used for energy crops, because they are a relatively small area of the UK essential for the production of food crops. The less productive land, graded as 3, 4 or 5, makes up the vast majority of the UK agricultural area. Applying the prohibition mask from the constraints, it was calculated that 8.5 Mha in the UK could be potentially considered for energy crops. The proportion of this that could be used without reducing the UK's capacity to produce food is a subject of much debate (see also Chapters 16 and 17). It is generally agreed that growing energy crops on 10% of this land (*i.e.* 850 kha) is a realistic target, which is unlikely to reduce food production capacity. Clearly, as *Miscanthus* is not always the appropriate choice of energy crop, due to factors such as thermal limitations on establishment, insufficient water availability, harvest logistics and available market, we project that it is likely to occupy less than half of the 850 Kha (*i.e.* ~400 Kha), which is in line with the 2007 UK Government target[26] for perennial energy crops of 350 Kha.

---

23 This accounts for N, P, K replacement of off-take amounts for rhizome planted *M. x giganteus* spring harvested over 15 years, with a discount for low yields during establishment in years 1 and 2.
24 Felten *et al.* (2013).
25 Lovett *et al.* (2014).
26 DEFRA (2007). http://www.biomassenergycentre.org.uk/pls/portal/docs/PAGE/RESOURCES/ REF_LIB_RES/PUBLICATIONS/UKBIOMASSSTRATEGY.PDF

The areas available vary country by country. In Ireland, grassland accounts for 60% the agricultural land. This is used largely for livestock production. It is possible in the future that a shift towards healthier diets with less meat and milk, could release significant pasture land available for energy crops. At present, most agree that there is enough land to make a substantial contribution to the renewable energy mix, bringing alternative income streams to farmers. In the event that yield improvements in food crops in the future do not match expectations and land under energy crops is needed for food, it is worth mentioning that it is relatively easy to remove *Miscanthus* with herbicides and to plough down the rhizomes. Where *Miscanthus* is grown on humus-depleted lower grade arable land, our research suggests that *Miscanthus* will have improved the soil characteristics for subsequent crops. These are complex issues for agricultural policy-makers and evidence needs to be weighed up carefully to produce policies that balance the demands on land use.

We also need to ask why has the uptake to date been so small in the EU, including Ireland and the UK, and in the USA? From the farmer's perspective, the obvious choice is to grow crops that provide the best return on their land. The decision to grow *Miscanthus* as a bioenergy crop depends on economic viability. In Ireland, the enthusiastic pioneer *Miscanthus* growers, stimulated by The Bioenergy Scheme[27], found themselves without a market. This has resulted in several businesses failing. In the UK, in 2014, a strong market for *Miscanthus* biomass has been established by the coal burning giant, Drax Power, which should give growers more confidence. Current work on developing seed-based hybrids propagated by novel agronomies will reduce the cost and speed up establishment. Selective breeding to produce the optimum traits for end use, such as lower chlorine and ash concentrations for biomass fuel and gasification or reduced lignin for ethanol fermentation and anaerobic digesters, will increase the value of the crop. Seed propagation will also enable a rapid increase in planted area, to be achieved in a timely fashion to make a meaningful contribution to Europe's energy needs. Nevertheless, the cessation at the end 2013 of the Rural Development Fund for England[28], funded by the European Union, may affect the willingness of some farmers to grow *Miscanthus*.

Ultimately, bioenergy's penetration of the market will depend on the awareness of the real costs of fossil fuel use and the value that humanity places upon a reduction in GHG emissions. It is hoped that the recent statements from the Inter-Governmental Panel on Climate Change (IPCC) that, in order to avoid disastrous global warming, use of fossil fuels must be completely phased out by the end of the century, will provide a strong enough incentive.

Selected References and Suggestions for Further Reading

Bond, C.E. *et al.* (2015) http://www.climatexchange.org.uk/reducing-emissions/
    life-cycle-assessment-ghg-emissions-unconventional-gas1/
Clifton-Brown, J.C. *et al.* (2004) *Global Change Biology*, **10**, 509–518.
Clifton-Brown, J.C., Renvoize, S.A., Chiang, Y.-C., Ibaragi, Y., Flavell, R. *et al.* (2011)
    Developing *Miscanthus* for bioenergy. In: Energy Crops (Halford, N.G. and Karp, A.
    eds), Royal Society of Chemistry, pp. 301–321.

---

27 http://www.agriculture.gov.ie
28 http://publications.naturalengland.org.uk/publication/46003?category=34022

Clifton-Brown, J.C., Stampfl, P. and Jones, M.B. (2004) *Miscanthus* biomass production for energy in Europe and its potential contribution to decreasing fossil fuel carbon emissions. *Global Change Biology*, **10**, 509–518.

Clifton-Brown, J.C., Breuer, J, and Jones, M.B. (2007) Carbon mitigation by the energy crop, *Miscanthus. Global Change Biology*, **13**, 2296–2307.

DEFRA (2007) UK Biomass Strategy. http://www.biomassenergycentre.org.uk/pls/portal/docs/PAGE/RESOURCES/REF_LIB_RES/PUBLICATIONS/UKBIOMASSSTRATEGY.PDF

Dohleman, F.G. *et al.* (2009) *Plant, Cell and Environment*, **32**, 1525–1537.

Dohleman, F.G., Heaton, E.A., Arundale, R.A. and Long, S.P. (2012) Seasonal dynamics of above- and below-ground biomass and nitrogen partitioning in *Miscanthus × giganteus* and *Panicum virgatum* across three growing seasons. *Global Change Biology: Bioenergy*, **4**, 534–544.

Dondini, M., Hastings, A., Saiz, G., Jones, M.B. and Smith. P. (2009) The potential of *Miscanthus* to sequester carbon in soils: comparing field measurements in Carlow, Ireland to model predictions. *Global Change Biology: Bioenergy*, **1**, 413–425.

Drewer, J. *et al.* (2012) *Global Change Biology Bioenergy*, **4**, 408–419.

Felten. D., Froba. N., Fries. J. and Emmerling, C. (2013) Energy balances and greenhouse gas-mitigation potentials of bioenergy cropping systems (*Miscanthus*, rapeseed and maize) based on farming conditions in Western Germany. *Renewable Energy*, **55**, 160–174.

Greef, J.M. and Deuter, M. (1993) *Angewandte Botanik.* **67**, 87–90.

Hastings, A., Clifton-Brown, J., Wattenbach, M., Mitchell, C.P., Stampfl, P. and Smith. P. (2009a) Future energy potential of *Miscanthus* in Europe. *Global Change Biology: Bioenergy*, **1**, 180–196.

Hastings, A., Clifton-Brown, J., Wattenbach, M., Mitchell, P. and Smith. P. (2009b) The development of MISCANFOR, a new *Miscanthus* crop growth model: towards more robust yield predictions under different climatic and soil conditions. *Global Change Biology: Bioenergy*, **1**, 154–170.

Hastings, A. *et al.* (2013) In: *Biofuel Crop Sustainability* (Singh, B.P. ed.), Wiley Online. doi: 10.1002/9781118635797.ch9781118635712.

Heaton, E., Clifton-Brown, J.C., Voigt, T., Jones, M.B. and Long. S.P. (2004a) *Miscanthus* for renewable energy generation: European Union experience and projections for Illinois. *Mitigation and Adaptation Strategies for Global Change*, **9**, 433–451.

Heaton, E., Voigt, T. and Long, S.P. (2004b) A quantitative review comparing the yields of two candidate C-4 perennial biomass crops in relation to nitrogen, temperature and water. *Biomass & Bioenergy*, **27**, 21–30.

Heaton, E. *et al.* (2008) *Global Change Biology*, **14**, 2000–2014.

Ibagari, Y. (2003) *Acta Phytotaxonomica et Geobotanica*, **54**, 109–125.

Kim, C., Lee, T.H., Guo, H., Chung, S.J., Paterson, A.H. *et al.* (2014) Sequencing of transcriptomes from two *Miscanthus* species reveals functional specificity in rhizomes, and clarifies evolutionary relationships. *BMC Plant Biology*, **14**, 134.

Lovett, A., Sunnenberg, G. and Dockerty, T. (2014) The availability of land for perennial energy crops in Great Britain. *Global Change Biology: Bioenergy*, **6**, 99–107.

Ma, X.F., Jensen, E., Alexandrov, N., Troukhan, M., Zhang, L.P. *et al.* (2012) High resolution genetic mapping by genome sequencing reveals genome duplication and tetraploid genetic structure of the diploid *Miscanthus sinensis. Plos One*, **7**, e33821.

Richter, G.M. *et al.* (2008) *Soil Use and Management*, **24**, 235–245.

Roth, B. *et al.* (2013) *Land*, **2**, 437–451.

Royal Society (2009) Reaping the benefits: Science and the sustainable intensification of Global Agriculture. Royal Society, London.

Scally, L., Hodkinson, T. and Jones, M.B. (2001). Origins and taxonomy of *Miscanthus*. In: *Miscanthus – for Energy and Fibre* (Jones, M.B. and Walsh. M, eds), London: James and James (Science Publishers), pp. 1–9.

Shield, I.F. *et al.* (2014) *Biomass & Bioenergy*, **68**, 185–194.

Styles, D. and Jones, M.B. (2008) *Energy Policy*, **36**, 97–107.

Zatta, A., Clifton-Brown, J., Robson, P., Hastings, A. and Monti, A. (2014) Land use change from C3 grassland to C4 *Miscanthus*: effects on soil carbon content and estimated mitigation benefit after six years. *Global Change Biology: Bioenergy*, **6**, 360–370.

8

Mangrove Palm, *Nypa fruticans*: '3-in-1' Tree for Integrated Food/Fuel and Eco-Services

C.B. Jamieson[1,2], R.D. Lasco[1] and E.T. Rasco[3]

[1] World Agroforestry Centre, Laguna, Philippines
[2] Next Generation, Kimpton, Herts, UK
[3] PhilRice, Philippnes Rice Research Institute, Munoz, Philippines

Summary

The surge in biofuel production in the first ten years of this century raised the issue of 'food *vs* fuel'. This is based on production of biofuels from crops normally used for animal or human nutrition or because of change of land use from food crops to biofuel crops. However, this competition between food and fuel can be avoided, in particular by using crops for which their use in the human food chain is not in conflict with their use as a sources of biofuels. Indeed, it is possible for food production and fuel production to be mutually reinforcing. Several agricultural systems in which this may be achieved are described and discussed. There is particular emphasis on sugar palms and especially on *Nypa fruticans*, which has great potential for use in integrated systems.

8.1 Introduction: The 'Food *vs* Fuel' and 'ILUC' Debates

Biofuels are not new. Henry Ford's original mass-produced cars were intended to run on ethanol made from agricultural crops and Rudolph Diesel's engine was originally designed to run on peanut oil (see Chapters 1, 5 and 6). There was much optimism in those early days that the demand created by the growing automobile industry would bring renewed opportunities and prosperity to the rural sector. Of course, this optimism was not fully realised as biofuels were unable to compete with cheap oil, which focused on the demand for fuels initially, and then developed refineries that produce the array of other products that have become foundational to modern-day life.

Such has become our dependence on oil that it threatens to collapse under its own weight, due to air pollution, climate change and price spikes leading to recessions, food insecurity and other serious knock-on effects. These threats led to a 'rediscovery' of biofuels in the late 20th/early 21st centuries – a hundred years after their promise was initially touted. The same reasons were given for this renewed optimism: a renewable, clean-burning fuel that required little modification to existing infrastructure and could

Biofuels and Bioenergy, First Edition. Edited by John Love and John A. Bryant.
© 2017 John Wiley & Sons Ltd. Published 2017 by John Wiley & Sons Ltd.

contribute to rural development. This time round, biofuels were expected to deliver solutions to climate change over and above the other benefits. Furthermore, during the intervening century, the world's human population had more than tripled and with it, demand for both food and fuel had increased exponentially. Concerns grew that the weight of expectation placed on biofuels was unrealistic and questions were asked about the scale of biofuel production required and its impacts on food production, since they shared common feedstocks[1].

The other side of that same 'food *vs* fuel' coin is that, if more food crops are grown because some are being used to produce fuels, the net effect is assumed to be an increased requirement for arable land. In economic terms, the increased demand for arable land increases its value, and this in turn increases the incentive to convert other land to arable use[2]. Fears mounted that biofuel production would increase deforestation, either directly – for example, through conversion of rainforest to oil palm plantations in Indonesia – or indirectly (see Chapters 16 and 17). The latter came to be known as Indirect Land Use Change (ILUC). The rationale is that even if a biofuel crop is grown on existing arable land, it is effectively displacing the food production that would have occupied that land. That food must then be produced somewhere else and, possibly after a complex series of knock-on effects, results in forest clearance or other land conversion to agriculture.

The ILUC issue has become the greatest barrier to biofuel expansion, because any loss of natural habitat and greenhouse gas (GHG) emissions from such land use change negate biofuels' 'green' credentials. Since the policies that encouraged the fledgling industry were partly motivated by environmental benefits that biofuels could bring, governments around the world were pressurised, especially by environmental NGOs, to reverse those policies and incentives.

By contrast, biofuel companies pleaded their innocence to the charges being made against them. Having made significant investments and expansion plans based on government-led targets for biofuels, they felt the rules were being adjusted against them unfairly.

In the UK, the Government commissioned an independent review to identify the indirect impacts of biofuel production. The landmark 'Gallagher Review'[3] concluded that there were indeed good biofuels as well as bad ones but, until reliable certification schemes were developed for the good ones, it was going to be difficult to distinguish between them. Without a guarantee of the biofuels' sustainability, many governments revised their escalating biofuel mandates, holding them at the level already attained and abandoning targets for increased levels of incorporation in the fuel mix. Few have been satisfied by this and the debate has become polarised.

8.2 Integrated Food-Energy Systems (IFES): A Potential Solution

8.2.1 Main Features of IFES

An important point, which was highlighted by the Gallagher Review, is that a hectare of biofuel crops does not necessarily mean a hectare lost to food production. Many biofuel systems produce co-products that can be used for animal feed, in addition to the biofuel

---

1 Brown (2006).
2 Searchinger *et al.* (2008).
3 Renewable Fuels Agency (2008).

itself. Therefore, both food and fuel is effectively produced from the same plot of land. The livestock feed can be said to save land dedicated to existing feeds production, and that saving attributed to the biofuel crop, thus reducing its land footprint.

An example of this is when wheat is used in bioethanol production. After fermentation, the protein element of the wheat grain remains and can be dried and sold as animal feed. Known as Dried Distiller's Grains and Solids (DDGS), it is a high protein feed that can substitute for soya beans in livestock rations (see also Chapters 3 and 5). In some situations, this has led to projected decreases in GHG emissions from soybean production in South America, reducing one of the drivers for rainforest clearance. Therefore, the biofuel can be said to have 'Negative ILUC' – that is to say, the GHG emissions from production of biofuel and animal feed combined are credited as being below zero. Thus, there is a net reduction of GHG emissions compared to the common practice[4].

Many biofuel production methods can be configured to produce food as well as fuel. Also, many food production methods can be configured to produce bioenergy as well as food. These together are known as 'Integrated Food-Energy Systems' (IFES). The UN Food and Energy Organisation (FAO) produced a report on IFES for small-scale farmers in developing countries that sought to identify and overcome obstacles to their wider implementation. It identified two broad types of IFES[5], named simply as Type 1 and Type 2 IFES.

8.2.1.1 Type 1 IFES

Type 1 IFES makes use of trees in agricultural systems. The trees may produce liquid biofuel feedstock or solid biomass (*e.g.* wood) for energy use. At the same time, they can modify the environment around the arable crop to enhance food production. Examples of this are where trees reduce soil erosion on slopes, increase nutrient recycling from the lower layers of soil or through nitrogen fixation, or provide protection from strong winds or strong sunshine. In that way, the energy crop and the food crop have a symbiotic relationship, where they produce more together than they would if either were in a monoculture. This is known as agroforestry and is particularly suited to combined food and fuel production.

8.2.1.2 Type 2 IFES

Type 2 IFES may just involve one crop, but the components of the crop are fractionated to produce food and fuel. This may be either simultaneously, as with rice husk being gassified for electricity at rice mills, or it may be in a 'cascading' system, such as when livestock are given feed and then their manure is fed into a digester for biogas production (Chapter 3).

Of course, the dividing line between type 1 and type 2 IFES can be crossed. Many agroforestry trees are multipurpose and most multipurpose arable crops can be grown in an agroforestry system. Generally, the more complex systems can have the highest productivity, but tend to be more difficult to manage, requiring some form of institutional support to help organise and train farmers. More demonstration of these innovative systems could help increase world food and bioenergy production sustainably.

---

4 Bauen *et al.* (2010).
5 Bogdanski *et al.* (2011).

The synergies between food and fuel can be further accentuated when the biofuel element is used within the food production chain. For example, 40% of farmers in South Asia lack access to reliable energy needed for development[6]. This can lead to reduced yields due to lack of inputs, or increased food spoilage after harvest because of lack of energy for storage or processing facilities, or lack of energy for transportation to market. At the same time, many rice farmers in the region are burning rice straw as a waste product of intensive irrigated rice production. If that straw could be turned to fuel, for example by anaerobic digestion with manure, the resulting biogas could help power the whole agricultural supply chain from production through processing to cooking (see Chapter 3). Even if that use of the rice straw does not directly increase food production, the use of its energy could, through powering farm machinery, help to dry and store rice or transport it to market. In the case of anaerobic digestion, the nutrients are also recycled to help feed the next crop and the cleaner air from reduced burning could benefit farmers' health and productivity. Therefore, bioenergy and food production can have natural synergies, both in their production and their processing or end use.

8.2.2 Baseline Productivity

There is an ideology that says no biomass should be used for energy purposes if it can be used for food production. There are a number of problems with this philosophy.

First, all biomass can theoretically be turned into food, either directly or indirectly. However, this 'theoretical maximum' food conversion target is not always practical or economical and bioenergy can be a useful alternative, which can itself help increase food production. This leads to the second point: without energy, food productivity levels are low, losses are high and processing or cooking is impossible. One of the most convenient, versatile and available forms of energy on farms is bioenergy. If produced well, it can be a natural companion to food production in integrated food-energy systems, providing the synergies mentioned above. As with any crop production, it has to be acknowledged that bioenergy has not always been produced well, and has sometimes had negative impacts rather than positive ones. Halting bioenergy altogether, even if this were practical, would be to miss out on the benefits it can bring for food production, climate change and the rural economy. Adopting policies to help ensure bioenergy is sustainably produced in the future would seem to be a more pragmatic and desirable alternative.

Therefore, rather than speaking of 'food *vs* fuel' as if they were distinct entities in conflict, it is proposed here that their potential to be mutually reinforcing should be recognised and encouraged. Rather than overlooking or seeking to halt bioenergy use, it should be supported where it is sustainable and has no negative impact on overall food production. This is most easily achieved where baseline food production is low and could benefit most from synergies with bioenergy. This 'optimised' option for integrated food-energy systems is represented by the middle section of Figure 8.1 below, which is proposed here as a sustainable option, by contrast with the two extremes of 100% theoretical food, which is not feasible, or 100% biofuels where food production is completely displaced and sacrificed.

---

6 International Energy Agency (2011).

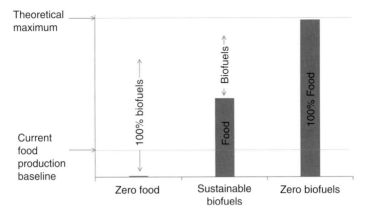

Figure 8.1 Sustainable food and fuel production.

Discussion 1
Producing food requires energy. Where that energy access is lacking, productivity decreases and losses increase. Where that energy is met by fossil fuels, it is subject to price fluctuations that cause food insecurity. On farms, biomass is often the cheapest and most convenient renewable energy resource: under what circumstances might using it for energy result in more food overall than if it were converted to food directly?

8.3   Land Use: The Importance of Forest Ecosystem Services

This chapter so far has mainly focused on the 'food *vs* fuel' controversy around biofuel production, demonstrating ways in which the two can be mutually supportive rather than in conflict. However, as touched upon earlier, ILUC is the other side of the same coin. While ILUC can refer to any change of land use, such as from pasture land to arable, and the GHG emissions incurred, we shall here focus on forests in particular. Forests supply broadly four 'ecosystem services'[7]:

1) provisioning, *e.g.* clean water;
2) regulating, *e.g.* carbon sequestration; regulation of local and global climate;
3) supporting, *e.g.* nutrient cycles and crop pollination; and
4) cultural, *e.g.* spiritual and recreational benefits

The issue of forest clearance for agriculture is not new and is driven by many factors beyond the scope or control of biofuels and bioenergy producers (Chapter 17). A major market failure is that there is no significant market for the services provided by forests, which are expected to be given without charge. In the absence of such markets, the main route for profiting from forests is to clear them, sell the timber and plant crops for which there are markets. However, there is growing interest in the concept of 'Payment for Environmental Services' (PES), whereby some of the forest's services – usually

---

7 Millennium Ecosystem Assessment (2005).

carbon storage and water provision – are valued and paid for[8]. Such payments often come through taxes on activities that have a negative environmental impact.

With policies that factor in environmental economics, there may be a route to maintaining ecosystem services; without them there is little hope, regardless of whether biofuels are produced or not. Hence, biofuels are part of a much larger framework, but their production can also play a positive role. One option would be to adopt more alternative feedstocks from tree species. The benefit of this is that indigenous tree crops, cultivated in the right way, can provide many of the environmental services of forests, while simultaneously having use for biofuels. Since biofuel processing plants can be adjusted to take multiple feedstocks, it may be feasible to have a mixed forest of domesticated tree species producing fuel and eco-services.

Although such a system still has to be developed, it is mentioned here as a concept to try to broaden thinking on ILUC with the following discussion:

Discussion 2

Controversy around biofuels typically pits food against forest as if they were competing land uses, thus questioning the place for biofuels on the assumption that they compete with food production and provide few ecological benefits. If tree-based biofuel systems are able to provide significant ecosystem services, does this exempt them from the ILUC debate, even if they produce no food?

8.4 Sugar Palms: Highly Productive Multi-Purpose Trees

An example of a highly productive tree crop is the sugar palm, a term given to a wide range of palm trees that are tapped for their sugary sap. Unlike sugar cane, the process does not result in waste biomass or cutting the plant to the ground. Therefore, the tree effectively becomes like a solar panel that continuously converts sunlight into sugar which is extracted. This makes it a very efficient process, with yields up to four times that of equivalent annual crops such as maize (see Figure 8.2). Palm tapping has been practised for centuries, producing products such as vinegar, wine, sugar or jaggery[9], and as an animal feed in areas with high population density. With rising world food demand and environmental concerns about the effects of modern agriculture, this has been proposed as a technique that could beneficially be adopted more widely[10]. One of the key sugar palm species, *Arenga pinnata*, has been shown to have good potential for bioethanol production in Southeast Asia[11]. This was as part of a sustainable mixed forest system, since this species is not suited to monoculture, unlike the oil palm that has come to be widely regarded as a threat to rainforests (see Chapter 17). Hence *A. pinnata* can be used in a non-destructive way to improve

---

8 Lasco *et al.* (2008).
9 Product made by concentrating palm sap without any separation of molasses from crystals.
10 Borin and Preston (1995); Dalibard (1997).
11 van de Staaij *et al.* (2011).

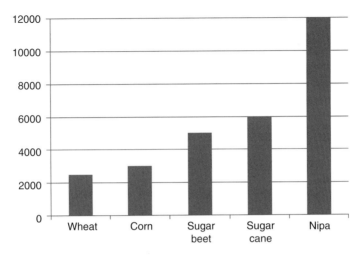

Figure 8.2 Ethanol production (litres/ha/yr).

livelihoods in the rainforest where it grows naturally, or in agroforestry systems where it could be cultivated in an agroforestry configuration.

With so many sustainability benefits, it is important to also be aware of the hindrances to sustainable biofuel production from sugar palms. A major factor is the laborious work involved in tapping, which involves cutting off the end of fruiting spikes and renewing the wound daily, collecting the sap twice daily and bruising or 'bashing' the stem to maintain sap flow. Therefore, labour is by far the largest cost of production[11]. This, combined with the climb necessary to reach the fruiting spikes of most sugar palms, makes it a tiresome and dangerous job.

Furthermore, anhydrous ethanol production favours economies of scale, which works against mixed cropping systems, since the distances to transport the sap become prohibitive. Possible solutions are to grow other sugar or starch crops along with it, or develop a feasible small-scale ethanol plant for the purpose, in conjunction with use of flex-fuel engines that can run on hydrous ethanol. This effort would be assisted if the sap is stabilised or pre-treated, to prevent it from deteriorating as soon as it leaves the

---

Discussion 3

If sugar palms are able to support up to four times as much livestock as 'conventional' crops such as maize, could half of the juice be used to make ethanol fuel instead, without being considered to compete with food production? If the food and the fuel components were in two geographically separate locations, could a mechanism be devised to credit the land saved by the livestock element to the biofuels? If the net result were an increase in food output overall, along with additional ecological services (carbon sequestration, etc.), would this be an example of negative ILUC, leading to reduced greenhouse gas emissions?

plant[12]. Finally, research is needed to find ways of shortening the period from sowing to production of first flowering shoots for tapping, which can take up to 15 years.

8.5 Nipa[13] (*Nypa fruticans*): A Mangrove Sugar Palm with Great Promise

In answering the challenges listed above, a focused research effort is required. A very promising species for that focus is the mangrove palm, *Nypa fruticans* (Nipa). As a biofuel crop, it has the distinction of being the only sugar palm that grows in saline environments, since it is the only mangrove species that is also a palm tree. This makes it suitable for areas where food crops would not be grown, avoiding one of the greatest arguments against liquid biofuel production. Among sugar palms, it is one of the earliest to produce, starting at around five years after sowing. It is also unique among sugar palms in that it does not require climbing – the trunk grows horizontally under the mud, which means that the inflorescences are produced at a convenient height for tapping. Finally, the trees can be tapped all-year-round and have high productivity[12].

In addition to these benefits, Nipa can provide many ecosystem services, such as shelter for coastal areas and habitats for wildlife. There may be potential for its use in reclamation of degraded, saline land and reforesting areas which have become denuded by mangrove clearance in its native Southeast Asia. However, care must be taken that such a productive crop does not simply replace fish-pond expansion as a threat to natural mangrove stands, not least because Southeast Asia is the most species-rich mangrove habitat on Earth. Even where clearance has been banned, such as in the Philippines, enforcement can be a challenge and many fish ponds are illegal. In its favour, the five-year establishment time for Nipa makes it less attractive for illegal activities where land title has not been secured. Policies would be required to incentivise planting in the places where it would bring the greatest social and environmental benefits.

As already mentioned, one of the greatest challenges for sugar palms is the need to reduce the cost of the tapping itself. To this end, a technology has been developed that achieves a similar effect to the stem 'bashing' but with only a fraction of the labour, by using ethylene applied to the base of the flowering spike. It appears to have been successful on coconut, chemically stimulating the sap to continue to flow, and the same approach is being transferred to Nipa palm tapping.

Nipa can grow not only in brackish water but also fresh water and has been demonstrated in what is believed to be the world's first rice-Nipa agroforestry system, by PhilRice in Mindanao, southern Philippines. It is in slightly brackish water there, along with salt-tolerant rice, making use of the flooded conditions favoured by both. The synergies do not end there: Asia produces 550 million tonnes of rice straw each year which, if incorporated back into the field, creates methane emissions in the flooded systems that predominate across the continent, and hinders establishment of the following crop. It is a poor-quality animal feed and, in the Philippines, 95% of it is piled up in the fields and burnt, creating a health and environmental hazard. It has been shown that sugar palm sap sprayed onto the straw makes it more palatable to ruminants and increases its energy value[14]. If the manure

---

12 Rasco (2012).
13 Also known as Nipa palm or Nipah palm.
14 Dalibard (1999).

from the livestock were mixed with additional straw and co-fed into an anaerobic digester to produce biogas, the Nipa could have a catalytic effect, turning a waste into a resource for additional food and fuel production. Furthermore, nutrients from the digester could be recycled to the rice paddy rather than lost to the air through burning (see Chapter 3). The fuel could in turn be used to power agricultural production and food processing on the farm. The Philippines has the second highest electricity prices in Asia, after Japan, making crop irrigation prohibitively expensive. With affordable bioenergy to pump water, many rice farmers could increase yields by 25% per crop and plant two crops a year instead of just one[15]: another example of the virtuous cycle of food and fuel co-production.

8.6   Conclusion

After initial optimism, the revival of biofuels as a substitute for mineral oil has become controversial, due to concerns about feedstock competition with food production and about damage to natural habitats. Such concerns have been emphasised very recently in new EU laws enacted by the European Parliament setting a limit on the quantity of crop-based biofuels that can be used to meet EU energy targets[16]. However, sugar palms offer a route to higher productivity whilst simultaneously reducing inputs and providing ecosystem services. Nipa palm appears to be a particularly promising species on which to focus a research effort. Being '3-in-1' multi-purpose trees, efforts to domesticate them could bring significant benefits not just for sustainable, **locally available** biofuels but also for food production, **job creation** and the environment. Following a century of neglect in favour of industrial monocultures, the time has surely now come to invest in these exciting, potentially transformational plants.

Selected References and Suggestions for Further Reading

Bauen, A., Chudziak, C., Vad, K. and Watson, P. (2010) *A Causal Descriptive Approach to Modelling the Greenhouse Gas Emissions Associated with the Indirect Land Use Impacts of Biofuels: A Study for the UK Department of Transport.* E4Tech.

Bogdanski, A., Dubois, O., Jamieson, C. and Krell, R. (2011) *Making Integrated Food-Energy Systems Work for People and Climate.* FAO, Rome.

Borin, K. and Preston, T.R. (1995) Conserving biodiversity and the environment and improving the well-being of poor farmers in Cambodia by promoting pig feeding systems using the juice of the sugar palm tree (*Borassus flabellifer*). *Livestock Research for Rural Development*, **7(2)**.

Brown, L. (2006) *Plan B 2.0: Rescuing a Planet under Stress and a Civilization in Trouble.* Earth Policy Institute/W.W. Norton & Company, New York and London.

Dalibard, C. (1997) The potential of tapping palm trees for animal production. In: *Livestock Feed Resources within Integrated Farming Systems.* FAO, Rome.

---

15 Personal Communication to the author by M. Regalado, Deputy Director for Research, Philippines Rice Research Institute (2013)

16 http://www.ipsnews.net/2015/04/european-biofuel-bubble-bursts/

Dalibard, C. (1999) Overall view on the tradition of tapping palm trees and prospects for animal production. *Livestock Research for Rural Development*, **11(1)**.

International Energy Agency (2011) *World Energy Outlook*. IEA, Paris.

International Energy Agency (2014) *World Energy Outlook*. IEA, Paris.

Lasco, R.D., Villamor, G., Pulhin, F., Catacutan, D. and Bertomeu, M. (2008) From principles to numbers: approaches in implementing Payments for Environmental Services (PES) in the Philippines. In: *Smallholder Tree Growing for Rural Development and Environmental Services* (Snelder, D.J. and Lasco, R.D. eds), Springer, Heidelberg, pp. 379–391.

Millennium Ecosystem Assessment (2005) *Ecosystems and Human Well Being: Synthesis*. Island Press, Washington DC.

Rasco, E. (2012) *Nipa: A Gift to Humankind from the Age of the Dinosaurs*. NAST, Taguig, Philippines.

Renewable Fuels Agency (2008) *The Gallagher Review of the Indirect Effects of Biofuels Production*. RFA/Department of Transport, London.

Searchinger, T., Heimlich, R., Houghton, R.A., Dong, F., Elobeid, A. *et al.* (2008) Use of US croplands for biofuels increases greenhouse gases through emissions from land-use change. *Science Express*, 319, 1238–1240.

Van de Staaij, J., van den Bos, A., Hamelinck, C., Martini, E., Roschetko, J. and Walden, D. (2011) *Sugar Palm Ethanol: Analysis of Economic Feasibility and Sustainability*. Ecofys, Utrecht and Winrock International, Brussels for NL Agency, The Netherlands.

9

The Use of Cyanobacteria for Biofuel Production

David J. Lea-Smith and Christopher J. Howe

Department of Biochemistry, University of Cambridge, Cambridge, UK

Summary

In this chapter, the use of blue-green algae, or cyanobacteria, for biofuel production is discussed. Relevant aspects of cyanobacterial biology are summarised, before reviewing the products that are obtained from cyanobacteria. Currently, these products may be described as high-value/low-volume, whereas biofuels, in order to compete economically with fossil fuels, are necessarily low-value/high volume products. The inherent advantages of cyanobacteria for large-scale culture and some of the problems presented by industrial production of cyanobacteria are described. The use of genetic modification techniques to engineer cyanobacteria is discussed, and examples where companies have begun commercial growth of genetically-modified cyanobacteria for biofuels are reviewed.

9.1 Essential Aspects of Cyanobacterial Biology

9.1.1 General Features

Cyanobacteria, alternatively known as blue-green algae, are a diverse group of photosynthetic bacteria found in almost every environment on the planet, from freezing Antarctic lakes to volcanic hot springs, and from wetlands, lakes and seas to desert sands and rock surfaces. They are probably the most abundant organisms on Earth and are particularly dominant in oceans where they are responsible for approximately half of all carbon fixation[1] and for the majority of nitrogen fixation[2]. Cyanobacteria are an important feature of the fossil record, with the earliest known examples dating back to 3 billion years or more. The impact on the Earth from cyanobacteria cannot be understated; over a period of a billion years the early cyanobacteria gradually changed the Earth's primal atmosphere from highly reducing with high levels of carbon dioxide

---

1 Zwirglmaier *et al.* (2008).
2 Galloway *et al.* (2004).

Biofuels and Bioenergy, First Edition. Edited by John Love and John A. Bryant.
© 2017 John Wiley & Sons Ltd. Published 2017 by John Wiley & Sons Ltd.

(CO_2) to one rich in oxygen with very low levels of CO_2. This crucial change was achieved by a (then) unique aspect of cyanobacterial biology, that of oxygenic photosynthesis[3], the conversion of CO_2 and water to oxygen and carbohydrates using energy derived from sunlight. The importance of the cyanobacteria to the Earth's ecosystem is further highlighted by the fact that the chloroplast, the organelle responsible for photosynthesis in eukaryotic (*i.e.* possessing a cell nucleus) algae and plants, likely evolved from endosymbiosis between ancient cyanobacteria and a non-photosynthetic eukaryotic host.[4]

Extant cyanobacteria can be broadly defined into two groups, unicellular (single-celled) species and filamentous (thread-shaped, colonial) species. Unicellular cyanobacteria include the *Synechococcus* and *Prochlorococcus* species which dominate marine environments, the nitrogen fixing *Cyanothece* species, freshwater *Synechococcus* species, *Cyanobacterium* UCYN-A, a marine species which can fix nitrogen but lacks key metabolic pathways common to most other species and oxygenic photosynthesis, and the model species *Synechocystis sp.* PCC 6803. Model species are particularly suited to the study of certain biological phenomena and are therefore a practical reference against which other observations may be made, experiments conducted or conclusions compared. Filamentous cyanobacteria consist of a series of cells connected in a chain and include *Arthrospira* species, *Trichodesmium* and *Crocosphaera* species responsible for a high proportion of marine nitrogen fixation and *Nostoc* species, which are found in freshwater environments or in symbiotic relationships with plants.

Compared to eukaryotic algae and plants, cyanobacteria are far simpler organisms. Cyanobacteria possess no nucleus; rather, their genetic material (genome) is harboured on circular DNA strands and they may possess several copies of the genome per cell. They are surrounded by a cell wall which, from the inside to the outside, consists of a lipid layer termed the cytoplasmic membrane, a layer consisting of a thick polymer of sugars and amino acids termed peptidoglycan, a second lipid layer termed the outer membrane which incorporates lipopolysaccharides (a lipid/polysaccharide molecule) on its outer sheaf and, in some species, a fourth 'S-layer' that is a thick envelope composed of proteins. The cyanobacterial cell wall is essential for survival and plays a major role in allowing cyanobacteria to accommodate varying environmental conditions. It also acts as a barrier against invasion from bacteriophages (viruses that infect bacteria) and unicellular of multicellular herbivores.

9.1.2 Photosynthesis and Carbon Dioxide Fixation

Inside the cyanobacterial cell, there is another series of lipid layers, termed the thylakoid membranes, where the light reactions of photosynthesis occur (Figure 9.1). Thylakoid membranes form characteristic layers and are found in all cyanobacteria, except in *Gloeobacter* species. The protein complexes that transform light into chemical energy (so-called photosystems I and II) are located in the thylakoids. In most cyanobacterial species, with the exception of *Prochlorococcus* species and *cyanobacterium* UCYN-A, the photosystem complexes are further associated with other peptides and prosthetic

---

3 Bendall *et al.* (2008).
4 Howe *et al.* (2008).
   See also Hodson and Bryant (2012).

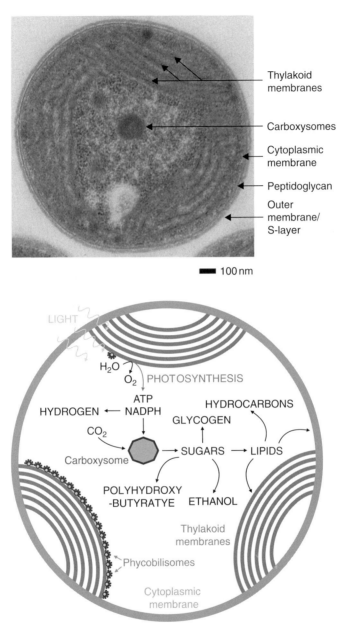

Figure 9.1 Above: transmission electron micrograph of a spherical cyanobacterium, with annotated structures. Below: Diagrammatic representation of the key metabolic pathways for the production of carbohydrates and biofuel molecules in cyanobacteria.

groups (pigments) in a large, antenna-shaped protein complex termed the phycobilisome. The main photosynthetic pigment in cyanobacteria, found in photosystems I and II, is chlorophyll a. Phycobilisomes contain a number of other pigments, generically termed 'phycobilins', that absorb light at different wavelengths from that of chlorophyll, thus allowing organisms to maximise harvesting across the entire visible spectrum. Thanks to their structure and localisation, phycobilisomes can transfer light energy to the site of photosynthesis with over 95% efficiency.

Despite their ancient origin, cyanobacteria remain the most efficient CO_2-fixing organisms on Earth due, in part, to an internal compartment termed the 'carboxysome' (Figure 9.1). Carboxysomes are present in all cyanobacteria with the exception of *Cyanobacterium* UCYN-A. The carboxysome contains RuBisCO[5], the enzyme complex that incorporates CO_2 into carbohydrates (sugars), which are then utilised in cellular processes. RuBisCO is a relatively inefficient enzyme, with a turnover rate of 3–10 molecules per second. Furthermore, RuBisCO is not specific for CO_2 and can also use O_2 to degrade carbohydrates in a process termed 'photorespiration'. The product of photorespiration, phosphoglycolate, is toxic and must be removed by the cell *via* an energetically expensive process. Therefore a high rate of photorespiration limits cellular growth. The carboxysome evolved as a highly efficient carbon dioxide concentrating compartment, which limits photorespiration and increases net levels of photosynthesis.

9.1.3   Nitrogen Fixation

Unlike plants, many cyanobacteria are capable of fixing nitrogen (N_2) to ammonia (NH_3), a form which can be used for cellular processes. Nitrogen is an essential component of DNA, proteins, metabolites and vitamins. The ability to fix N_2 directly allows some cyanobacterial species to exploit nitrogen-deprived environments such as the open ocean. Nitrogen fixation occurs *via* the enzyme nitrogenase, which is oxygen sensitive. N_2 fixation cannot therefore occur simultaneously with photosynthesis, which generates O_2 as a by-product. Different cyanobacteria overcome this limitation by employing a range of strategies. Unicellular *Cyanothece* species fix nitrogen when it is dark (*i.e.* when photosynthesis is at a standstill), using energy stored as carbohydrates accumulated during the day. *Trichodesmium* fix nitrogen during the day, but separate the competing processes of photosynthesis and nitrogen fixation both spatially and temporally. Some filamentous cyanobacteria, including *Nostoc* spp., are capable of cellular differentiation and fix nitrogen in oxygen impermeable, non-photosynthetic cells termed heterocysts. In *Nostoc*, the energy for N_2 fixation is provided by the other, photosynthetic cells in the filament.

From an industrial perspective, cultivating nitrogen-fixing cyanobacteria limits the need to add (expensive) nitrogen-based fertiliser to the culture. However, there is a trade-off; nitrogen fixation is an energetically expensive process that results in slower growth and lower biomass accumulation, overall. For example, in *Cyanothece sp.* PCC 51472, a transition to a nitrogen fixing state results in the time taken for cell numbers to double increasing from 12 to 18 hours[6].

9.2   Commercial Products Currently Derived from Cyanobacteria

Cyanobacteria, notably *Arthrospira* species, have been exploited as a food source for over a millennium in the region of Lake Chad, Africa. The same species are commercially cultivated worldwide for production of Spirulina, a high protein food source

---

5 Ribulose bis-phosphate carboxylase-oxygenase.
6 Reddy *et al.* (1993).

marketed as a dietary supplement. At approximately US\$ 50 per 500 g (Nutrex-Hawaii), food grade cyanobacteria is classified as a high-value product. Other high-value products derived from cyanobacteria include phycocyanin, a blue dye used as a food colouring and scientific reagent (~£80 per mg), and a variety of cyanobacteria-derived metabolites that are under investigation as potential anti-cancer compounds[7].

Unlike these speciality items, fuel is a low-value product that must be produced in very high volumes. A barrel of oil costing US\$ 100 equates to approximately \$US 1 per kg for jet fuel, gasoline and diesel, although the price of each of these commodities varies depending on availability and demand. An alternate fuel source, hydrogen, costs approximately US\$ 5 for 1 kg (energy equivalent to US\$ 2 a kg of gasoline), excluding delivery, when derived from natural gas. Due to the cost of installation, running and processing algal cultures (see Chapter 12), cyanobacteria-derived biofuels are not currently economically viable compared to the alternatives. Consequently, research tends to focus particularly on identifying and/or generating cyanobacterial strains that produce industrial chemicals that have a substantially higher market value than biofuels. Polyhydroxybutyrate (PHB) is a carbon-based polymer used to make biodegradable plastics and is accumulated in a wide range of cyanobacteria under conditions of physiological stress. However, plastics derived from oil products cost between US\$ 1 and 2 per kg, so the same issues in generating a low value product at a reasonable cost exist.

9.3   Cyanobacteria Culture

Under ideal conditions, cyanobacteria grow extremely fast compared to other photosynthetic microbes or plants. Under appropriate light and temperature, and with plentiful nutrients, many cyanobacterial species can divide every 3 to 4 hours. Doubling times of cyanobacteria increase when cells are cultured in the presence of extra glucose, rather than relying only on photosynthesis. For example, the doubling time of *Synechocystis sp.* PCC 6803 was reduced from 22.2 hours to 8.2 hours when the culture medium contained glucose[8]. However, although the addition of glucose improves growth rate, the chances of contamination of the cyanobacterial culture from undesirable, non-photosynthetic microbes are also increased by this practice.

Obtaining predictably high yields of cyanobacterial biomass from cultures over an extended period of time is a major challenge. Typically, microalgae, including cyanobacteria, are cultured using contrasting methods; open ponds or photobioreactors. Open ponds (or 'raceways') are generally rectangular-shaped, closed-loop systems lined with either plastic or cement. Raceways are approximately 20–35 cm in depth to ensure adequate exposure to sunlight and the cultures are mixed by a large paddle wheel. As their name suggests, raceways are open to the atmosphere; contamination by other organisms is therefore an issue, as is exposure to extreme climate conditions (freezing or heating) and loss of water due to evaporation. Conversely, photobioreactors (PBRs) are closed systems made of either transparent plastic or glass and illuminated either by natural sunlight or by artificial lights. While ideal for controlling environmental conditions and minimising contamination, PBRs are more expensive to build and maintain than raceways (see Chapters 12 and 16).

---

7 Tan (2007).
8 Bhaya *et al.* (2006).

Cyanobacteria can be cultured in a variety of media ranging from fresh to salty or brackish water and in installations located on arid, marginal or otherwise unproductive land. This versatility makes cyanobacteria appropriate for large-scale production, as significant regions of desert land adjacent to oceans are located on every continent, excluding Europe. As for all algal culture, nitrogen input is required, typically ammonia derived from fertiliser, for growth of strains unable to fix N_2. Smaller amounts of phosphorus, also derived from fertiliser, are required. Other trace elements essential for growth include sodium, chloride, magnesium, sulphur, boron, manganese, zinc, molybdenum, iron, copper, potassium and cobalt. All these elements with the possible exception of iron are found in sufficient quantities in seawater.

Different cyanobacteria have adapted to a wide range of environmental conditions, notably extremes of heat, pH or salinity that are fatal to other organisms. For example, growth under alkaline (pH > 8.5) conditions is typical for species of *Arthrospira*. *Synechocystis* sp. PCC 6803 can tolerate a very high pH (up to pH 11, the maximum being pH 14) and grow at similar rates whether in alkaline or neutral (pH 7) media[9]. The culture of cyanobacteria that are adapted to 'extreme' conditions minimises contamination from other organisms. Similarly, growth of marine or hypersaline-tolerant cyanobacteria in land-based raceways will also limit contamination from organisms that are sensitive to salt. Other methods such as controlling contaminants by antibiotics or pesticides may be effective but not economical in generating a low-value product.

A major challenge for biofuel production is that, unlike many eukaryotic algal species, cyanobacteria do not accumulate significant amounts of lipids that can be converted into biofuels (see Chapter 6). Instead, they accumulate either polyhydroxybutyrate or cyanophycin, an amino acid based polymer. Consequently, in order to produce biofuels from cyanobacteria, the biomass itself must be converted to biofuels, either chemically or through digestion by another organism, or strains required must be specifically engineered to produce biofuels molecules through genetic modification.

9.4   Cyanobacterial Genomes and Genetic Modification for Biofuel Production

As at May 2015, 103 cyanobacterial genomes had been entirely sequenced. These sequenced genomes range in size from approximately 1.44 million nucleotide base pairs (Mbp) and 1,199 genes in *Cyanobacterium* UCYN-A to approximately 9.06 Mbp and 6,689 genes for *Nostoc punctiforme* PCC 73102 and 8.65 Mbp and 8,294 genes for *Microcoleus chthonoplastes* PCC 7420. For comparison, the human genome is approximately 3,200 Mbp and contains approximately 21,000 genes, that of the gut bacterium *Escherichia coli* is 4.6 Mbp (4,377 genes) and that of the Norway spruce (or 'Christmas tree': *Picea abies*) is 19,600 Mbp (just under 28,500 genes). As stated previously, *Cyanobacterium* UCYN-A lacks carboxysomes and phycobilisomes and is also deficient in many key pathways for amino acid and vitamin biosynthesis. However, this species is capable of nitrogen fixation and exists in symbiosis with other organisms that can provide the nutrients it lacks. *Prochlorococcus* genomes range in size from 1.66 to 2.68

---

9 Summerfield and Sherman (2008).

Mbp, containing approximately 1,900 to 3,000 genes. As free-living organisms, these species are highly adapted to nutrient-poor, tropical ocean waters where environmental conditions are relatively stable. Many cyanobacteria are also adapted to life at low light levels, and can grow at depths of 200 m. Marine *Synechococcus* species typically have larger genomes, ranging in size from 2.3 to 3.01 Mbp, containing approximately 2,500 to 3,010 genes. *Synechococcus spp.* dominate temperate and polar latitudes and coastal regions where conditions are typically more variable than in the open ocean. Some cyanbacteria are adapted to more variable, even extreme environments, such as the conditions that prevail in arid land ecosystems or freshwater, estuarine or hypersaline water. Typically, cyanobacteria that are found in such environments have larger genomes and more genes than those found in more stable environments. In general, the filamentous species possess the largest genomes.

One advantage in using cyanobacteria for biofuel production over other photosynthetic microbes is the ease with which many species can be genetically manipulated. Genetic manipulation is defined as the direct human manipulation of an organism's genome using molecular biology techniques. In the case of biofuel production, genetic engineering is performed either to produce or increase production of a compound of interest, or to improve growth of the engineered cyanobacteria under industrial conditions (or, ideally, both). Genetic manipulation can involve either the insertion of genes from foreign species in order to synthesise a novel product; the alteration of (a) native (*i.e.* genes that are already present in the cyanobacterium) gene(s), in order to increase or decrease production of a native compound; the insertion of additional copies of (a) native gene(s) in order to increase production of a native compound; or the deletion of genes, typically those encoding proteins responsible for the synthesis of undesirable products or for the degradation of desirable products, in order to direct the organisms metabolism towards biofuel production.

In order to understand the physiology and biochemistry of cyanobacteria, genetic manipulation has predominantly been performed on model cyanobacteria, including *Synechocystis* sp. PCC 6803, *Synechococcus* sp. PCC 7002, *Synechococcus elongatus* and *Nostoc* sp. PCC 7002. Largely because of this history, these species are now investigated for industrial applications, including production of biofuels. Genetic transformation of more widespread marine *Synechococcus* species[10] and *Prochlorococcus* strain MIT 9313[11] has been conducted, but the tools required for the genetic engineering are not as advanced as for the former species of cyanobacteria. For example, in *Synechocystis* sp. PCC 6803 and *Synechococcus elongatus*, it is possible to generate so-called 'unmarked mutants', that are genetically modified strains containing no foreign DNA, except when desirable.

To generate unmarked mutant cyanobacterial strains (Figure 9.2), artificial DNA constructs must first be generated. These constructs are synthesised using small, circular DNA molecules called plasmids. Plasmids contain an origin of replication, which allows the bacteria to make additional copies of the plasmid that can be passed on to daughter cells when bacteria divide. The origin of replication is typically specific for only some laboratory strains of bacteria, mainly *E. coli*. By the use of molecular biology techniques it is possible to join various fragments of DNA together in the desired

10 Brahamsha (1996).
11 Tolonen *et al.* (2006).

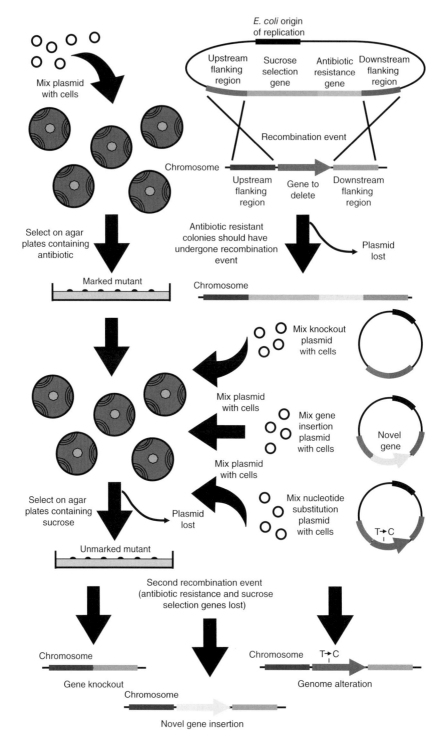

Figure 9.2 Diagrammatic representation of the process of generating unmarked knockout mutations for cyanobacteria.

combination, in *E. coli*. The resulting plasmid construct, containing the assembled DNA, is purified from the *E. coli* and then inserted into the cyanobacterium. To generate knockout mutations in cyanobacteria, that is strains in which genes are deleted, two DNA fragments that are homologous to specified regions in the cyanobacterial chromosome and flank the gene to be deleted (termed the upstream and downstream flanking regions) are identified and engineered into the plasmid to be used. In the case of *Synechocystis* sp. PCC 6803, two genes are inserted between these flanking regions; one of these genes encodes a protein which confers resistance to an antibiotic and the second encodes a protein which is lethal to many bacteria when cells are grown in the presence of sucrose.

In the first stage of the process, the plasmid construct is mixed with the cyanobacteria cells. In the case of *Synechocystis* sp. PCC 6803, *Synechococcus* sp. PCC 7002 and *Synechococcus elongatus*, the DNA is naturally taken up by the cells. Cells are then cultured on agar plates, a solid medium which contains the required nutrients for growth, and also in the presence of the antibiotic to which the plasmid confers resistance. By growing cells on agar plates containing the antibiotic, only cells that have successfully absorbed the plasmid construct will survive; it is therefore possible to select those cells that have incorporated the antibiotic resistance gene into the chromosome, as the plasmid cannot replicate in cyanobacteria and is rapidly lost. Incorporation of the antibiotic resistance gene from the plasmid into the cyanobacterial chromosome is only possible if a recombination event occurs. This is a natural process whereby two similar sequences of DNA are exchanged, hence the requirement to select the homologous, flanking regions with care.

By insertion of the antibiotic resistance gene between the two flanking regions, it is possible to target a specific region of the chromosome, since the recombination event will only occur at the part of the chromosome where the flanking regions are located. The result of this genetic event is the production of a cyanobacterium in which a specific gene has been replaced by one that confers antibiotic resistance. However, that marker can be subsequently removed to yield 'unmarked mutants'. The mutant produced is dependent on what plasmid is inserted into the cell. If a deletion mutant is required, whereby part of the chromosome is removed, a plasmid containing only the flanking regions is inserted. In the unmarked mutant, the chromosomal DNA between these flanking regions will be removed. If a mutant containing a novel gene (*i.e.* a gene not normally present in the recipient organism's genome) is required, for example one encoding an enzyme to produce biofuel compounds, the novel gene is inserted between the flanking regions on the plasmid. In the unmarked mutant, not only will the chromosomal DNA between these flanking regions be removed, but the novel gene will be inserted.

Even more subtle alteration of chromosomal DNA can be performed. For example, if a single base pair change in a gene is required (*e.g.* from a thymine (T) to a cytosine (C)) the native region between the two flanking regions is altered accordingly in the plasmid. Upon generation of the unmarked mutant, the only difference between the wild-type and mutant strains will be at this nucleotide. In each case, after the marked mutant cells are mixed with the plasmid, the cells are grown on agar plates containing sucrose. As sucrose is lethal to cells when the *sacB* gene product is expressed, only cells in which a second recombination event occurs, whereby the sucrose sensitivity gene in addition to the antibiotic resistance gene are recombined out of the chromosome and onto the plasmid, will survive. In exchange, the flanking regions and the DNA between them are inserted into the chromosome.

There are a number of advantages to generating targeted, unmarked mutants by genetic manipulation. By removing the antibiotic resistance or sucrose sensitivity genes from the unmarked mutant, the entire engineering process can be repeated in the same strain to generate strains with multiple alterations in metabolic pathways. Moreover, the elimination of antibiotic resistance genes in the engineered organism obviates the possibility of transfer of these genes to other microbes in the environment.

9.5   Industrial Production of Biofuels from Cyanobacteria

Cyanobacteria naturally produce biofuel compounds, albeit in vanishingly small amounts. Two pathways for hydrocarbon production have been identified in cyano-bacteria[12,13]. The first pathway involves a two-step process whereby fatty acids are converted to alkanes, predominantly heptadecane (a straight-chain alkane containing 17 carbons; C17). Heptadecane, a major component of diesel, can be readily used in diesel vehicles without modification to engine components. The second pathway is a one-step process involving conversion of fatty acids to alkenes, predominantly nona-decene (C19 alkene). Alkenes are not as suitable as alkanes for biofuels, but are a useful feedstock for polymer production. Cyanobacteria can also produce hydrogen (H_2). However the enzyme involved in hydrogen production, termed the hydrogenase, is inactivated by O_2. Consequently, H_2 production is impaired when photosynthesis is active, and H_2 production by cyanobacteria requires the use of very sophisticated culture systems.

Several commercial companies currently use cyanobacteria to generate biofuels. However, it is difficult to assess their progress and the exact process by which they generate biofuels, due to the lack of publications in the peer reviewed literature. Consequently, the details for most of the remainder of this chapter are derived from company websites and patent applications.

Algenol is a Florida-based company, founded in 2006, that has patented a technology they call 'DIRECT TO ETHANOL®'. According to their website[14], the company claims to produce ethanol for less than US$ 1 per US gallon (3.8 litres) and are targeting a production level of 6,000 US gallons per acre per year/9,200 litres per hectare per year. As ethanol is not produced in significant quantities in cyanobacteria, two genes from the bacterium *Zymomonas mobilis* encoding pyruvate decarboxylase and alcohol dehydrogenase enzymes were inserted into *Synechococcus elongatus*[15]. These enzymes convert pyruvate, a key metabolite produced in large quantities in bacteria, to ethanol. As ethanol is a small compound, it readily diffuses out of the cell. Therefore cells constantly produce ethanol and do not have to be harvested and broken apart to collect the biofuel product. Algenol cultures are conducted in closed polyethylene

---

12   Schirmer *et al.* (2010).
13   Mendez-Perez *et al.* (2011).
14   http://www.algenolbiofuels.com/
15   Deng and Coleman (1999).

photobioreactors. Ethanol and water from the culture media condense in the top of the bioreactors at night when the temperature drops. The ethanol/water mix then runs down the side of the photobioreactor where it is collected. Ethanol is then concentrated to biofuel concentrations (99.7% pure). This is the most energy intensive and complicated part of the process. The advantage of the technology is the low cost of the photobioreactors and the relative simplicity of the process. This process has been tested at lab-scale and the company is currently building a 36 acre/14.6 hectare pilot plant, which they claim will produce approximately 100,000 US gallons/(≈379,000 litres of ethanol per year[14].

Joule Unlimited was founded in 2007 in Massachusetts and holds a number of key patents for engineering cyanobacteria, notably *Synechococcus* sp. PCC 7002, to produce and secrete C13–C17 alkanes by insertion of two genes encoding the alkane biosynthesis enzymes originally identified in another species of cyanobacteria (US Patent No. 7794969); for engineering cyanobacteria to produce and secrete ethanol by insertion of the genes from *Zymomonas mobilis* encoding pyruvate decarboxylase and alcohol dehydrogenase (Patent WO 2010 062707 A1); for engineering cyanobacteria to produce C18–C19 alkenes *via* over-expression of a polyketide synthase, which is produced at low levels in some cyanobacteria species (US Patent No. 2010/0330642); and for the design of a novel photobioreactor capable of culturing photosynthetic organisms to a concentration of 14 g dry cell weight per litre of culture (Patent WO 2010 068288 A3R4). It should be noted that the process to engineer cyanobacteria to produce ethanol is similar to the process first patented by Algenol. Moreover, the patent to synthesise alkanes in cyanobacteria is based on a biosynthetic pathway first identified and patented by another company, LS9, based in California (Patents WO 2010 0249470 and 2010 0199548) and published in *Science*[12]. LS9 and Joule Unlimited were both founded by the same venture capital firm, Flagship Ventures. According to the Joule Unlimited website[16], the company claims they can produce 15,000 US gallons/57,000 litres of diesel (C17 alkanes) and 25,000 US gallons per acre/210,000 litres per hectare of ethanol per year at a cost of US$ 50–60 a barrel (≈120 litres) and US$ 1.28 per US gallon (3.785 litres) respectively. The company is currently operating a pilot plant in Texas and a 5-acre/2-hectare demonstration plant in New Mexico.

It should be noted that other biofuel companies such as Amyris, which uses yeast to convert sugars to small hydrocarbons such as farnesene, have encountered significant difficulties when scaling their process up to industrial production. The Amyris facility in Brazil was originally targeted to produce 9 million litres of hydrocarbons in 2011 and 44 to 50 million litres of hydrocarbons in 2012, but this target was later reduced to 1 to 2 million litres in 2011 and the company has now scaled down plans to produce low value, high volume products such as fuels[17]. The company did not put forward the exact reasons for the difficulties. However, when organisms are engineered so that their metabolism is diverted towards high levels of production of a single metabolite and away from growth, this results in a significant drop in 'fitness'. Therefore any contaminant with fast growth may quickly dominant a culture, especially when a ready energy source, such as sugar, is available. This may be less of an issue for photosynthetic organisms cultured in the absence of sugar. Another issue is that any engineered organism

---

16 http://www.jouleunlimited.com/about/overview
17 http://www.amyris.com/

that can divert its energy from metabolite production back to growth, most likely due to inactivation of genes inserted into the organism, will have a significant advantage and quickly dominate the culture. Obviously this will cause a significant drop in metabolite production. Algenol is approaching this issue by growing cyanobacteria in small-scale bioreactors. In the event that the engineered organism losses dominance in the culture, the bioreactor can be quickly shut down and the process restarted.

9.6 Conclusion

In this chapter we have focused on the potential use of cyanobacteria for biofuel production. We noted that the cyanobacteria possess numerous desirable properties for industrial deployment, including the direct fixation of CO_2, extremely fast growth rates, the possibility for culture in extreme environments that reduce or preclude contamination risks and the possibility of targeted and subtle genetic engineering. In certain sectors, cyanobacteria are a commercial reality. However, biofuels must be cheap and produced in high volumes; certain companies have taken on that challenge and, although the volumes of biofuel produced may be small, the knowledge acquired through practice at scale is critical for any wider deployment and long-term improvements to the existing technologies.

Selected References and Suggestions for Further Reading

Bhaya, D. *et al.* (2006) Phototaxis and impaired motility in adenylyl cyclase and cyclase receptor protein mutants of *Synechocystis* sp. strain PCC 6803. *Journal of Bacteriology*, **188**, 7306–7310.

Bendall, D.S., Howe, C.J., Nisbet, E.G. and Nisbet, R.E.R. (2008) Introduction: Photosynthetic and atmospheric evolution. *Philosophical Transactions of the Royal Society B-Biological Sciences*, **363**, 2625–2628.

Brahamsha, B. (1996) A genetic manipulation system for oceanic cyanobacteria of the genus *Synechococcus*. *Applied and Environmental Microbiology*, **62**, 1747–1751.

Deng, M.D. and Coleman, J.R. (1999) Ethanol synthesis by genetic engineering in cyanobacteria. *Applied and Environmental Microbiology*, **65**, 523–528.

Galloway, J.N., Dentener. F.J., Capone, D.G., Boyer, E.W., Howarth, R.W. *et al.* (2004) Nitrogen cycles: past, present, and future. *Biogeochemistry*, **70**, 153–226.

Hodson, M.J. and Bryant, J.A. (2012) *Functional Biology of Plants.* Wiley-Blackwell, Chichester, UK.

Howe, C.J. *et al.* (2008) The origin of plastids. *Philosophical Transactions of the Royal Society B-Biological Sciences*, **363**, 2675–2685.

Mendez-Perez, D., Begemann, M.B. and Pfleger, B.F. (2011) Modular synthase-encoding gene involved in alpha-olefin biosynthesis in *Synechococcus* sp. strain PCC 7002. *Applied and Environmental Microbiology*, **77**, 4264–4267.

Reddy, K.J. *et al.* (1993) Unicellular, aerobic nitrogen-fixing cyanobacteria of the genus *Cyanothece*. *Journal of Bacteriology*, **175**, 1284–1292.

Schirmer, A., Rude, M.A., Li, X., Popova, E. and del Cardayre, S.B. (2010) Microbial biosynthesis of alkanes. *Science*, **329**, 559–562.

Summerfield, T.C. and Sherman, L.A. (2008) Global transcriptional response of the alkali-tolerant cyanobacterium *Synechocystis* sp. strain PCC 6803 to a pH 10 environment. *Applied and Environmental Microbiology*, **74**, 5276–5284.

Tan, L.T. (2007) Bioactive natural products from marine cyanobacteria for drug discovery. *Phytochemistry*, **68**, 954–979.

Tolonen, A.C., Liszt, G.B. and Hess, W.R. (2006) Genetic manipulation of *Prochlorococcus* strain MIT9313: green fluorescent protein expression from an RSF1010 plasmid and Tn5 transposition. *Applied and Environmental Microbiology*, **72**, 7607–7613.

Zwirglmaier, K. *et al.* (2008) Global phylogeography of marine *Synechococcus* and *Prochlorococcus* reveals a distinct partitioning of lineages among oceanic biomes. *Environmental Microbiology*, **10**, 147–161.

10

Third-Generation Biofuels from the Microalga, *Botryococcus braunii*

Charlotte Cook[1], Chappandra Dayananda[2], Richard K. Tennant[1] and John Love[1]

[1] *College of Life and Environmental Sciences, University of Exeter, Exeter, UK*
[2] *Central Food Technological Research Institute, Mysore, India*

Summary

By 2030, the global demand for transport fuel is likely to surge by 45%, requiring the production of approximately 500 billion litres of biofuel per year (see Chapter 2). While a significant component of that demand will be filled by first- and second-generation biofuels (sugar- and lignocellulosic-derived ethanol and biodiesels from different sources), oleaginous microbes, notably heterotrophic yeasts (Chapter 14) and microscopic algae (Chapters 11 and 12) have the potential to produce cost-effective, biologically-derived, long-chain hydrocarbons and other platform chemicals for industrial uses. We have discussed at length the issues that make algal biofuels difficult to implement economically, yet the idea of third-generation biofuels from microalgae remains compelling and stimulates the imagination of researchers and entrepreneurs alike. A possible reason for that ongoing interest is the particular properties of *Botryococcus braunii*, a green, colonial microalga that produces and secretes large amounts of liquid hydrocarbons that can be converted into combustible fuel oils. In this chapter, we review the potential and the constraints of biofuel production from *B. braunii*.

10.1 *Botryococcus braunii*

Botryococcus braunii is a planktonic, green, colonial microalgae (Figure 10.1) and has attracted particular interest as a potential source of biofuel for two main reasons. First, and uniquely among microalgae, *B. braunii* synthesises and secretes large quantities (up to 80% of its dry mass) of long-chain (C_{20}–C_{40}), liquid hydrocarbons that can be subsequently treated to generate gasoline, diesel, aviation fuel or lubrication oil by catalytic cracking[1]. These hydrocarbons are generically termed 'botryococcenes' and comprise a variety of molecular structures (Figure 10.2). Second, most sources of crude

---

1 Hillen *et al.* (1982).

Biofuels and Bioenergy, First Edition. Edited by John Love and John A. Bryant.
© 2017 John Wiley & Sons Ltd. Published 2017 by John Wiley & Sons Ltd.

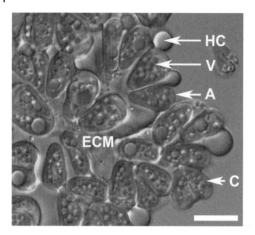

Figure 10.1 Brightfield Micrograph of *Botryococcus braunii*. *B. braunii* is a colonial microalga. Colonies are formed of adjoined, pyriform algal cells (A). The green colour is from the single, lobed chloroplast (C) that is present is each cell. Cells also contain a number of cytoplasmic vesicles (V) that contain the characteristic hydrocarbons synthesised by the algae. The hydrocarbons are secreted into the extracellular matrix (ECM) and, when placed under a microscope slide, ooze out of the colony in droplets (HC). The scale bar represents 10 μm.

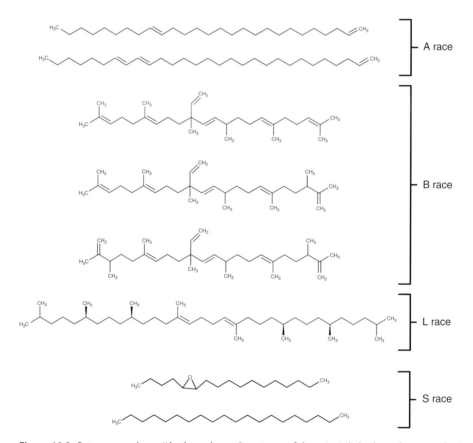

Figure 10.2 *Botryococcus braunii* hydrocarbons. Structures of the principle hydrocarbons synthesised by the different 'chemical races' of *B. braunii*.

oil are particularly rich in the fossilised (hydrogenated) form of botryococcenes, termed 'botryococcanes', or contain high levels of *Botryococcus* micro-fossils. For example, botryococcane comprises 1% of the crude oil from the Duri oil field and Sumatran crude contains 1.4% botryococcane. While these figures might appear low at first, it must be noted that there are no other reported occurrences of a single fossil marker in crude oil anywhere near the 1% level. Considering that botryococcane originates from botryococcene, the contribution of ancient *Botryococcus* to current oil fields is remarkable. As well as a direct contribution to crude oil, *Botryococcus* constitutes a large proportion of worldwide oil shale deposits[2]. *Botryococcus* fossils are particularly prevalent in Torbanite, one of the richest oil shales in the world, as well as in the Eocene Green River shales of Colorado and Utah, and in the Ordovician Kukersite of Estonia. *Botryococcus* has therefore had a huge impact on present-day oil reserves, and the potential of this alga to generate quantities of biofuels and platform chemicals is compelling.

Extant *Botryococcus* are distributed throughout the globe, and found in freshwater lakes, reservoirs, ponds, brackish waters and ephemeral lakes. Colonies vary in size from 30 μm to 2 mm in diameter and consist of single or multiple cell clusters united by transparent strands (Figure 10.1). The cells are pyriform (pear-shaped), approximately 13 μm in length and 7 μm wide, and are embedded within an extracellular matrix (ECM) that is impregnated with polymerised and liquid hydrocarbons. The cells possess no flagellae. Ultrastructural investigations show that the ECM consists of remnants of outer cell walls originating from successive cellular divisions. A polysaccharidic fibrillar sheath, rich in arabinose and galactose, extends into the medium from the outside surface of the colony. Inside, each *B. braunii* cell has a single, lobed chloroplast, an apical (*i.e.* close to the more pointed end of the cell) Golgi body and a cortical, fenestrated endoplasmic reticulum (ER) that is thought to participate in the secretion of the retaining wall and fibrillar sheath components. In the non-apical (*i.e.* blunt end of the cell) domain of cells of the B-race, inclusion bodies containing lipid and botryococcenes are present and associate with the ER, nuclear and chloroplast envelopes. The non-apical ER is in close contact with the cell membrane from where botryococcenes are secreted from the cell.

Until recently, the *Botryococcus* genus was classified under the *Chlorophycaeae*, which is one of three monophyletic groups of the Chlorophyta, or green algae. However, closer phylogenetic analyses using the 18S rRNA sequences of *B. braunii* suggested that its closest relatives are, in fact, in the genus *Choricystis* (*Trebouxiophyceae*)[3]. At least 13 *Botryococcus* species have been described morphologically. Reproduction in *B. braunii*, at least in laboratory culture, is typically asexual and occurs by mitosis and subsequent fragmentation of colonies into two or more parts, each one becoming a new colony. Zoosporogenesis, *via* the formation of four autospores per cell, has also been observed in *Botryococcus*. The degree to which *Botryococcus* is dependent on vegetative (asexual) propagation is unknown but, to date, sexual reproduction in *B. braunii* remains undocumented.

The most distinctive characteristic of *B. braunii* is the production and secretion of long-chain, liquid hydrocarbons[4]. This particularity is so unique among microalgae that, despite the ambiguities surrounding precise phylogenetic relationships, *B. braunii*

---

2 Volkman *et al.* (2015).
3 Senousy *et al.* (2004).
4 Metzger and Largeau (2005).

is conventionally classified into three 'chemical races', termed A, B and L, based on the type of hydrocarbons synthesised (see below and Figure 10.2). Recently a fourth race, S, that is similar to the Race L. has been tentatively identified[5]. It has been suggested that the different races of *B. braunii* are, in fact, different species[6], but definitive confirmation has been complicated by morphological variation attributed to culture age and that induced by varying culture conditions of *B. braunii*; it thus remains unclear as to whether the numerous strains of *B. braunii* belong to a single species, to three species or to several sub-species.

10.2   Microbial Interactions

Algae exist in the natural habitat as part of a varied consortium of microbes whereby the species may exert considerable influence over one another to exploit unique biological functions or to exchange metabolites. *B. braunii* have been found to associate with varied bacterial partners, including *Arthrobacter sp., Cornynebacterium sp., Pseudomonas sp., Erwinia sp., Alcaligenes sp., Flavobacterium sp., Azatobacter sp.* and *Streptomyces*. So-called 'pure' cultures of *B. braunii* may contain only one strain of alga, but are often contaminated with bacteria, which associate with the exopolysaccharide matrix of algal colonies (Figure 10.3). Association of bacterial consortia with the extracellular matrix of *B. braunii* colonies renders the removal of contamination from the culture without damaging microalgal cells challenging. The difficulty of obtaining and maintaining axenic cultures of *B. braunii* indicates that the associations are the product of long-standing mutualistic relationships, as opposed to a contamination event[7].

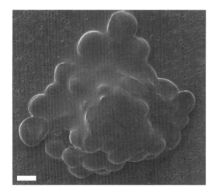 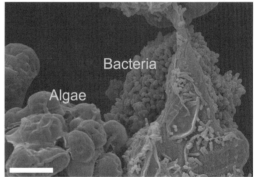

Figure 10.3 Association between *Botryococcus braunii* and bacteria. The left panel is a scanning electron micrograph of a single *B. braunii* colony. The algal cells are barely visible and appear shrouded by the layer of secreted hydrocarbons. The right panel shows a scanning electron micrograph of a *B. braunii* colony from which the hydrocarbons have been removed. The algae appear as flattish pear-shaped cells; note a number of rod-shaped bacterial cells that are closely associated with the surface of the algae. The scale-bars represent 5 μm.

---

5 Kawachi *et al.* (2012).
6 Dayananda *et al.* (2010).
7 Tanabe *et al.* (2012).

The impact bacterial associations have upon *B. braunii* cultures are dependent on the bacterial species involved[8]. *B. braunii* cultured with *Flavobacterium aquatile* produced a larger yield of algal biomass and hydrocarbon[9], whereas co-culture with *Pseudomonas oleovorans* caused a considerable reduction in algal biomass and hydrocarbon production, possibly attributable to bacterial degradation of the hydrocarbons. Association with *Bacillus, Cornynebacterium* and *Rhizobium* species can stimulate hydrocarbon production in *B. braunii*; indeed, extrapolation of results suggested an increase of approximately 50% in associations with *Rhizobium sp*[10]. Conversely, co-culture with *Acinetobacter sp.* was detrimental to algal growth and resulted in reduced culture density. These antagonistic effects may occur through the release of toxic compounds or competition for identical resources.

The exact basis for the apparently beneficial relationships between *B. braunii* and the potentially ectosymbiotic bacteria is not established, although *B. braunii* may benefit from a bacterial consortium by capture of vitamins and other essential micronutrients, B_{12} (riboflavin) in particular. *De novo* synthesis of B_{12} is so far undiscovered in algae, although in many species it is known to be required as a cofactor for methionine synthase[11] (Croft *et al.*, 2006). In addition, there may be benefits to maintenance of bacterial associations, such as the exclusion of other harmful contaminating organisms by co-culture with *Rhizobium sp*[7], as well as provision of CO_2 from respiring bacteria or bio-available nitrogen, phosphate and other inorganic compounds.

The research community is yet to fully comprehend the mechanism(s) by which association of bacterial consortia with *B. braunii* may lead to increased biomass and hydrocarbon production. Nonetheless, it is clear that maintenance of stable microalgal-bacterial relationships could have a profound impact on the commercial viability of biofuels from microalgae.

10.3 *Botryococcus braunii* as a Production Platform for Biofuels or Chemicals

10.3.1 Hydrocarbons, Lipids and Sugars

B. braunii synthesises a number of potentially interesting compounds that may be used as fuel or platform chemicals, most notably long-chain hydrocarbons (highly reduced organic compounds), but also lipids and polysaccharides. There are a number of reviews that describe hydrocarbon synthesis in *Botryococcus*, so only a broad outline of these biochemistries is included here; readers are referred to Banerjee *et al.* (2002) and to Metzger and Largeau (2005) for further details.

Race A of *B. braunii* produces C_{23}–C_{33} odd numbered n-alkadienes and mono-, tri-, tetra- and pentaenes, derived from fatty acids. These linear olefins can constitute up to 61% of the dry cell mass of the green active state colonies. In addition, Race A cells may produce ether lipids such as alkadieenyl-o-alkatrienyl ether, alkenyl-o-botryalyl ether

---
8 Chirac *et al.* (1985).
9 Banerjee *et al.* (2002).
10 Rivas *et al.* (2010).
11 Croft *et al.* (2006).

and resorcinolic ether that can constitute 5–42% of the biomass on a dry weight basis. Race B typically produces C_{30} to C_{37} hydrocarbons derived from the terpenoid metabolic pathway and referred globally as botryococcenes and C_{31} to C_{34} methylated squalenes. Certain strains of the B race may also synthesise cyclobotryococcenes. Although Race B strains of *B. braunii* can produce ether lipids such as diepoxy-tetramethylsqualene, botyolin-A and braunixanthin-1 (general name botryolins), these are present in minor amounts. In contrast to the metabolic complexity of Races A and B, Race L cells produce only a single hydrocarbon, lycopadiene, a $C_{40}H_{78,}$ tetraterpene which constitutes 2–8% of the dry biomass[12]. There are about 12 other ether lipids called lycopanerols isolated from L race strains, which can accumulate up to 10% on a dry weight basis depending on culture conditions.

In nature, strains of *B. braunii* Race B can accumulate 20–86% of hydrocarbons on a dry weight basis[13]. These compounds are readily converted to starting materials for industrial chemical manufacturing and high-quality fuels under standard hydrocracking/distillation conditions in yields approaching 97%. As a proof-of-principle, hydrocarbons extracted from a natural bloom of *B. braunii* produced a distillate comprising of 67% gasoline fraction, 15% aviation fuel, 15% diesel fuel fraction and 3% residual oil when hydrocracked[14]. These figures are used throughout the industry to illustrate the potential of *Botryococcus* as a source of biofuel on a par with that produced from oilseed crops, although commercial production of *B. braunii*-derived hydrocarbons has never been trialled at scale.

In addition to terpenoid-derived hydrocarbons and ether-lipids, *B. braunii* also produces large amounts of fatty acids that are suitable for the production of biofuel, including palmitic (16:0), oleic (18:1), linoleic (18:2) and linolenic acids (18:3). As with most microalgae (see Chapters 11, 12 and 14), the physiological conditions in the culture dictate lipid yields, which may vary from 10 to 60% (w/w).

All of the three *B. braunii* races are known to produce exopolysaccharides, comprising heterogeneous polymers of galactose, glucose, arabinose, rhamnose, fucose, uronic acids and some unusual sugars such as 3-o-methyl fucose and 3-o-methyl rhamnose. These polysaccharides could be used in the production of adhesives, paper, paint, textiles and in the food industry or for microbial fermentation to produce ethanol, hydrogen or methane. *B. braunii* UC58, isolated in Portugal, has been reported to produce 4–5.51 kg m^3 exopolysaccharide.

Another important bio-polymer is algaenan, a highly aliphatic, non-hydrolysable and insoluble macromolecule that is common in the outer cell wall of all three chemical races of *B. braunii*. Algaenan is highly resistant to degradation, allowing it to be selectively preserved during fossilisation, which may go some way to explain the presence of fossilised *B. braunii* as a major component of several high oil potential sediments. It is also regarded as a major 'player' in the Earth's carbon cycle[15].

Despite decades of research focusing on hydrocarbons from *Botryococcus*, it remains unclear why these algae should synthesise and secrete such complex and energetically expensive metabolic products on such a large scale. Several possible explanations exist

---

12 Metzger and Casadevall (1987).
13 Brown *et al.* (1969).
14 Hillen *et al.* (1982).
15 Hedges and Kell (1995); Volkman *et al.* (2015).

and any one, or any combination, may be true. It is possible that the presence of hydrocarbons within the colony matrix allows the colonies to float and therefore increase exposure to sunlight, maximising photosynthetic potential. However, it must be noted that all that is required to float is a cellular density less than that of water; to achieve this, the algae do not need to secrete the oils, but simply retain them within intracellular compartments. Consequently, the oily matrix probably serves multiple functions, including protecting *B. braunii* from desiccation and extreme changes in salinity, pH, temperature and light intensity. Resistance to environmental changes enable the dispersal of microalgae (*e.g.* by wind or as hitchhikers on the feathers or legs of waterfowl) and it has been implied that the extra-cellular matrix has thus facilitated the global dispersal of *B. braunii*. Other, equally plausible drivers for the prolific level of hydrocarbon production in *B. braunii* include protection from pathogens or as a role in maintaining commensal or symbiotic interactions with other microbes.

10.3.2 Controlling and Enhancing Productivity

Like all photosynthetic organisms, *B. braunii* requires light, CO_2, water and inorganic nutrients (nitrogen, phosphates and trace minerals) to grow. Other factors, including temperature, pH and salinity also have marked effect on *B. braunii* growth and productivity. Changing these factors may have a drastic effect on the type and amount of hydrocarbons produced[16]. For biofuel production, a judicious selection of *Botryococcus* strain and culture conditions is therefore important to ensure success. *Botryococcus* growth rate is slow compared to other microalgae, with doubling times measured in days to weeks, rather than hours. The slow growth rate is explained by the channelling of photosynthate towards hydrocarbon production rather than to cell division. Hydrocarbon production occurs at all stages of growth, and therefore appears to be a normal feature of *B. braunii*. Consequently, in most *B. braunii* cultures, hydrocarbon yields correlate closely with biomass.

In addition to the impact of the growth parameters that *B. braunii* is exposed to, hydrocarbon production has been demonstrated to be growth-associated under varying culture conditions and maximum productivity occurs during the exponential and early stationary growth phases[17]. Qualitative differences have also been noted in the lipids accumulated by *B. braunii* throughout the growth stages, with a greater proportion of ether lipids produced during the exponential and early stationary growth phases. Unsaponifiable lipid (up to 80% w/w) and exopolysaccharides are synthesised during the stationary growth phase[18] and colonies turn from green to orange-red in colour as carotenoids accumulate[19].

B. braunii is found in the tropics as well as in higher latitudes, and can grow under a wide range of light levels, from 40 to over 800 $\mu E\ m^{-2}\ s^{-1}$. A number of studies have been performed to identify the optimal irradiance levels for supporting both growth and hydrocarbon production in *B. braunii*. High light intensity increased the carotenoid (photoprotectant and antioxidant) to chlorophyll ratio and resulted in changes of the

---

16 Wake and Hillen (1980).
17 See, *e.g.* Dayananda *et al.* (2007).
18 Brown *et al.* (1969); Banerjee *et al.* (2002).
19 Metzger *et al.* (1985).

colour of algal colonies. A decrease in concentrations of carbohydrate, cellular nitrogen and phosphate in cultures has been observed[20] following extended exposure to a light intensity of 25–75 µE m^{-2} s^{-1}. It was also reported that cultures adapted to high irradiance (>140 µE m^{-2} s^{-1}) during a pre-culture stage could attain a higher biomass (7 kg m^{-3}) and hydrocarbon content (50% w/w) than those adapted to a low irradiance of 42 µE m^{-2} s^{-1}. A two-fold increase in biomass and four-fold increase in hydrocarbon content were reported in cultures under continuous illumination when compared to cultures illuminated with a diurnal light cycle (Lupi *et al.*, 1994), although the long-term effects of constant illumination on the algae or on the composition of the hydrocarbons was not assessed.

Under laboratory conditions, with optimal nutrient concentrations and increased levels of CO_2, faster growth can be achieved for which hydrocarbon productivity is best during the exponential phase of growth[21]. Similarly, ether lipid production was shown to be maximal during the exponential and early deceleration stages of growth. A modified Chu-13 medium[21] is commonly used for culturing of all three *B. braunii* races. In this culture medium, phosphate concentration is not a limiting factor for growth. Continuous absorption of phosphate was observed throughout the initial active growth phase, during which hydrocarbon production is at its peak; whilst under the stationary growth phase, uptake of phosphate ceases and instead phosphate is gradually released into the medium[22]. Increased concentrations of phosphate and sulphate lead to the production of greater biomass and thence larger lipid yields. The levels of nitrogen in the algal growth medium also has a significant influence on both hydrocarbon and lipid production in *B. braunii*[22]. Nitrogen sources like urea, ammonia and different salts of nitrites and nitrates, may be utilised for growth and a higher concentration of available nitrogen in the growth medium extends the exponential growth phase of *B. braunii*, ultimately attaining a greater biomass, but leads to a decrease in hydrocarbon production relative to biomass[22]. Micronutrients also have a notable influence on *B. braunii* culture characteristics, with increased biomass and hydrocarbon yields associated with optimised trace element concentration.

Supplementation with carbon dioxide has considerable influence on both growth and hydrocarbon production in *B. braunii*. Addition of CO_2 to 20% has little effect on the pH of the Chu medium. Air enriched to only 1% CO_2 considerably enhances *B. braunii* growth with a mean doubling time of 2 days compared with that of 7 days when cultured without CO_2 enrichment, and faster growth rates can be achieved with enrichment to 2% CO_2. Hydrocarbon production increased 5-fold when the algae were cultured in 1% CO_2 compared to cultures grown in ambient air with normal CO_2 concentrations[23].

The optimal growth temperature for the majority of *B. braunii* strains is around 25°C, although the alga can grow at temperatures from approximately 10°C to 32°C, albeit more erratically towards the extremes[24]. Hydrocarbon biosynthesis is affected by temperature of the ambient medium, and ceases when *B. braunii* is cultured below 23°C or above 32°C. Fatty acid production is also affected by ambient temperature of *B. braunii* cultures, with a marked decrease above and below 25°C noted in the B race strain, Kutz.

---

20 Oh *et al.* (1997).
21 Largeau *et al.* (1980).
22 Casadevall *et al.* (1985).
23 See footnote 8.
24 Lupi *et al.* (1991).

B. braunii tolerates a range of pH from 6 to 8.5. In closed cultures, the pH of the medium increases during the active growth phase followed by a slight decline. This increase in pH is partially due to the consumption of dissolved carbon dioxide for photosynthesis. A similar trend occurs in CO_2 supplemented cultures.

Although *B. braunii* is often quoted as tolerant of brackish habitats, and can adapt to NaCl concentrations of up to 3M, the effects of salinity on the algae remain unclear; *B. braunii* lipid and hydrocarbon production appear to increase when the algae are cultured in salinities between 15 and 85 mM NaCl[25], but are hampered at concentrations above 150 mM.

10.3.3 Alternative Culture Systems

As for most algae, the leitmotiv for *B. braunii* culture is as a photoautotroph in a mixed, water-based medium (see Chapter 12). Industrial production of microalgae depends on improvements in mass cultivation technology, strain improvements, species control, microbial consortia and process engineering. The US Department of Energy's Aquatic species programme identified the need for adequately funded, long-term R&D programmes to achieve the high productivities necessary for sustainable biodiesel production from microalgae[26]. These studies demonstrated the potential use of brackish and saline water, marginal land and CO_2 resources for mass cultivation of microalgae and estimated that around 0.4% of global arable land would be sufficient to meet global fuel demands from algae-based biofuel plants. However, such infrastructure is, as yet, non-existent and would require significant investment to construct.

Alternative culture techniques for *B. braunii* have also been explored, often with intriguing results. In parallel with other algal species, protocols for heterotrophic culture of *B. braunii* using organic carbon sources such as simple sugars, amino acids, free fatty acids and glycerol, have been developed, and *B. braunii* has been successfully employed to clarify secondarily treated sewage[27]. Immobilised algal cells, rather than planktonic cultures, offer the potential to reduce energy and water requirements, while providing a platform that can remove pollutants from the ambient environment. Compared to culture in open ponds, *B. braunii* grown as a biofilm in a photobioreactor required 45% less water and 99% less energy for the post-harvest dewatering process[28]. Alternatively, microalgae may be immobilised in calcium alginate beads, which can enhance photosynthetic activity and increase hydrocarbon content by up to 20%, despite a decrease in the rate of biomass production. Reduced biomass is a drawback to the use of immobilised algal systems, particularly for slow growing species such as *B. braunii*, although more recent studies are encouraging and have reported enhanced biomass yields for *B. braunii* in attached growth systems[29]. Although technological developments have progressed the field of immobilised or attached algal culture systems, in the case of *B. braunii*, the problem of extracting hydrocarbons easily and cost-effectively from immobilised or encapsulated *B. braunii* cells remains to be addressed.

---

25 Ranga Rao *et al.* (2007).
26 Sheehan *et al.* (1998).
27 Sawayama *et al.* (1992).
28 Ozkan *et al.* (2012).
29 Cheng *et al.* (2013).

10.3.4 Harvesting *Botryococcus* Biomass and Hydrocarbons

Downstream processing of microalgae involves several steps including harvesting, de-watering and processing of algal metabolites for commercial applications and therefore determines the economics of biofuel production from these organisms (see also Chapter 16). Careful engineering of downstream process design and inputs from intensive R&D is therefore fundamental in the development of commercially viable products from microalgae.

Harvesting is one of the major hurdles for industrially relevant microalgal species, mostly because of their cell size and morphology. The harvesting procedure for *B. braunii* is atypical for microalgae, in that the species does not settle or flocculate but floats. Wild *B. braunii* blooms have been reported to form a rubbery mat that, once dry, is easily collected for processing[30]. The main issue however remains, as for all algae, that of de-watering by a cost-effective process (Chapters 12, 14 and 16). Skimmers, strainers, filters, centrifuges or hydrocyclones may all be used to harvest *Botryococcus*. However, with *B. braunii*, complete de-watering may not be as important as the method of hydrocarbon harvest. Biocompatible solvents can recover up to 70% of liquid hydrocarbons from *B. braunii* [31] and thus offer the possibility of hydrocarbon extraction at defined periodic intervals (also termed 'milking'), or even continuous extraction of hydrocarbon from the ECM. Furthermore, harvesting oils several times from the same (slow-growing) biomass may result in more favourable process economics and reduce the overall cost of production of biofuel molecules from *B. braunii*.

10.3.5 Processing *Botryococcus* into an Alternative Fuel

Algal biomass may be converted into energy by chemical, thermochemical or biological processes, or by combination of these. The chemical composition of the algal biomass dictates the quality of the end-products, such as gasoline or diesel fuels. Hydrocracking has been used to process crude *B. braunii* extracts[32] into gasoline (67% of total), jet fuel (15% of total) and automotive gas-oil (or diesel, 15% of total). Thermo-chemical liquefaction (TCL) involves subjecting the dry or wet biomass to high temperatures and pressures with a suitable catalyst. The reaction temperature and catalysts dictate yields of gasoline and other fractions. TCL eliminates the need for an initial solvent extraction and has been successfully employed to extract oil from *B. braunii*. The overall oil yield from liquefied biomass is, however, lower than the yield obtained by solvent extraction.

Fast pyrolysis is used for direct conversion of biomass into liquid oils (often called 'bio-crude'), but can also be adjusted to favour char, liquid or gas production. Fast pyrolysis is a high temperature process in which biomass is rapidly heated in the absence of oxygen. The essential features of a fast pyrolysis process are reaction temperatures of approximately 500°C, very high heat transfer rates to the biomass, short vapour residence times of less than 2 seconds and rapid cooling of the pyrolysis vapour. The total liquid products after pyrolysis comprise aqueous and oil phases that can be separated easily. The remaining solid char and gas can also be collected. Oil from fast pyrolysis of

30 Hillen *et al.* (1982).
31 Frenz *et al.* (1989).
32 Hillen *et al.* (1982).

microalgae has low oxygen content and thus offers suitable storage stability (see Chapter 2), and may be used in many applications as a direct substitute for conventional fuels. Despite these advantages, and possibly due to the difficulties in obtaining sufficient initial biomass, fast pyrolysis has not yet been used to process *B. braunii.*

10.4 Improving *Botryococcus*

The future demand for biofuel far exceeds production capacity (see Chapters and 16), without the development of alternative strategies including all those described in other chapters of this volume. Consequently, there is significant interest in the development of microalgae for future biofuel production[33].

Compared to other microalgae, *B. braunii* has a number of advantages that make it especially suitable to meet the requirements for the production of alternative transport fuels. As previously mentioned, *Botryococcus* algae are present throughout the globe and therefore may, in theory at least, be cultured anywhere. The size of *Botryococcus* colonies and the fact that they float allows simple filtration with surface skimmers to harvest the algal biomass. *B. braunii* synthesise petroleum-like hydrocarbons, but not, as do other algae, plant-oils in the form of triacylglycerides (TAGs) that require significant processing to make a useable transport fuel additive (see Chapters 2, 6 and 11). Importantly, *B. braunii* secretes these hydrocarbons into the colonial, ECM, thereby protecting the algae from predation, maintaining it in the photic zone at the surface of a pond and, as mentioned earlier, offering the possibility of continuous oil extraction from living biomass. This latter point is critical, considering the slow growth rates of *B. braunii.* In any industry, time is money, so harvesting the product without killing the biomass may be a way of circumventing the inherent limitations of *B. braunii* growth rates.

Although a significant amount of research has focused on *B. braunii* products, comparatively little has actually aimed to understand the alga *per se.* This lack of fundamental, biological knowledge is critical, as it impedes moving beyond wild-type algae to improved strains that are better suited to the specific demands of aquaculture. Indeed, it is unlikely that any one *B. braunii* strain will display all the characteristics that industry demands, but the implementation of molecular technologies (strain selection and/or genetic modification; see Chapter 11) could maximise the capabilities of *B. braunii,* making it commercially competent. To date, 30 different microalgal species have been genetically engineered. Several methods for DNA delivery into algal cells have been developed, including biolistic ('gene-gun') direct DNA transfer, *Agrobacteruim* mediated gene transfer, cell agitation with DNA-coated micro- or macro-particles, polyethylene glycol-mediated direct gene transfer and electroporation. Despite these advances, and possibly due to both its colonial growth habit and to the oily coats that shrouds its cells, *B. braunii* remains genetically intractable. Moreover, very little detail is known about the *B. braunii* life cycle; these neglected areas of fundamental biology must be addressed in order to progress along the route of strain improvement.

---

33 Wijffels and Barbosa (2010); Georgianna and Mayfield (2012).

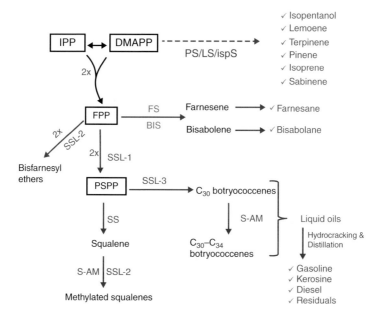

Figure 10.4 Potential synthetic biology of *Botryococcus braunii* hydrocarbon synthesis. The *B. braunii* isoprenoid biosynthetic pathway of *B. braunii* may be leveraged to produce advanced biofuels through the application of synthetic biology approaches. Enzyme functions are indicated in boxes, and conversions are indicated by arrows. SSL, squalene synthase like enzymes 1, 2 and 3; S-AM, S-adenosyl methonine transferase; FS, farnesene synthase; BIS, bisbolene synthase; PS, pinene synthase; LS, limonene synthase; ispS, isoprene synthase.

In the absence of a *B. braunii* 'horticulture', the molecular resource of *B. braunii* may be mined to transfer important pathways into more tractable algal species. Several research groups are involved in sequencing the genome of *B. braunii* to increase our understanding of cellular processes at a molecular level. Genome analysis may shed light on the identity of key genes involved in cell division and botryococcene biosynthesis, facilitating enhancement of overall productivity.

Up to the mid-2015, only a handful of microalgal genomes had been sequenced and annotated. However, many algal genome analysis projects are in progress. The *B. braunii* race B genome is estimated to be 166.2 Mbp, larger than all other available microalgal genomes. Hence, sequencing the *B. baunii* genome would augment the understanding of microalgae more generally, as well as being an end in itself[34].

B. braunii Race B produces large amounts of triterpenes *via* the isoprenoid pathway and genome mining may therefore offer the possibility for producing advanced isoprenoid-based biofuels using synthetic biology in other algal or microbial species (Figure 10.4; see also Chapter 15). Isopentenyl diphosphate (IPP) and dimethylallyll diphosphate (DMAPP) are the starting precursor molecules required for the production of advanced isoprenoid-based biofuels such as farnesene, bisbolene, terpinene, lemoene,

---

34 Weiss *et al.* (2011).

sabainene and pinene and their associated alcohols isopentanol, terpineol, geraniol and farnesol. Farneseyl pyrophosphate (FPP) can be converted into farnesene by farnesene synthase, which is then chemically hydrogenated to produce a valuable diesel fuel substitute, farnesane[35]. Similarly, FPP can be converted into bisabolane, another good source of biofuel, by introducing heterologous bisbolene synthase into the *B. braunii* isoprenoid pathway to produce bisbolene, followed by chemical reduction into bisabolane.

B. braunii Race A strains have the ability to produce very long chain alkenes from fatty acids and as they are able to accumulate relatively large amounts of neutral lipids, they offer the possibility to produce advanced fatty acid derived alkanes, alkenes, fatty aldehydes, fatty acid methyl esters, fatty alcohols and fatty acid ethyl esters using synthetic biology approaches[36].

10.5 Future Prospects and Conclusion

Botryococcus braunii is clearly a versatile alga and adaptable to a wide range of conditions. Recent business evaluations[37] suggest that by the mid-21st century, *B. braunii* could come to share much of the oil market with fossil-derived transportation fuel. Regardless of the availability of fossil fuels, it is for society to consider the environmental and health hazards that are associated with these fuels and to source our energy accordingly. Increased reliance on fossil fuels has contributed to the increase in concentration of atmospheric CO_2 from 280 to 400 parts per million (ppm) since the Industrial Revolution, the probable cause for the 0.7°C rise in the average global surface temperature over the past century. If left unchecked, the concentration may exceed 500 ppm by the end of this century. As populations increase and economies surge, global energy demand is predicted to rise by 37% between 2015 and 2040, with about 60% of the increase arising from developing and less-developed countries[38].

However, despite the prolific oil production, the growth rate of *B. braunii* is not competitive with that of several microalgae currently being cultivated for commercial applications. Continuous extraction of hydrocarbons while maintaining the biomass may result in more favourable process economics and reduce the overall cost of production of biofuel molecules from *B. braunii*, but such a process has yet to be tested at scale. To date, *B. braunii* remains genetically intractable and very little is known about the *B. braunii* life cycle; research must address these neglected areas to enhance the natural capacities of *B. braunii*. However, it is also possible that *B. braunii* itself will not be used in the future as a source of biofuel, rather that the *B. braunii* genome will be mined for new metabolic pathways, regulators and enzymes that will be improved and transferred, using synthetic biology, into species more tractable for industry. Whatever the route, *B. braunii* may well be a foundation for future, sustainable transport fuels.

---

35 Renninger and McPhee (2008).
36 Georgianna and Mayfield (2012); Wang *et al.* (2012).
37 Shiho *et al.* (2012).
38 http://www.iea.org/textbase/npsum/weo2014sum.pdf

Selected References and Suggestions of Further Reading

Banerjee, A., Sharma, R., Chisti, Y. and Banerjee, U.C. (2002) *Botryococcus braunii*: A renewable source of hydrocarbons and other chemicals. *Critical Reviews in Biotechnology*, **22**, 245–279.

Brown, A.C. *et al.* (1969) *Phytochemistry*, **8**, 543–547.

Casadevall, E., Dif, D., Largeau, C., Gudin, C., Chaumont, D. and Desanti, O. (1985) Studies on batch and continuous cultures of *Botryococcus braunii*: hydrocarbon production in relation to physiological state, cell structure, and phosphate nutrition. *Biotechnology and Bioengineering*, **27**, 286–295.

Cheng, P., Ji, B., Gao, L., Zhang, W., Wang, J. and Liu, T. (2013) The growth, lipid and hydrocarbon production of *Botryococcus braunii* with attached cultivation. *Bioresource Technology*, **138**, 95–100.

Chirac, C. *et al.* (1985) *Journal of Phycology*, **21**, 380–387.

Croft, M.T. *et al.* (2006) *Eukaryotic Cell*, **5**, 1175–1183.

Dayananda, C., Sarada, R., Usha Rani, M., Shamala, T.R. and Ravishankar, G.A. (2007) Autotrophic cultivation of *Botryococcus braunii* for the production of hydrocarbons and exopolysaccharides in various media. *Biomass and Bioenergy*, **31**, 87–93.

Dayananda, C., Venkatachalam, L., Bhagyalakshmi, N. and Ravishankar, G.A. (2010) Assessment of genetic polymorphism among green microalgae *Botryococcus* of distinct origin by RAPD. *Genes, Genomes and Genomics*, **2010, Special Issue 1**, 65–69.

Frenz, J., Largeau, C., Casadevall, C., Kollerup, F. and Daugulis, A.J. (1989) Hydrocarbon recovery and biocompatibility of solvents for extraction from cultures of *Botryococcus braunii*. *Biotechnology and Bioengineering*, **34**, 755–762.

Georgianna, D.R. and Mayfield, S.P. (2012) Exploiting diversity and synthetic biology for the production of algal biofuels. *Nature*, **488**, 329–335.

Hedges, J.I. and Kell, R.G. (1995) Sedimentary organic matter preservation: an assessment and speculative synthesis. *Marine Chemistry*, **49**, 81–115.

Hillen, L.W. *et al.* (1982) *Biotechnology and Bioengineering*, **24**, 193–205.

Hillen, L.W., Pollard, G., Wake, L.V. and White, N. (2004) Hydrocracking of the oils of *Botryococcus braunii* to transport fuels. *Biotechnology and Bioengineering*, **24**, 193–205.

Kawachi, M., Tanoi, T., Demura, M., Kaya, K. and Watanabe, M.M. (2012) Relationship between hydrocarbons and molecular phylogeny of *Botryococcus braunii*. *Algal Research*, **1**, 114–119.

Largeau, C. *et al.* (1980) *Phytochemistry*, **19**, 1043–1051.

Lupi, F.M. *et al.* (1991) *Applied Phycology*, **3**, 35–42.

Metzger, P. *et al.* (1985) *Phytochemistry*, **24**, 2305–2312.

Metzger, P. and Casadevall, E. (1987) *Tetrahedron Letters*, **28**, 3931–3934.

Metzger, P. and Largeau, C. (2005) *Botryococcus braunii*: a rich source for hydrocarbons and related ether lipids. *Applied Microbiology and Biotechnology*, **66**, 486–496.

Oh, H.M. *et al.* (1997) *Misaengmul Hakhoe chi (Korean Journal of Microbiology)*, **25**, 339–343

Ozkan, A., Kinney, K., Katz, L. and Berberoglu, H. (2012) Reduction of water and energy requirement of algae cultivation using an algae biofilm photobioreactor. *Bioresource Technology*, **114**, 542–548.

Ranga Rao, A., Dayananda, C., Sarada, R., Shamala, T.R. and Ravishankar, G.A. (2007) Effect of salinity on growth of green alga *Botryococcus braunii* and its constituents. *Bioresource Technology*, **98**, 560–564.

Renninger, N.S. and Mcphee, D.J. (2008) Fuel Compositions Comprising Farnesane and Farnesane Derivatives and Method of Making and Using Same. US patent No. 7,846,222.

Rivas, M.O., Vargas, P. and Riquelme, C.E. (2010) Interactions of *Botryococcus braunii* cultures with bacterial biofilms. *Microbial Ecology*, **60**, 628–635.

Sawayama, S. *et al.* (1992) *Applied Microbiology and Biotechnology*, **38**, 135–138.

Senousy, H.H. *et al.* (2004) *Journal of Phycology*, **40**, 412–423.

Sheehan, J., Terri, D., John, B. and Paul. R. (1998) A Look Back at the US Department of Energy's Aquatic Species Program – Biodiesel from Algae. Close-out report. NREL/TP-580-241901617. National Renewable Energy Laboratory Cole Boulevard Golden, Colorado, pp. 1–294.

Shiho, M., Kawachi, M., Horioka, K., Nishita, Y., Ohashi, K. *et al.* (2012) Business evaluation of a green microalgae *Botryococcus braunii* oil production system. *Procedia Environmental Sciences*, **15**, 90–109.

Tanabe, Y., Kato, S., Matsuura, H. and Watanabe, M.M. (2012) A *Botryococcus* strain with bacterial ectosymbionts grows fast and produces high amount of hydrocarbons. *Procedia Environmental Sciences*, **15**, 22–26.

Volkman, J.K., Zhang, Z., Xie, X., Qin, J. and Borjigin, T. (2015) Biomarker evidence for *Botryococcus* and a methane cycle in the Eocene Huadian oil shale, NE China. *Organic Geochemistry*, **78**, 121–134.

Wake, L.V. and Hillen, L.W. (1980) *Biotechnology and Bioengineering*, **22**, 1637–1656.

Wang, B., Wang, J., Zhang, W. and Meldrum, D.R. (2012) Application of synthetic biology in cyanobacteria and algae. *Frontiers in Microbiology*, **344**, 1–15.

Weiss, T.L., Johnston, J.S., Fujisawa, K., Okada, S. and Devarenne, T.P. (2011) Genome size and phylogenetic analysis of the A and L races of *Botryococcus braunii*. *Journal of Applied Phycology*, **23**, 833–839

Wijffels, R.H. and Barbosa, M.J. (2010) An outlook on microalgal biofuels. *Science*, **329**, 796–799.

11

Strain Selection Strategies for Improvement of Algal Biofuel Feedstocks

Leyla T. Hathwaik and John C. Cushman

Department of Biochemistry and Molecular Biology, University of Nevada, Reno, USA

Summary

Sustainable production of renewable fuels is needed as reserves of fossil petroleum become depleted and its damaging impact on the environment becomes more apparent (see Chapter 1). Over the last decade, microalgae have attracted a great deal of interest as feedstocks for biofuel production, because they are able to produce large amounts of lipids and starch, and can serve as a non-seasonal renewable energy crop that can be grown in fresh, waste, brackish or salt water on marginal lands currently considered unusable for traditional agricultural applications. Although a number of genetic engineering approaches have been developed to manipulate the lipid or starch production of selected algal strains, environmental release of genetically-modified strains raises concerns. Reiterative, transgressive selection strategies provide an alternative, non-transgenic approach for manipulating microalgae strains to enhance lipid or starch production. In this chapter, the various strategies that have been used to select and isolate algal strains with altered feedstock characteristics are discussed, including manipulation of growth conditions and genetic, chemical and physical mutagenesis strategies with an emphasis on reiterative (and transgressive) selection methods based upon flow cytometry, fluorescence-activated cell sorting, and buoyant density centrifugation.

11.1   Introduction

Microalgae have become increasingly prominent as potential biofuel feedstocks due to their high rates of annual productivity and their unique capacity to offset, in part, the production limitations of conventional biofuel feedstocks such as terrestrial crops and waste streams. Microalgae are a large and highly diverse group of microscopic auto-trophic, mixotrophic or heterotrophic organisms that can be grown under conditions unfavourable to terrestrial feedstocks on marginal lands with fresh, waste, brackish or saline water[1] (Gouveia and Oliveira, 2009). With over 40,000 species already identified,

---

1 Gouveia and Oliveira (2009).

Biofuels and Bioenergy, First Edition. Edited by John Love and John A. Bryant.
© 2017 John Wiley & Sons Ltd. Published 2017 by John Wiley & Sons Ltd.

microalgae have been classified into multiple major groupings including, but not limited to, the following families: cyanobacteria (Cyanophyceae), green algae (Chlorophyceae), diatoms (Bacillariophyceae), yellow-green algae (Xanthophyceae), golden algae (Chrysophyceae), red algae (Rhodophyceae), brown algae (Phaeophyceae), dinoflagellates (Dinophyceae) and 'pico-plankton' (Prasinophyceae and Eustigmatophyceae)[2]. Microalgae are considered a promising source of renewable biofuel due to their ability to accumulate large amounts of lipid and starch, their high photosynthetic efficiency that results in high rates of biomass production and their higher annualised growth rates compared with most terrestrial crops (see also Chapters 10, 12, 14 and 16). Many strategies have been developed to increase microalgal productivity by improving their solar energy capture or conversion efficiency[3]. However, novel strategies are needed to enhance lipid or starch production in order to optimise the economic viability of the microalgae-based biofuel industry[4] (Chisti, 2013; Day *et al.*, 2012). Here, lipid and starch biosynthetic pathways and their interconnection are reviewed. Then, various screening and selection strategies to isolate microalgae strains with improved lipid and starch biosynthesis or accumulation are discussed, including manipulation of growth conditions, genetic engineering and mutagenesis, flow cytometry (FCM), fluorescence activated cell sorting (FACS) and buoyant density centrifugation.

11.2   Lipids in Microalgae

Microalgae have been the focus of considerable interest because they produce a wide range of hydrocarbon molecules including, but not limited to, neutral lipids, polar lipids, wax esters, sterols and hydrocarbons, along with prenyl derivatives such as tocopherols, carotenoids, terpenes, quinones and phytylated pyrrole derivatives such as the chlorophylls. The lipids of interest for biofuels have essential roles in microalgae, including energy storage, structural support and intercellular signalling. Storage lipids exist largely in the form of triacylglycerols (TAGs) or esters derived from glycerol and three fatty acids (FAs). Structural lipids (phospholipids) and sterols function as physical constituents of cell membranes, which act as selectively permeable barriers between subcellular compartments and the interior of the cell or the outside environment. In addition, lipids can act as ligands or intermediates that activate signal transduction pathways[5]. Lipids in the form of TAGs are one of the most energy-dense and abundant forms of reduced carbon in nature that represent an ideal, renewable energy source. Under artificial chemical or environmental stress, many microalgal species are able to produce significant amounts (20–50%) of TAGs that can be used to produce biodiesel or green diesel as alternative fuel sources (see also Chapters 2, 6 10 and 14).

Based on *in silico* analysis, the basic pathways of TAG biosynthesis in microalgae are generally believed to be analogous to those described in higher plants[6]. TAG biosynthesis (described in detail in Chapters 6 and 14) begins with the *de novo* synthesis of FA in the

---

2   Hu *et al.* (2008).
3   Stephenson *et al.* (2011); Simionato *et al.* (2013).
4   Day *et al.* (2012); Chisti (2013).
5   Eyster (2007).
6   See, *e.g.* Nguyen *et al.* (2011) and Rismani-Yazdi *et al.* (2011).

plastids and involves the cyclic condensation of two carbon units in which acetyl-CoA is the precursor. The first committed step in the pathway is the synthesis of malonyl-CoA from acetyl-CoA and CO_2 by the enzyme acetyl-CoA carboxylase. The malonyl group is transferred to acyl carrier protein (ACP) to give rise to malonyl-ACP. The malonyl moiety formed is used subsequently for the elongation of the acyl group as the central carbon donor. Subsequent condensation reactions, which are catalysed by 3-ketoacyl ACP reductase, 3-hydroxyacyl ACP dehydrase and enoyl ACP reductase, occur to form butyryl-ACP. This cycle continues until 16 or 18 carbons have been added (see diagrams in Chapters 6 and 14). The resultant FAs are then either used directly in the plastids or exported into the cytosol in the endoplasmic reticulum (ER), to acylate glycerol-3-phosphate to produce phosphatidic acid (PA), and, following phosphate removal, diacylglycerol (DAG). As a final step, a diacyglycerol acyltransferase (DGAT), which uses acyl CoA as an acyl donor, converts DAG to TAG.

TAGs are stored mainly in lipid bodies (LB) composed of a single layer of phospholipids. The synthesis of TAGs and their storage in LB appears to be a protective mechanism by which microalgae deal with stress conditions, typically under nutrient limitation or salinity stress[7]; however, little is known about the molecular regulation of TAG and LB formation[8]. Under nitrogen-replete conditions, starch-forming and starchless *Chlamydomonas reinhardtii* cells store TAGs in LB in the cytoplasm[9]. Under nitrogen-depleted conditions, starch-forming and starchless cells augment cytoplasmic LB production; however, starchless cells produce chloroplast LBs that are far larger than plastoglobules and are typically enclosed within one or more thylakoids[10]. Unlike traditional terrestrial oilseed crops, pressure extraction of lipids from algae is typically inefficient, presumably due to their complex cell walls. Therefore, combined, simultaneous chemical extraction and transesterification of lipids can often improve biodiesel quality or quantity[11].

11.3 Starch in Microalgae

Starch is another major carbon and energy storage compound in plants and microalgae that can be used as a biofuel feedstock. Starch produced by microalgae and higher plants is composed of two polysaccharide fractions: amylopectin and amylose. Amylopectin is the major polymer, typically making up 70–80% of the starch granule, and is responsible for the granular nature of starch. Amylopectin is composed of glucose units that are linked into linear chains by α-1,4 bonds and branched chains arising from α-1,6 bonds. Amylose is a lesser component of the granule with very few branched points. As the main assimilatory product of photosynthesis, many microalgae are able to produce large amounts (8–57%) of starch as storage materials in their plastids under stress conditions, such as during inhibition of cytoplasmic protein synthesis or sulphur limitation[12], or under nitrogen or iron depletion[13]. The accumulated starch provides

7 Siaut *et al.* (2011).
8 See, *e.g.* San Pedro *et al.* (2013).
9 Wang *et al.* (2009); Li *et al.* (2010).
10 Goodson *et al.* (2011).
11 See, *e.g.* Dong *et al.* (2013).
12 Brányiková *et al.* (2011).
13 Dragone *et al.* (2011).

an ideal substrate for the production of various biofuels, including ethanol, butanol and hydrogen.

Starch biosynthesis has been well studied in microalgae, especially in the model algae *C. reinhardtii*. As in higher plants, the biosynthesis of starch begins in the chloroplast with triose phosphate as the starting substrate. This is dimerised to form fructose-1, 6-bisphosphate, which is converted to fructose-6-phosphate, glucose-6-phosphate and glucose-1-phosphate through three reactions catalysed by fructose-1, 6-bisphosphotase, hexose phosphate isomerase and phosphoglucomutase, respectively. The glucose-1-phosphate is converted to ADP-glucose *via* ADP-glucose pyrophosphorylase in a reaction that requires ATP and generates pyrophosphate (PPi). PPi is hydrolysed into two orthophosphate (Pi) molecules using a specific inorganic pyrophosphatase (PPi), thereby driving the reaction towards ADP-glucose synthesis. Lastly, the glucose moiety of ADP-glucose is transferred to the non-reducing end of the terminal glucose of a growing starch chain.

Starch is stored in the plastids in the form of an insoluble crystalline granule. The synthesis of starch granules is a defence mechanism in which microalgae cease cell division and switch photosynthetic carbon partitioning towards starch synthesis under macro-element (*e.g.* nitrogen, sulphur or phosphorus) limitation[14]. Nevertheless, the initiation of starch granules and the mechanisms that determine granule size, number and morphology are still not well understood[15]. In general, starch accumulation is favoured in a plurality of microalgae as it is energetically less costly to synthesise than lipids, although larger surveys should be conducted.

11.4   Metabolic Interconnection Between Lipid and Starch Biosynthesis

In plants and microalgae, lipid and starch biosynthesis share a common carbon precursor; however, the regulation of and possible interactions between the two pathways are less understood[16]. In *Chlamydomonas reinhardtii* starchless mutants, the disruption of genes encoding either ADP-glucose pyrophosphorylase or isoamylase blocks starch biosynthesis, and can result in an up to ten-fold increase in lipid accumulation under nutrient deprivation (see previous section). Studies in *Chlorella pyrenoidosa* have also demonstrated that cells deficient in starch production can synthesise higher amounts of polyunsaturated FAs and produce 50% more lipids[17]. In addition, studies in *Pseudochlorococcum sp.* have shown that, under nutrient-replete conditions, starch was the primary carbon storage product. However, under nutrient-deprived conditions, carbon storage shifts into neutral lipids, leading to decreased starch content[18].

Starch and lipid biosynthesis serve as competing pathways for carbon storage, with starch rather than lipids serving as the dominant sink in *C. reinhardtii*. Lipid accumulation under nutrient deprivation or high acetate conditions depends upon *de novo* FA

---

14 Yao *et al.* (2012).
15 Zeeman *et al.* (2010).
16 Rawsthorne (2002); Weselake *et al.* (2009).
17 Ramazanov and Ramazanov (2006).
18 Li *et al.* (2011).

synthesis, rather than catabolism of membrane lipids[19]. Therefore, the starch and lipid biosynthetic pathways compete for the same carbon source. Lipid accumulation under nutrient deprivation conditions actually lags behind that of starch, so that rapid oil synthesis occurs only when carbon reserves exceed the capacity for starch biosynthesis[20]. In addition, the pathways of starch and lipid synthesis not only compete for carbon precursors, but also for space in the chloroplast for storage of end products. As previously mentioned, LBs can be found in the chloroplast, in addition to in the cytosol. In *C. reinhardtii*, 40% of the cell is occupied by the chloroplast which, in the wild type strain, is crowded with starch granules under nutrient deprivation. In the starchless mutant STA6, part of this 40% of otherwise empty space is used to store lipids under nutrient-limited conditions[21]. Taken together, these results indicate that lipid accumulation under nutrient deprivation is limited by carbon supply and that the increased lipid accumulation in various starchless mutants is attributable to increased carbon supply in the absence of starch synthesis. Therefore, improved understanding of the regulation of lipid and starch biosynthesis and the triggers that redirect carbon flux from starch to lipid biosynthesis, form the basis for designing effective tactics to enhance lipid production.

11.5   Strategies for the Selection of Microalgae Strains with Enhanced Biofuel Feedstock Traits

11.5.1   Manipulation of Growth Conditions

The manipulation of nutrient or cultivation conditions is often used to enhance lipid or starch production in microalgae. Nutrient deprivation is the most widely-used approach for directing carbon flux to either starch or, more often, lipid biosynthesis (see previous section). Under nutrient deprivation, micro-algal cells accumulate lipid or starch as a means of chemical energy storage when the energy source (light) and carbon source (CO_2) are abundant for photosynthesis. While many nutrients, such as acetate, ammonium nitrate, iron, phosphorous, potassium, silicon, sulphur and urea, have been reported to increase lipid or starch production in microalgae, nitrogen limitation is the most commonly reported factor for activating lipid or starch accumulation. Manipulation of cultivation conditions, including temperature, light, salinity, pH and heavy metal concentrations can also induce either lipid or starch biosynthesis[22]. Screening and selection of specific microalgal species under such diverse growth conditions can lead to the identification of species capable of substantial lipid or starch accumulation. Once identified, growth conditions for these species can be further optimised and their ability to grow under a wide range of environmental parameters can be tested.

11.5.2   Genetic Mutagenesis

Genetic mutagenesis is an effective strategy to generate microalgae with altered lipid or starch productivity. Insertional, chemical or UV mutagenesis are used to create random

---

19 Li *et al.* (2010).
20 Wang *et al.* (2009); Work *et al.* (2010).
21 Goodson *et al.* (2011).
22 Sharma *et al.* (2012).

mutations in nuclear DNA. Once created, the mutagenised population is then subjected to high-throughput screening and selection of specific microalgae cells or clonal populations with either altered lipid or starch content. Screening is done typically in 96-well plates using a spectrophotometer or a FCM in combination with lipid-specific dyes or starch-specific stains. The exposure of microalgae to mutagens can result in the hyper-accumulation of lipids or starch by blocking metabolic pathways that lead to other energy-rich compounds. For example, STA6 and STA7 insertional *C. reinhardtii* mutants preferentially accumulate lipids under nitrogen deprivation due to the blockage of starch synthesis *via* disruption of either ADP-glucose pyrophosphorylase (STA6) or isoamylase activity (STA7) as mentioned previously. Another example is the Δ5 desaturase-deficient mutant of *Parietochloris incise*, which can shift the production of arachidonic acid to dihomo-γ-linolenic acid[23].

In addition, genetic mutagenesis has led to the isolation of microalgae with enhanced photosynthetic productivity. Microalgae are efficient at converting sunlight energy into biomass; however, this process can be limited by shading effects within dense cultures. Photosynthetic productivity can be improved in mutants with small or truncated chlorophyll antenna obtained *via* insertional mutagenesis[24] or UV light-induced mutagenesis[25]. Small chlorophyll antennae offer several advantages for biofuel production, such as improved light penetration properties of the culture, reduced photo-damage under high light, improved bioreactor efficiency and higher yield. The results of these various studies using genetic mutagenesis to alter micro-algal feedstock characteristics are summarised in Table 11.1.

11.5.3 Flow Cytometry

Lipid and starch accumulation can be manipulated by the application of environmental stress; however, optimisation of culture conditions and subsequent selection of algal species can be a slow and labour-intensive process. In order to speed up the selection process, researchers have begun to exploit FCM as a rapid, high-throughput screening method to monitor not only lipid and starch production, but also cell viability of algal cultures[26]. Collection of information about cell growth, biomass, lipid or starch production over a time course can permit the optimisation of culturing conditions and harvest time in order to maximise biomass, and lipid or starch production.

FCM is used to analyse multiple physical and chemical properties of thousands of microalgae cells per second as they flow in a fluid stream through a beam of light. Assessed cell properties include the relative size (from 0.2 to 150 μm) of the algal cells, their internal complexity and their fluorescence intensity. These traits are determined using an optical-to-electronic detection apparatus that records how the microalgae scatter incident laser light and emit fluorescence. FCM can perform multi-parameter analyses on the algal cell properties by measuring light scatter signals along with chlorophyll *a* autofluorescence, after excitation at 488 nm. Shifts in forward angle light scatter (<15°) and side-angle light scatter (15–85°) can indicate differences in cell size, shape or morphology (*e.g.* intracellular structures with specific chemical compositions, such

---

23 Iskandarov *et al.* (2011).
24 Polle *et al.* (2003); Tetali *et al.* (2007); Mitra and Melis (2010); Kirst *et al.* (2012).
25 Nakajima and Ueda (2000).
26 See, *e.g.* da Silva *et al.* (2009) and Hyka *et al.* (2013).

Table 11.1 Summary of studies for microalgae biofuel improvement *via* genetic mutagenesis.

Mutagenesis method	Species	Disrupted gene (enzyme)	Altered trait	References
Insertional	*Chlamydomonas reinhardtii*	ADP-glucose pyrophosphorylase	Starchless mutant with up to 10-fold increase in TAG content	Li *et al.* (2010a,b)
	Chlamydomonas reinhardtii	ADP-glucose pyrophosphorylase	Starchless mutant with 15-fold increase in lipid bodies	Wang *et al.* (2009)
	Chlamydomonas reinhardtii	ADP-glucose pyrophosphorylase and isoamylase	Starchless mutants with 2-fold increase in lipid accumulation	Work *et al.* (2010)
	Chlorella pyrenoidosa	Unknown	Starchless mutant with 22% higher growth rate and 50% increase in lipid content	Ramazanov and Ramazanov (2006)
	Parietochloris incise	Delta-5 desaturase	Production shift of arachidonic acid to dihomo-γ-linolenic acid	Iskandarov *et al.* (2011)
	Chlamydomonas reinhardtii	Truncated light-harvesting antenna 1 (TLA1)	Increased photosynthetic productivity due to an antenna reduction	Polle *et al.* (2003)
	Chlamydomonas reinhardtii	Truncated light-harvesting antenna 1 (TLA1)	Increased photosynthetic productivity due to 50% antenna reduction	Tetali *et al.* (2007) Mitre and Melis (2010)
	Chlamydomonas reinhardtii	Truncated light-harvesting antenna 2 (TLA2)	Increased photosynthetic productivity due to a 65% antenna reduction	Kirst *et al.* (2012)
UV-light	*Chlorella vulgaris*	Unknown	2.5-fold increase in lipid content	Deng *et al.* (2011)
	Pavlova lutheri	Unknown	32.8% and 32.9% increase of eicosapentaenoic and docosahexaenoic acids	Miereles *et al.* (2003)
	Chlorella sorokiniana	Unknown	Increase in total lipid and TAG content and decrease in starch content	Vigeolas *et al.* (2012)
	Scenedesmus obliquus	Unknown	Increase in total lipid and TAG content	Vigeolas *et al.* (2012)
	Chlamydomonas perigranulata	Light-harvesting complex pigment (LHC-1)	1.5-fold increase in photosynthetic productivity due to less light-harvesting pigment	Nakajima and Ueda (2000)
EMS	*Nannochloropsis sp.*	Unknown	21% increase total lipid productivity	Anandarajah *et al.* (2012)
	Nannochloropsis sp.	Unknown	1.5- to 2.1-fold increase in total lipid content	Doan and Obbard (2012)

as starch granules), respectively. Similarly, chlorophyll *a* fluorescence, detected at 660–700 nm, can be used to measure cell concentration and to discriminate cells from heterotrophic organisms and non-living particles.

FCM can be used to measure lipid accumulation in conjunction with lipophilic fluorescent dyes such as Nile red (NR; 9-diethylamino-5H-benzo[alpha]phenoxazine-5-one) or BODIPY 505/515 (4,4-difluoro-1,3,5,7-tetramethyl-4-bora-3a,4a-diaza-s-indacene). NR has been used extensively for over 20 years to detect lipid droplets and measure lipid content in algal cells. NR is a relatively photo-stable hydrophobic probe that is intensely fluorescent in organic solvents and hydrophobic environments. Several studies comparing gravimetric and gas chromatography (GC) analytical methods have shown a correlation between fluorescence of NR, measured as yellow-gold (585/40 nm) or red fluorescence (635/20 nm), with neutral or polar lipid content, respectively. Although this correlation has been reported routinely, caution in interpreting the NR fluorescence has been recently recommended[27]. NR staining can be used to monitor the accumulation of neutral and polar lipids, and the ratio of polar/neutral lipids, under different cultivation conditions. However, a major drawback is that NR cannot be used to adequately determine the concentration of lipids in several green microalgae with thick, rigid cell walls that prevent the stain from effectively entering the cells and binding with the lipids. In addition, the utility of NR fluorescence is hampered because of its spectral similarity to that of chlorophylls. These problems can be partially overcome using DMSO as a stain carrier at 20–30% and elevated temperatures. However, the high concentration of DMSO or elevated temperatures can impair cell viability and hinder their usefulness for subsequent algal cultures.

BODIPY 505/515 has also been used to monitor the storage of lipids within live algal cells. BODIPY 505/515 offers a useful alternative to NR because of its high oil/water coefficient, which allows it to successfully cross membranes with the addition of 2% DMSO; it is also more resistant to photo-bleaching. The use of low concentrations of DMSO (2%) allows the cells to remain viable to use for further studies. BODIPY 505/515 also has a narrower emission spectrum (530 ± 15 nm) than NR, which is spectrally distinguishable from chlorophyll auto-fluorescence, and allows for fluorescent enhancement and three-dimensional imaging of LBs within living cells, in addition to its use in staining for FACS and microplate-based isolation of lipid-rich microalgae[28].

In addition to lipids, FCM can be applied to quantify starch accumulation within microalgae using fluorescent probes or by measuring the changes in cell granularity. Safranin O has been an effective fluorescent probe to determine the starch concentration of *Chlamydomonas reinhardtii* and *Scenedesmus subspicatus* under nitrogen-limited growth conditions when using 435 nm excitation and 480 nm emission wavelengths[29]. In addition, starch content can be quickly and roughly measured by monitoring the change in intensity in the side scatter channel, which increases with increasing cell granularity. However, increasing cell granularity is also influenced by starch accumulation, therefore starch content can only be measured in cells when starch is relatively abundant. In addition, this assay should be complemented with conventional enzymatic methods to determine the absolute amount of starch within a cell culture.

In addition, FCM can be used to monitor algal culture viability using membrane impermeant fluorescent probes that bind to nucleic acids. For example, propidium

---

27 Elsey *et al.* (2007); Doan and Obbard (2011b).
28 Periera *et al.* (2011).
29 Dean *et al.* (2010).

iodide (PI), PicoGreen, SITOX Green I or II, TOTO-1, TOPRO-1, YOYO-1 and YOPRO-1 are probes that are used routinely to assess cell viability by identifying cells with permeabilised membranes (dead cells) in an algal culture. However, no universal staining procedure exists for assessing cell viability within the wide taxonomic diversity of microalgae. The use of PI requires the extraction of algal pigments with ethanol or methanol, because PI emits fluorescence in the same spectral range (red) as algal pigments. In addition, PI, PicoGreen, SITOX Green I or II, TOTO-1, TOPRO-1, YOYO-1 and YOPRO-1 require cell fixation before staining. These probes provide useful tools for monitoring cell viability throughout the growth phases of an algal culture. This is important because a high proportion of dead cells can decrease process yields because they do not accumulate lipid or starch, nor do they reproduce as do living cells. Such information is useful for determining when to harvest cultures at optimum cell concentrations in order to achieve maximal lipid or starch yields.

11.5.4   Fluorescence-Activated Cell Sorting

Fluorescence-activated cell sorting (FACS) is a major application of FCM that allows sorting and collecting of cell populations based upon the specific light scattering and fluorescence characteristics of individual cells. To sort cells, the FCM first identifies each cell of interest, then separates the individual cells from one another based on selected criteria. Once a population of interest has been identified on a data plot, a gate is drawn around it, which identifies the cells of interest to be sorted out of the stream. After the sample is loaded, each cell is probed with a beam of light as it flows past a detector (Figure 11.1). The scattered light and fluorescence criteria are then compared to that of the sorting gate characteristics. If a cell matches the selection criteria, then the cells (single, thousands or millions) can be collected and subsequently grown to use for further assays.

FACS is an efficient method that enables the isolation from heterogeneous cultures or environmental isolates of microalgae with intrinsic characteristics suitable for biofuel production, based on the morphology (side- or forward-scatter channel) and autofluorescence of the cells. Furthermore, the ability to sort single cells provides the advantages of obtaining subsequent cultures, which are clonal as well as potentially axenic, and of accelerating the discovery of new algal strains.

FACS can be also used in conjunction with lipophilic fluorescent dyes to evaluate, sort and select cells based upon their intracellular lipid content. The concept of using FACS to reiteratively select for cell populations is summarised in Figure 11.1. FACS selection can be unidirectional or bidirectional (or transgressive) through the selection for cells with either increased or decreased lipid (or starch) content, depending on the desired trait in the selected cell population. For example, NR staining in combination with three rounds of FACS was used to select for daughter *Nannochloropsis sp.* cells with enhanced lipid accumulation estimated at approximately 55% on a dry weight basis[30]. The sorted cells maintained the trait of enhanced intracellular lipid content for over approximately 100 generations. Similarly, two rounds of FACS selection of NR-stained wild-type *Tetraselmis suecica* cells resulted in the isolation of rapidly growing cells with an enhanced mean fluorescence trait that remained stable for 30 cell generations[31]. FACS was also used for the high-throughput screening for intracellular LBs in wild-type *C. reinhardtii* and the

---

30 Doan and Obbard (2011a).
31 Montero *et al.* (2011).

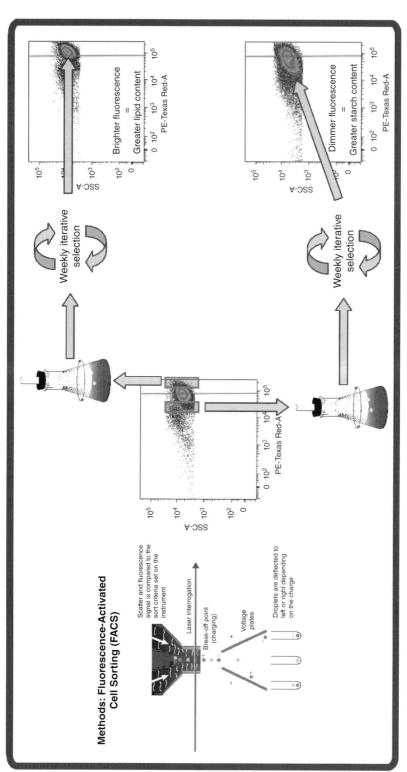

Figure 11.1 Reiterative and transgressive selection by fluorescence-activated cell sorting (FACS). Fluorescently stained microalgae cell populations are collected based upon their fluorescence intensity using FACS. After collection, cells with either greater or lesser fluorescence intensity are collected, re-cultured in an aerated flask, and iteratively selected on a weekly basis until significant differences in fluorescence are evident. Cultures are analysed at regular intervals for starch and lipid content. FACS figure adapted from Rahman (2006).

cell wall-less *sta6-1* mutant[32]. After a single cell-sorting event, cells stained with NR failed to recover, whereas cells stained with BODIPY 505/515 continued to grow after selection. The use of glycerol instead of dimethyl sulfoxide (DMSO) is recommended when performing FACS, as glycerol does not inhibit the growth of the selected cells. In addition, FACS has been used to select lipid-rich BODIPY 505/515-stained microalgal cells from environmental samples, followed by microplate-based isolation.

To demonstrate the utility of the FACS reiterative, transgressive selection approach, NR- or BODIPY 505/515-stained cells can be sorted and selected for either higher or lower fluorescence intensity, then tested for either increased or decreased lipid content, respectively. For example, one of us (LH) has shown that reiterative, transgressive selection of wild-type *Dunaliella salina* cells gave cells with a 200% (DW) increase in total TAG and FFA, which were isolated after 20 rounds of selection[33]. The selected trait remained stable for at least seven months. Because starch and lipid synthesis are two competing pathways for carbon storage, starch-accumulating strains can also be identified on the basis of reduced fluorescence intensity, then sorted to isolate cells with greater starch content. It was also shown that 30 rounds of FACS selection of *Dunaliella salina* cells with overall reduced NR fluorescent intensity resulted in a population of cells with a 71% increase in starch accumulation[33]. Alternatively, Safranin O staining combined with FACS can be used to isolate cells with increased starch accumulation by reiteratively selecting the cells with greater fluorescence intensity.

Although FACS allows the efficient isolation of wild-type microalgal cells, the use of UV-light or chemical mutagenesis in conjunction with FACS has the potential to further improve desirable biofuel feedstock traits. A two-step method based upon UV-C light irradiation of *Isochrysis affinis galbana* cells followed by NR-FACS-based selection resulted in cells with enhanced lipid accumulation[34]. Following the first round of mutation-selection, neutral lipid FAs increased by 20%, whereas phospholipid FA content declined by 40%. After the second round of mutation-selection, total FA content increased by 80% with an FA distribution similar to that found after the first round of mutation selection. EMS mutagenesis followed by FACS selection can result in the isolation of mutants with enhanced intracellular lipid content. For example, a mutant strain of *Nannochloropsis sp.* with up to a four-fold increase in total FA was isolated using this approach after a single round of selection[35]. EMS mutagenesis and 10 rounds of FACS selection of *Dunaliella salina* resulted in a 32% increase in NR fluorescence[36].

11.5.5 Buoyant Density Centrifugation

Buoyant density centrifugation (BDC) separates different cells, sub-cellular organelles and macromolecular complexes on the basis of buoyant force, which exploits the differences in the relative density of the biological constituents within a gradient medium. The method involves the formation of continuous or discontinuous (step) gradient using cesium chloride (CsCl), sucrose or Percoll® in analytical or preparative scale centrifuge tubes. Samples are then layered on top of the gradient and the tubes are centrifuged with sufficient

---

32 Velmurugan *et al.* (2013).
33 Hathwaik, L.T., unpublished data.
34 Bougaran *et al.* (2012).
35 Doan and Obbard (2012).
36 Hathwaik, L.T., unpublished data.

centrifugal force so that samples or sample components sediment to a position within the gradient that is in equilibrium with their buoyant density (*e.g.* equilibrium density). Thus, sample components that sediment towards the bottom of the gradient have a higher density, whereas components that migrate towards the top of the gradient have a lower density.

The technique of BDC can be applied to microalgae to successfully separate axenic microalgae from heterogeneous cultures. Percoll® (colloidal silica coated with polyvinylpyrrolidone (PVP)) has become the density medium of choice for the isolation and selection of microalgae, because of its low viscosity, its low osmolarity, lack of permeation of biological membranes and of toxicity towards living cells (and viruses) compared with other gradient materials. The use of Percoll® has been effective for axenic separation and isolation of several species of marine and freshwater microalgae from heterogeneous samples based upon their buoyant densities by density centrifugation.

BDC can also be used to perform *in situ* determination of lipid, hydrocarbon or starch content of microalgae using continuous sucrose and CsCl gradients based on direct density equilibrium measurements. Lipid and hydrocarbon accumulation results in lesser overall density of the microalgae cells compared with typical cell density. In contrast, starch accumulation results in greater cell density. Thus, discontinuous or continuous buoyant density gradient centrifugation (BDGC) is an extremely valuable tool to monitor and quantify microalgae cells with altered lipid, hydrocarbon, or starch contents under different environmental stress or culture conditions. BDGC can also be used to reiteratively select microalgae with enhanced traits for biofuel production (Figure 11.2). Weekly reiterative selection of wild-type or EMS-mutagenised *Dunaliella salina* cells towards either lesser or greater density was effective in either increasing

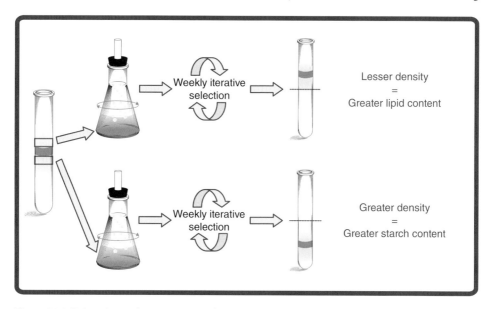

Figure 11.2 Reiterative and transgressive selection by continuous buoyant density gradient centrifugation (BDGC). Microalgae cells are placed atop a continuous Percoll® density gradient and centrifuged until cells reach their buoyant density equilibrium. After centrifugation, cell populations with either greater or lesser buoyant densities are removed, re-cultured in an aerated flask, and iteratively selected on a weekly basis until significant differences in buoyant density are evident. Cultures are analyzed at regular intervals for starch and lipid content.

lipid or starch content, respectively. For example, 60 rounds of such transgressive selection resulted in the isolation of cell populations with significantly lesser or greater buoyant densities. Lipid measurements using NR indicated that the wild-type and EMS-mutagenised population with lesser density exhibited 26% and 45% (DW) increases in NR fluorescence, respectively; however, changes in total TAGs and FFAs in these populations were not statistically significant. Conversely, starch analysis indicated that the wild-type and EMS-mutagenised population with higher density had 165% and 78% (DW) increases in starch content, respectively. In addition, the alterations in lipid and starch produced during the reiterative selections remained stable for at least 70 weeks[37].

11.6 Conclusions

Microalgae represent a diverse group of micro-organisms that are considered to be one of the most promising biofuel feedstocks, because of their high lipid and starch productivity, high photosynthetic efficiency and rapid growth rates. Understanding the factors affecting the shift of carbon flux between lipid and starch biosynthesis is crucial for manipulating microalgae for the production of the desired biofuel. The coupling of genetic mutagenesis with FCM, FACS and discontinuous or continuous BDGC offers useful technologies for the screening and isolation of diverse species of microalgae with improved traits for biofuel production. These approaches avoid the need to resort to genetic engineering strategies which, while extremely useful, often raise public concerns about the release of such genetically-modified organisms into the environment. Furthermore, once selected, numerous studies now indicate that the selected trait remains stable for many cell generations. Overall, the reiterative (and transgressive) selection technologies outlined here should prove very useful for improving the quality traits of selected microalgae species for use as biofuel feedstocks.

Acknowledgements

The authors would like to thank Lori Kunder (Kunder Design Studio, LLC) for assistance with preparation of the figures and Mary Ann Cushman for critical review of and clarifying comments on the manuscript. The authors would like to acknowledge funding support from the Sun Grant Initiative, Western Regional Center (U.S. Department of Transportation), NIH funding (P20 GM103440 & P30 GM110767) to support flow cytometry at the University of Nevada, Reno, and the University of Nevada Agricultural Experiment Station (NEV-00342). The authors would like to dedicate this review to the memory of Professor Carl A. Price and his expertise in buoyant density centrifugation.

Selected References and Suggestions for Further Reading

Ahmad, A., Yasin, N., Derek, C. and Lim, J. (2011) Microalgae as a sustainable energy source for biodiesel production: A review. *Renewable and Sustainable Energy Reviews*, **15**, 584–593.

---

37 Hathwaik L.T., *et al.* 2015.

Amaro, H., Guedes, A. and Malcata, F. (2011) Advances and perspectives in using microalgae to produce biodiesel. *Applied Energy*, **88**, 3402–3410.

Anandarajah, K. *et al.* (2012) Characterization of microalga *Nannochloropsis sp.* mutants for improved production of biofuels. *Applied Energy*, **96**, 371–377.

Araujo, G., Matos, L., Gonçalves, L., Fernandes, F. and Farias, W. (2011) Bioprospecting for oil producing microalgal strains: evaluation of oil and biomass production for ten microalgal strains. *Bioresource Technology*, **102**, 5248–5250.

Beckmann, J., Lehr, F., Finazzi, G., Hankamer, B., Posten, C. *et al.* (2009) Improvement of light to biomass conversion by de-regulation of light-harvesting protein translation in *Chlamydomonas reinhardtii. Journal of Biotechnology*, **142**, 70–77.

Bougaran, G., Rouxel, C., Dubois, N., Kaas, R., Grouas, S. *et al.* (2012) Enhancement of neutral lipid productivity in the microalgae *Isochrysis affinis Galbana* (T-Iso) by a mutation-selection procedure. *Biotechnology and Bioengineering*, **109**, 2737–2745.

Brányiková, L. *et al.* (2011) Microalgae—novel highly efficient starch producers. *Biotechnology and Bioengineering*, **108**, 766–776.

Chisti, Y. (2007) Biodiesel from microalgae. *Biotechnology Advances*, **25**, 294–306.

Chisti, Y. (2013) Constraints to commercialization of algal fuels. *Journal of Biotechnology*, **167**, 201–214.

Cooper, M.S., Hardin, W.R., Petersen, T.W. and Cattolico, R.A. (2010) Visualizing 'green oil' in live algal cells. *Journal of Bioscience and Bioengineering*, **109**, 198–201.

Courchesne, N.M.D., Parisien, A., Wang, B. and Lan, C.Q. (2009) Enhancement of lipid production using biochemical, genetic and transcription factor engineering approaches. *Journal of Biotechnology*, **141**, 31–41.

da Silva, T. *et al.* (2009) Multi-parameter flow cytometry as a tool to monitor heterotrophic microalgal batch fermentations for oil production towards biodiesel. *Applied Biochemistry and Biotechnology*, **159**, 568–578.

Day, J.G., Slocombe, S.P. and Stanley, M.S. (2012) Overcoming biological constraints to enable the exploitation of microalgae for biofuels. *Bioresource Technology*, **109**, 245–251.

Dean, A.P. *et al.* (2010) Using FTIR spectroscopy for rapid determination of lipid accumulation in response to nitrogen limitation in freshwater microalgae. *Bioresource Technology*, **101**, 4499–4507.

Deng, X. *et al.* (2011) Effects of selective medium on lipid accumulation of Chlorellas and screening of high lipid mutants through ultraviolet mutagenesis. *African Journal of Agricultural Research*, **6**, 3768–3774.

Dismukes, G., Carrieri, D., Bennette, N., Ananyev, G. and Posewitz, M. (2008) Aquatic phototrophs: efficient alternatives to land-based crops for biofuels. *Current Opinion in Biotechnology*, **19**, 235–224.

Doan, T. and Obbard, J. (2011a) Enhanced lipid production in *Nannochloropsis sp* using fluorescence-activated cell sorting. *Global Change Biology Bioenergy*, **3**, 264–270.

Doan, T. and Obbard, J. (2011b) Improved Nile Red staining of *Nannochloropsis sp. Journal of Applied Phycology*, **23**, 895–901.

Doan, T. and Obbard, J. (2012) Enhanced intracellular lipid in *Nannochloropsis sp. via* random mutagenesis and flow cytometric cell sorting. *Algal Research*, **1**, 17–21.

Dong, T., Wang, J., Miao, C., Zheng, Y. and Chen, S. (2013) Two-step *in situ* biodiesel production from microalgae with high free fatty acid content. *Bioresource Technology*, **136**, 8–15.

Dragone, G., Fernandes, B.D., Abreu, A.P., Vicente, A.A. and Teixeira, J.A. (2011) Nutrient limitation as a strategy for increasing starch accumulation in microalgae. *Applied Energy*, **88**, 3331–3335.

Durre, P. (2007) Biobutanol: an attractive biofuel. *Biotechnology Journal*, **2**, 1525–1534.

Durrett, T.P., Benning, C. and Ohlrogge, J. (2008) Plant triacylglycerols as feedstocks for the production of biofuels. *Plant Journal*, **54**, 593–607.

Elsey, D. *et al.* (2007) *Journal of Microbiological Methods*, **68**, 639–642.

Eroglu, E., Okada, S. and Melis, A. (2011) Hydrocarbon productivities in different *Botryococcus* strains: comparative methods in product quantification. *Journal of Applied Phycology*, **23**, 763–775.

Eyster, K.M. (2007) The membrane and lipids as integral participants in signal transduction: lipid signal transduction for the non-lipid biochemist. *Advances in Physiology Education*, **31**, 5–16.

Goodson, C., Roth, R., Wang, Z.T. and Goodenough, U. (2011) Structural correlates of cytoplasmic and chloroplast lipid body synthesis in *Chlamydomonas reinhardtii* and stimulation of lipid body production with acetate boost. *Eukaryotic Cell*, **10**, 1592–1606.

Gouveia, L. and Oliveira, A.C. (2009) Microalgae as a raw material for biofuels production. *Journal of Industrial Microbiology and Biotechnology*, **36**, 269–274.

Greenwell, H.C., Laurens, L.M., Shields, R.J., Lovitt, R.W. and Flynn, K.J. (2010) Placing microalgae on the biofuels priority list: a review of the technological challenges. *Journal of the Royal Society Interface*, **7**, 703–726.

Harwood, J.L. and Guschina, I.A. (2009) The versatility of algae and their lipid metabolism. *Biochimie*, **91**, 679–684.

Hathwaik, L.T., Redelman, D., Samburova, V., Zielinska, B., Shintani, D.K., Harper, J.F., Cushman, J.C. (2015) Transgressive, reiterative selection by continuous buoyant density gradient centrifugation of *Dunaliella salina* results in enhanced lipid and starch content. *Algal Research*, **9**, 194–203.

Hu, Q., Sommerfeld, M., Jarvis, E., Ghirardi, M., Posewitz, M. *et al.* (2008) Microalgal triacylglycerols as feedstocks for biofuel production: perspectives and advances. *Plant Journal*, **54**, 621–639.

Hyka, P., Lickova, S., Pribyl, P., Melzoch, K. and Kovar, K. (2013) Flow cytometry for the development of biotechnological processes with microalgae. *Biotechnology Advances*, **31**, 2–16.

Iskandarov, U., Khozin-Goldberg, I. and Cohen, Z. (2011) Selection of a DGLA-producing mutant of the microalga *Parietochloris incisa*: I. Identification of mutation site and expression of VLC-PUFA biosynthesis genes. *Applied Microbiology and Biotechnology*, **90**, 249–256.

Johnson, M. and Wen, Z. (2009) Production of biodiesel fuel from the microalga *Schizochytrium limacinum* by direct transesterification of algal biomass. *Energy Fuels*, **23**, 5179–5183.

Jones, C. and Mayfield, S. (2012) Algae biofuels: versatility for the future of bioenergy. *Current Opinion in Biotechnology*, **23**, 346–351.

Kirst, H., Garcia-Cerdan, J.G., Zurbriggen, A. and Melis, A. (2012) Assembly of the light-harvesting chlorophyll antenna in the green alga *Chlamydomonas reinhardtii* requires expression of the TLA2-CpFTSY gene. *Plant Physiology*, **158**, 930–945.

Li, P., Miao, X., Li, R. and Zhong, J. (2011) *In situ* biodiesel production from fast-growing and high oil content *Chlorella pyrenoidosa* in rice straw hydrolysate. *Journal of Biomedicine and Biotechnology*, **2011**, 1–8.

Li, Y., Han, D., Hu, G., Dauvillee, D., Sommerfeld, M., Ball, S. and Hua. Q. (2010a) *Chlamydomonas* starchless mutant defective in ADP-glucose pyrophosphorylase hyper-accumulates triacylglycerol. *Metabolic Engineering*, **12**, 387–391.

Li, Y. *et al.* (2010b) Inhibition of starch synthesis results in overproduction of lipids in *Chlamydomonas reinhardtii. Biotechnology and Bioengeneering*, **107**, 258–268.

Li, Y., Han, D., Sommerfeld, M. and Hu, Q. (2011) Photosynthetic carbon partitioning and lipid production in the oleaginous microalga *Pseudochlorococcum sp.* (Chlorophyceae) under nitrogen-limited conditions. *Bioresource Technology*, **102**, 123–129.

Li, Y., Horsman, M., Wu, N., Lan, C.Q. and Dubois-Calero, N. (2008) Biofuels from microalgae. *Biotechnology Progress*, **24**, 815–820.

Mata, T.M., Martins, A.A. and Caetano, N.S. (2010) Microalgae for biodiesel production and other applications: A review. *Renewable and Sustainable Energy Reviews*, **14**, 217–232.

Miereles, L.A. *et al.* (2003) Increase of the yields of eicosapentaenoic and docosahexaenoic acids by the microalga *Pavlova lutheri* following random mutagenesis. *Biotechnology and Bioengineering*, **81**, 50–55.

Miller, R., Wu, G., Deshpande, R.R., Vieler, A., Gartner, K. *et al.* (2010) Changes in transcript abundance in *Chlamydomonas reinhardtii* following nitrogen deprivation predict diversion of metabolism. *Plant Physiology*, **154**, 1737–1752.

Mitra, M. and Melis, A. (2010) Genetic and biochemical analysis of the TLA1 gene in *Chlamydomonas reinhardtii. Planta*, **231**, 729–740.

Moellering, E.R., Miller, R. and Benning, C. (2010) Molecular genetics of lipid metabolism in the model green alga *Chlamydomonas reinhardtii. Lipids in Photosynthesis*, **30**, 139–155.

Montero, M.F., Aristizabal, M. and Reina, G.G. (2011) Isolation of high-lipid content strains of the marine microalga *Tetraselmis suecica* for biodiesel production by flow cytometry and single-cell sorting. *Journal of Applied Phycology*, **23**, 1053–1057.

Nakajima, Y. and Ueda, R. (2000) The effect of reducing light-harvesting pigment on marine microalgal productivity. *Journal of Applied Phycology*, **12**, 285–290.

Nguyen, H.M. *et al.* (2011) Proteomic profiling of oil bodies isolated from the unicellular green microalga *Chlamydomonas reinhardtii*: with focus on proteins involved in lipid metabolism. *Proteomics*, **11**, 4266–4273.

Pereira, H., Barreira, L., Mozes, A., Florindo, C., Polo, C. *et al.* (2011) Microplate-based high throughput screening procedure for the isolation of lipid-rich marine microalgae. *Biotechnology for Biofuels*, **4**, 61.

Polle, J.E., Kanakagiri, S.D. and Melis, A. (2003) tla1, a DNA insertional transformant of the green alga *Chlamydomonas reinhardtii* with a truncated light-harvesting chlorophyll antenna size. *Planta*, **217**, 49–59.

Ramazanov, A. and Ramazanov, Z. (2006) Isolation and characterization of a starchless mutant of *Chlorella pyrenoidosa* STL-PI with a high growth rate, and high protein and polyunsaturated fatty acid content. *Phycological Research*, **54**, 255–259.

Rawsthorne, S. (2002) Carbon flux and fatty acid synthesis in plants. *Progress in Lipid Research*, **41**, 182–196.

Rismani-Yazdi, H., Haznedaroglu, B.Z., Bibby, K. and Peccia, J. (2011) Transcriptome sequencing and annotation of the microalgae *Dunaliella tertiolecta*: pathway description and gene discovery for production of next-generation biofuels. *BMC Genomics*, **12**, 148.

Rosenberg, J.N., Oyler, G,A., Wilkinson, L. and Betenbaugh, M.J. (2008) A green light for engineered algae: redirecting metabolism to fuel a biotechnology revolution. *Current Opinion in Biotechnogy*, **19**, 430–436.

San Pedro, A., González-López, C., Acién, F. and Molina-Grima, E. (2013) Marine microalgae selection and culture conditions optimization for biodiesel production. *Bioresource Technology*, **134**, 353–361.

Schenk, P.M., Thomas-Hall, S.R., Stephens, E., Marx, U.C., Mussgnug, J.H. *et al.* (2008) Second generation biofuels: High-efficiency microalgae for biodiesel production. *Bioenergy Research*, **1**, 20–43.

Sharma, K.K., Schuhmann, H. and Schenk, P.M. (2012) High lipid induction in microalgae for biodiesel production. *Energies*, **5**, 1532–1553.

Siaut, M., Cuine, S., Cagnon, C., Fessler, B., Nguyen, M. *et al.* (2011) Oil accumulation in the model green alga *Chlamydomonas reinhardtii*: characterization, variability between common laboratory strains and relationship with starch reserves. *BMC Biotechnology*, **11**, 7.

Simionato, D., Basso, S., Giacometti, G. and Morosinotto, T. (2013) Optimization of light use efficiency for biofuel production in algae. *Biophysical Chemistry*, **182**, 71–78.

Stephenson, P.G., Moore, C.M., Terry, M.J., Zubkov, M.V. and Bibby, T.S. (2011. Improving photosynthesis for algal biofuels: toward a green revolution. *Trends in Biotechnology*, **29**, 615–623.

Tetali, S.D. *et al.* (2007) Development of the light-harvesting chlorophyll antenna in the green alga *Chlamydomonas reinhardtii* is regulated by the novel Tla1 gene. *Planta*, **225**, 813–829.

Velmurugan, N., Sung, M., Yim, S., Park, M., Yang, J. and Jeong, K. (2013) Evaluation of intracellular lipid bodies in *Chlamydomonas reinhardtii* strains by flow cytometry. *Bioresource Technology*, **138**, 30–37.

Vigeolas, H., Duby, F., Kaymak, E., Niessen, G., Motte, P. *et al.* (2012) Isolation and partial characterization of mutants with elevated lipid content in *Chlorella sorokiniana* and *Scenedesmus obliquus*. *Journal of Biotechnology*, **162**, 3–12.

Wang, Z.T., Ullrich, N., Joo, S., Waffenschmidt, S. and Goodenough, U. (2009) Algal lipid bodies: stress induction, purification, and biochemical characterization in wild-type and starchless *Chlamydomonas reinhardtii*. *Eukaryotic Cell*, **8**, 1856–1868.

Weselake, R.J., Taylor, D.C., Rahman, M.H., Shah, S., Laroche, A. *et al.* (2009) Increasing the flow of carbon into seed oil. *Biotechnology Advances*, **27**, 866–878.

Wong, D. and Franz, A. (2013) A comparison of lipid storage in *Phaeodactylum tricornutum* and *Tetraselmis suecica* using laser scanning confocal microscopy. *Journal of Microbiological Methods*, **95**, 122–128.

Work, V.H., Radakovits, R., Jinkerson, R.E., Meuser, J.E., Elliott, L.G. *et al.* (2010) Increased lipid accumulation in the *Chlamydomonas reinhardtii* sta7-10 starchless isoamylase mutant and increased carbohydrate synthesis in complemented strains. *Eukaryotic Cell*, **9**, 1251–1261.

Yao, C., Ai, J., Cao, X., Xue, S. and Zhang, W. (2012) Enhancing starch production of a marine green microalga *Tetraselmis subcordiformis* through nutrient limitation. *Bioresource Technology*, **118**, 438–444.

Yu, W.L., Ansari, W., Schoepp, N.G., Hannon, M.J., Mayfield, S.P. and Burkart, M.D. (2011) Modifications of the metabolic pathways of lipid and triacylglycerol production in microalgae. *Microbial Cell Factories*, **10**, 91.

Zeeman, S.C. *et al.* (2010) Starch: its metabolism, evolution, and biotechnological modification in plants. *Annual Review of Plant Biology*, **61**, 209–234.

12

Algal Cultivation Technologies

Alessandro Marco Lizzul[1] and Michael J. Allen[2]

[1] University College London, London, UK
[2] Plymouth Marine Laboratory, Plymouth, UK

Summary

This chapter reviews the different methods for micro-algal cultivation. The choice for micro-algal culture is often between open ponds or enclosed systems, and each have their advantages and disadvantages. To understand the potential of microalgae for the production of biofuels, it is imperative to have a realistic assessment of these systems. Consequently, here we focus particularly on the productivity, cost and scalability of the most typical installations for algal culture.

12.1   Introduction

Anyone with a pool, garden pond or even bird bath will know how easy it can be to grow microalgae, indeed often wilful neglect might be all that is necessary to turn clear standing water green with algae. Uncontrolled or unplanned growth of microalgae is a common blight of water bodies (be they ponds, pools, canals, rivers, coastal regions and estuaries), which creates the misconception that controlled microalgae culture is therefore a straightforward process. In fact the reliable, reproducible and controlled growth of microalgae over prolonged periods of time poses significant problems. Put simply, larger, industrial-scale algal cultures are harder to control and maintain. For some applications there can be more leeway; for example, the use of microalgae for remediation will not usually require the culture of a single species, indeed a diverse, robust and resilient mixed population may have significant advantages with regards to stability and performance. However, for specific processes such as nutraceutical, pharmaceutical or other high-value compound production, a single species is often essential to maintain a high degree of quality control. Therefore, in general, high-volume, lower-value products from algal biomass typically utilise open style systems (*i.e.* raceway ponds) where contamination is either not an issue or an easily implemented selection for the species cultured is available, such as high salt or temperature, whilst for the newer generation of lower-volume, higher-quality products from microalgal biomass, cultivation within closed systems known as photobioreactors (PBRs) is often favoured.

Biofuels and Bioenergy, First Edition. Edited by John Love and John A. Bryant.
© 2017 John Wiley & Sons Ltd. Published 2017 by John Wiley & Sons Ltd.

The design and construction of an algal cultivation system is a multi-parametric problem, in that many different factors must be considered to ensure success. This means that designing an optimum for each condition cannot be satisfied under most circumstances without incurring effects on the other factors. Consequently, algal cultivation platforms are a compromise between biotic, abiotic and economic factors, with the designs very much tailored to the desired end application. Many of the conventional designs described in the literature attempt to maximise the efficiency of as many of the parameters as possible, within practical design limits. The end result is an attempt to operate the system within a multi-parametric 'sweet spot'. In terms of practical designs, this philosophy has resulted in numerous small-scale approaches to cultivation alongside several general configurations for larger-scale platforms. Most algal culture systems in operation today can be categorised as having characteristics based on the following designs, which include tubular, column, plate, membrane and pond reactors[1].

The design considerations for PBRs are relatively well established[2] and can be simplified into several major categories. First, exposure to a suitable source of light is of great importance, as without adequate irradiance the photosynthetic processes cannot occur. The next most important factor is to make provision for the addition and containment of media containing the appropriate components for algal growth. This often includes an appropriate carbon source, usually bubbled into the culture as carbon dioxide (CO_2) as well as a source of soluble nitrogen, phosphorous and trace elements. Within this bulk fluid, temperature control is a very important factor in the maintenance of optimal growth and must be kept within set constraints according to the preference of the algal species cultured. Likewise, the ability to remove inhibitory waste products from the process is imperative, especially when growing the culture to higher densities. This waste removal is particularly true for the level of dissolved oxygen, which is known to inhibit photosynthesis. A final factor of considerable importance is the way in which these biotic and abiotic factors are combined; this is achieved by mass transfer of all these components under well-mixed (or turbulent) conditions within the reactor. To date, many different approaches have been employed within PBR design, and all successful systems give some degree of consideration to the afore-mentioned factors.

12.2   Lighting

Access to sufficient light is imperative for phototrophic algal cultivation, and its maximisation within biological and economic constraints is desirable at all times. Algal cultivation can be undertaken using either natural or artificial light, provided they have sufficient quanta at the correct wavelengths. Traditional open ponds often rely solely on natural sunlight, and growth in these systems is subject to the seasonal and diurnal cycles[3]. Reliance on natural light has obvious drawbacks, not least because a sizable proportion of the day will have insufficient light, meaning photosynthesis will be less efficient. Given these considerations, it is likely that vast tracts of the Northern Hemisphere would be unsuitable for algal cultivation with natural light for a considerable part of the year. Another factor to consider is that many algal strains respond differently to diurnal and seasonal cycles, which can alter the final biomass composition, thereby affecting the end-product[4]. Despite the

---

1 Ugwu *et al.* (2008).
2 Molina *et al.* (2001); Pulz (2001); Ugwu *et al.* (2008); Weissman *et al.* (1988).
3 Park *et al.* (2001).
4 Chen (2011).

inherent problems with natural light, the one notable benefit is the significant energy cost reduction that is imparted on the overall cultivation process. This simple fact means that to date a vast majority of commercially grown algae are produced with natural lighting[5].

Artificially illuminated systems have the benefit of offering consistent light regimes, but at considerable energetic cost. Conventional systems for lighting include fluorescent bulbs, which are often deployed at laboratory scale due to a favourable wavelength profile as well as a reasonable cost per Watt (Figure 12.1a). However, fluorescent lights often lack the power output required for larger, and more densely growing, cultures. For larger systems, metal halide lights (Figure 12.1b) are favoured, due to a spectral output that closely matches natural light and hence is ideal for photosynthesis. Recent developments in light emitting diode (LED) technology, which can now be manufactured to specific wavelengths at reasonable cost, have made them an attractive alternative to fluorescent or metal-halide lighting systems. LEDs allow the cultivator to tailor the incident wavelengths to suit the algal strain or process. For example, red LEDs with wavelengths between 620 and 700 nm and blue LEDs with wavelengths between 455 and 492 nm can be incorporated into lighting in varying proportions (Figure 12.1c). The result is that wavelengths not used during photosynthesis can be excluded, leading to an increase in total energy efficiency[6].

(a) (b)

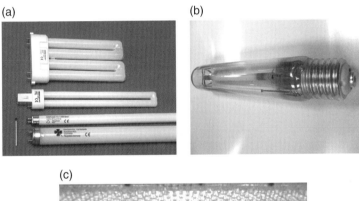

(c)

Figure 12.1 Examples of common lighting arrangements. (a) fluorescent light tubes[7]; (b) a Metal Halide lamp[8]; and (c) an LED lighting block with red/blue lighting[9].

5 Chaumont (1993); Chen *et al.* (2011).

6 Lee and Palsson (1996).

7 Christian Taube, CC BY-SA 2.0 de, https://commons.wikimedia.org/w/index.php?curid=102772.

8 Mike Allen (2015).

9 By NASA Marshall Space Flight Center - http://www.msfc.nasa.gov/news/news/photos/2000/photos00-336.htm, Public Domain, https://commons.wikimedia.org/w/index.php?curid=39544881.

12.3 Mixing

Adequate mixing is required to maintain homogenous culture conditions within a heterogeneous mixture. For algal cultures, mixing maintains access to sufficient nutrients, whilst allowing for gaseous exchange and access to sufficient quanta of light. Mixing can occur *via* convective or intensive methods; within most algal applications intensive and direct fluid displacement is preferred for the afore-mentioned reasons. The displacement required for fluid mixing can be undertaken in one of many ways, but conventional systems include impeller or paddlewheel agitation, direct liquid displacement or airlift systems[10].

Impellers and paddle wheels may be used to move an aqueous dispersion around predictable circulatory patterns by introducing kinetic energy into the fluid. Impellers

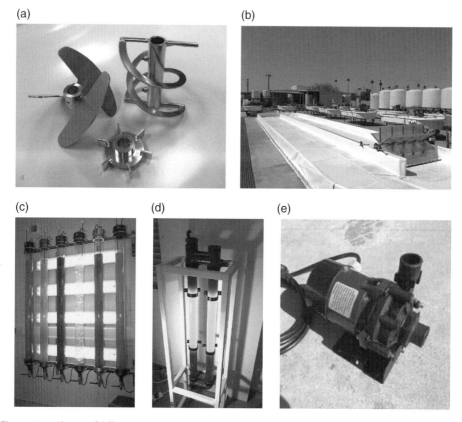

(a)　　　　　　　　　　(b)

(c)　　　　　　(d)　　　　　　(e)

Figure 12.2 Photos of different mixing systems, (a) Turbines are often deployed to create radial mixing within stirred tank reactors[11]. (b) Paddlewheel, used to create directional fluid mixing[12]. (c) Bubble column reactors, where turbulence is created by direct bubbling of the culture[13]. (d) ALR, where air injection creates directional liquid circulation. (e) Velocity pump, clearly showing the inlet and outlet[14].

10　Chaumont (1993).
11　Allen (2015).
12　Allen (2015).
13　Allen (2013).
14　Reef Central Online Community (2013).

tend to be used within smaller laboratory-scale systems, or within classical enclosed fermenters (Figure 12.2a). This method of mixing has the benefit of transferring a large amount of kinetic energy into the fluid, but can also have some negative attributes, such as high-energy consumption and the creation of considerable shearing effects. It is also challenging to cultivate photosynthetic algae with impellers on a larger scale, due to poor light penetration in larger volume fermentation vessels[15]. Consequently, impeller mixed systems have had more success when deployed for heterotrophic cultivation of algae, with Solazyme producing algal oil in this way in the USA[16]. Paddlewheel systems differ from impellers in that they are only partially submerged and positioned in a horizontal plane, and run at a much lower rpm (Figure 12.2b). This mode of mixing is preferred for larger-scale algal production, due to its lower running costs and comparative ease of maintenance[17]. However, the low liquid velocities induced by this type of mixing means that the algal culture is sub-optimally mixed, and dead zones are often created within certain areas of the pond (Figure 12.3). Some of these problems can be overcome to an extent by introducing baffles and maintaining certain depth and width to length relationships, or utilising them within rotating contactor style systems[18].

Other prominent mixing methods deployed within closed systems include the use of liquid or air pumps. Conventional pumps can be characterised into three main groups, based on how they create the actual mixing. The first type of pump is the impulse or airlift pump, which creates a density difference in the fluid circulation pathway, thereby forcing liquid circulation to occur. These reactors are often described as having a bubble column or airlift configuration (Figures 12.2 c and d), and are discussed further later. The second type of pumping system is the positive displacement pump, which creates movement by trapping a fixed volume of fluid and moving it into a discharge pipe by creating a driving motion through reciprocating or rotary motion. The final category can be described as velocity driven pumps, which operate by adding kinetic energy into the

Figure 12.3 AlgaeConnect control system and pH probe. The probes can interface with control systems such as AlgaeConnect produced by AlgaeLab Systems[19].

---

15 Singh and Sharma (2012).
16 Franklin *et al.* (2012).
17 Terry and Raymond (1985).
18 Christenson and Sims (2011); Hadiyanto *et al.* (2013).
19 Lee (2012); Champion Lighting and Supply Co. (2005)

fluid, increasing the pressure and flow rate around a set pathway (Figure 12.2e). These types of pump form a broad category, which include rotodynamic and centrifugal drives.

12.4   Control Systems and Construction Materials

Larger-scale algal applications have a fundamental requirement for process control to allow for maintenance of optimal operational parameters and conditions, whilst minimising the workload for operators. The actual control systems usually take the form of a series of sensor and control relays that can measure, relay and then adjust parameters to a set point. More complex systems allow for the ability of dynamic control, including turbidostat culture where the culture density is maintained. Several common parameters are measured and controlled during algal cultivation. These include the light intensity, temperature, pH, dissolved oxygen and cell density. Once these variables are detected by the sensors they are passed onto a computer which relays the information to the control loop, which can act to alter the conditions back to the set point (Figure 12.4).

PBRs are often constructed from cheap, durable and readily available materials. In the case of open pond systems, this often means levelling and compacting the ground and, cost permitting, using concrete or neoprene/plastic under-lining to contain the culture[20]. In externally illuminated PBRs, the choice of construction material is often decided by the requirements for transparent materials with high optical clarity, such as durable polymers like acrylic or polycarbonate. Glass is also deployed, and can have many benefits in comparison to plastics, especially in terms of overall longevity. Other less optically important parts of the reactor are often constructed from cheaper materials such as polyvinyl chloride (PVC)[21]. The final decision on the building material is often driven by a trade-off in lower upfront expenditure *vs* longer-lasting and more expensive building materials. Modern trends in reactor design include an increasing focus on the environmental impact of construction materials, achieved *via* detailed life-cycle assessment[22] (see also Chapter 16).

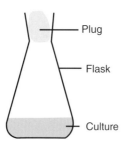

Figure 12.4 Example of a flask culture, and shaker arrangement. Cotton wool is used to allow for gaseous exchange whilst preventing the entrance of contaminants. The flasks are placed on rotational shakers to provide adequate mixing for growth.

---

20 Weissman *et al.* (1988); Tredici (2004).
21 Chaumont (1993).
22 Soratana and Landis (2011).

12.5   Algal Production Systems at Laboratory Scale

The need to cultivate algae for scientific study within the laboratory has led to the development of a variety of techniques to grow sufficient quantities of biomass from millilitre to litre (ml – l) scales. For these purposes, many laboratories deploy simple systems such as incubator shakers, which operate by rotational (or orbital) mixing of flasks. The flasks are usually sterilised and the algae grown within a controlled growth chamber that can maintain stable biotic and abiotic factors. This method of algal production is ideal when cost constraints are taken into account, and is particularly suitable for the small-scale requirements of many molecular biology and physiological laboratories. Using these smaller 'starter' cultures can also provide sufficient biomass for inoculation of larger-scale systems (Figure 12.4).

Other small-scale production methods tend to resemble simplified versions of larger-sized systems. A pertinent example of this is the conversion of standard laboratory bottles into miniaturised bubble columns. This is achieved by introducing air into the bottom of the reactor to create turbulence. Employing these systems within the laboratory has the benefit of keeping costs down, whilst also allowing for more flexibility than a simple conical shake flask[23]. Other laboratories deploy conventional small-volume stirred tank reactors (with volumes between 5 and 10 litres) for growing algal cultures (Figure 12.5). However, caution must be observed when choosing the mixing speed, as the impeller can cause considerable levels of shear in more sensitive strains, such as those with thinner cell walls, non-spherical shapes or flagella[24].

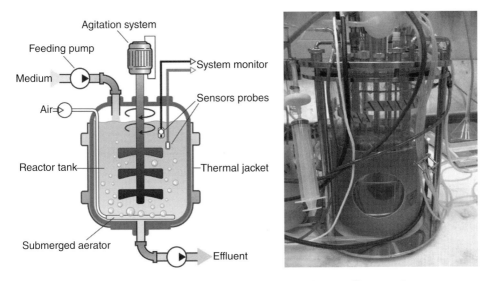

Figure 12.5 Example of a flask culture, and shaker arrangement. Diagram illustrates the components of a stirred fermenter, note submerged aerator and impeller. Lights can be arranged around the reactor or submerged inside, as long as they do not interfere with the circulation of the impeller[25].

---

23 Lizzul *et al.* (2103).
24 Joshi *et al.* (1996).
25 Mrabet (2009).

There is currently increasing interest in the use of microscale technologies within alga culture that reflects a wider trend within the biotechnological sciences of scaling down experiments for higher throughput[26]. The results from these faster and smaller studies can then be combined with the appropriate statistical analysis and used to predict parameters suitable for scale up within many applications. Growing algae in this way can allow for the rapid exploration of multiple biotic or abiotic parameters at a cost that is considerably lower than other scaled down systems. Another developing trend is the development of suspended cultures. These can be enclosed in the form of alginate beads or distributed as a biofilm onto suitable membrane surfaces. Suspended cultures have opened up a number of novel applications, particularly within the field of bioremediation[27].

12.6   Algal Production in Open Systems

12.6.1   Pond-Based Systems

The design of high yield, open algal ponds can be traced back to work initiated in the latter part of the 1940s and into the early 1950s[28]. Such systems are also described as 'high rate algal ponds' (HRAPs) within the literature. To date, open ponds have remained the most widespread and historically successful of large-scale production systems. There are several common variations in the design of an open system, with most examples taking the form of concrete or plastic lined channels that form a raceway loop system[29]. Most ponds tend to be fairly shallow (10–30 cm deep), to allow for maximal solar penetration of the culture[30]. Mixing is usually achieved by paddle wheel agitation, with nutrients added either in batch or continuously. Although most open ponds are mixed with paddle wheels, there have also been some attempts to design airlift ponds. However, the findings from these studies have shown that they compare unfavourably with paddle wheel mixed systems[31]. CO_2 is often introduced to the pond *via* submerged aerators located within sumps[32], although adequate gaseous retention and distribution within the medium can be hard to maintain due to the large contact areas involved. Harvesting methods are dependent on the desired product and can occur in either continuous or batch unit operations downstream. As the name suggests, most ponds are open to the elements, although some coverings have been employed in smaller pilot type projects. Typical arrangements are shown in Figure 12.6.

There are several process benefits that can be realised by the adoption of an open pond system, which helps to explain why it remains the preferred production system for many algal culture applications. The single most important factor is the lower energy input that is required to maintain a sufficient level of mixing in comparison to other

---

26 Ojo *et al.* (2014); van Wagenen *et al.* (2014).
27 Naumann *et al.* (2013); Shi *et al.* (2014).
28 Borowitzka (1999).
29 Jiménez *et al.* (2003).
30 Oswald (1995).
31 Chaumont (1993).
32 Weissman *et al.* (1988).

(a)

(b)

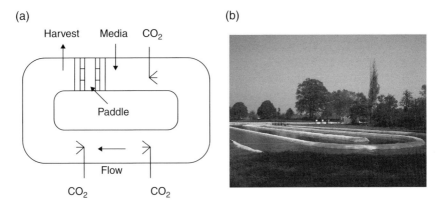

Figure 12.6 Schematic and photograph showing the typical arrangement of a raceway pond. (a) Shows an aerial view, indicating how the pond is mixed and sparged with CO_2[33]; and (b) Typical raceway pond[34].

production systems[35]. This low energy requirement makes most production processes more economical within open pond systems. Other benefits include cheap and simple constituent parts, as well as easy access to the whole system. This means that fouling can be cleaned relatively easily, whilst the lack of internalised parts also makes the systems relatively flexible should operational changes be required. There are also some biotic benefits to the relatively low liquid velocity within open ponds, notably the reduction in the shear levels encountered within the system, which can allow for the cultivation of more sensitive algal species. Another important factor that should not be understated is that raceway ponds have been in existence for a number of years and there is therefore a considerable body of operational knowledge on successful commercial operating procedures (see also Chapter 16).

Limitations of open pond designs include the fact that they achieve relatively poor levels of biomass productivity[36], on average $0.18–0.25\,g\,l^{-1}\,day^{-1}$. These lower yields can be attributed to the suboptimal mixing conditions found within raceway systems, as well as the inconsistent and variable light profiles found through the reactor. This constraint in available light is caused by a combination of low liquid velocity and relative culture depth, which means that much of the sunlight is blocked from all but the uppermost part of the culture, slowing down the average growth rate, especially in denser cultures. Other issues with the use of open ponds concern their exposed nature, making contamination with competing organisms or predator species particularly problematic. These problems of contamination can be partially overcome by using growth conditions that favour the cultivated strain, for example the use of extremophilic or extremotolerant organisms is a preferred option to avoid contamination[37]. However, such selections are not always feasible. Other more

33 Based on a diagram in Chisti (2007).
34 By JanB46 - Own work, CC BY-SA 3.0, https://commons.wikimedia.org/w/index.php?curid=16863305
35 Stephenson *et al.* (2010).
36 Brennan and Owende (2010); Ugwu *et al.* (2008).
37 Schenk *et al.* (2008).

general problems with open ponds include the relatively large areas of land required to establish a production facility. Due to their large size, open ponds are also particularly vulnerable to changes in temperature and light fluctuations, making most open ponds unsuitable above certain latitudes. Finally, from an environmental perspective, there are also some concerns regarding the evaporative losses that can occur from open ponds, as well as the high potential of environmental contamination by the cultured microalgae.

Due to their cost and versatility, open pond systems are particularly suited to the low to middle value range of biomass products, including biofuels and feed production. Ponds have also found considerable usage in the production of some higher value pigments[38] (Borowitzka, 1992). Several companies successfully produce commercially viable algae in open ponds, including Cyanotech, who cultivate *Haematococcus pluvialis* to produce astaxanthin[39] in a two-stage system, with the first vegetative stage within closed PBRs and the second maturation stage within HRAP systems. Other companies such as Seambiotic[40] produce algae in open ponds grown using flue gas and wastewater as input streams. The use of algae in wastewater treatment infrastructure in this way is a promising avenue for HRAPs, particularly within warmer climates, and past research has considered the integration of such systems[41]. It is likely that any serious attempt to produce large quantities of lower value products such as algal biomass for biofuel will require the large-scale deployment of HRAP type systems due to their comparably favourable operational cost.

12.6.2 Membrane Reactors

The use of membrane technology is increasingly finding its way into the wastewater treatment sector. This is due to the ability of membrane reactors to increase biological retention within high flow systems, thereby decreasing the hydraulic retention time and improving system efficiency. Adaptations of membrane designs are increasingly being trialled as algal production systems (Figure 12.7). Most of the designs have a liquid mobile phase sandwiched between two semi-permeable membranes. These membranes allow for the liquid and nutrients to pass through specific pore sizes, but the algae remain attached as a biofilm to the other side of the membrane (Naumann *et al.* 2013; Shi *et al.* 2014). Although much work is still needed to characterise and develop these systems at scale, the potential advantages could be numerous, most notably a short light path to enable high levels of algal growth and a lack of requirement for downstream de-watering. Their potential *in situ* deployment in the open ocean, without the requirement for large tracts of land, is particularly appealing. Foreseeable problems with membrane reactors include labour-intensive harvesting, which would resemble conventional agricultural practices. Other potential problems centre on membrane fouling, which could lead to a considerable performance drop within the system.

---

38 Borowitzka (1992).
39 www.cyanotech.com/bioastin.html
40 www.seambiotic.com
41 Oswald (1988); Sheehan *et al.* (1998).

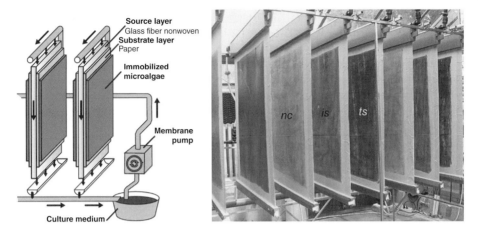

Figure 12.7 Schematic diagram and photograph of a membrane PBR. The photograph on the right shows the strains *Nannochloropsis* (*nc*), *Isochrysis* (*is*) and *Tetraselmis* (*ts*) grown within drip fed membrane reactors[42].

12.7 Algal Production in Closed Systems

12.7.1 Introduction

Enclosed vessels for algal cultivation are commonly described as PBRs and are used to culture algal biomass under more stringent and optimised conditions than open systems. Although many different PBR configurations exist, three main geometries dominate; these include horizontal/vertical tubular, plate and single column configurations. The actual categorisation and distinction between these three system types can become more complicated in practice, as many reactors display hybridised geometries and configurations, with the mixing method and configuration often having greater impact than the reactor geometry.

12.7.2 Plate or Panel Based Systems

In theory, flat panel (or plate) reactors are the most efficient enclosed PBR systems[43] due to their large surface area to volume ratio, which results in very little culture-induced self-shading. Flat panel reactors therefore display a particularly high conversion efficiency of incident sunlight and can achieve some of the highest reported daily levels of biomass production, with averages reaching $2.4\,g\,l^{-1}\,day^{-1}$. In addition, a panelled array can easily be tilted towards the Sun to maximise solar penetration[44] (Figure 12.8a). Dependent upon configuration, plate reactors can benefit from lower levels of dissolved oxygen build-up than many other types of closed reactor. This advantage is due to the relatively short circulatory path, meaning the culture is able to rapidly de-gas. The operational costs of panelled systems are comparable to tubular or column reactors, and are heavily dependent on the selected mixing mode and intensity. Indeed, the mixing within plate systems tends not to follow idealised patterns, and could be

---

42 Naumann *et al.* (2013).
43 See Hu *et al.* (1996).
44 Richmond and Hu (2013).

(a)   (b)

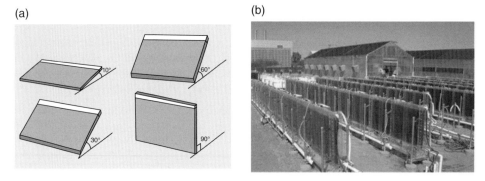

Figure 12.8 Diagram showing the potential arrangements of flat panelled reactors. (a) Gives an indication of the many different tilt angles that are possible to optimise solar penetration. (b) Shows an array of flat panel bioreactors.[45]

described as generally very turbulent, in a spectrum between bubble column and airlift flow patterns, depending on internal structures. Like other closed PBRs, external contamination can be kept to a minimum, with an ability to be sterilised more effectively than many other configurations. Internal fouling can also be kept lower in flat panel reactors than in tubular systems, due to a less convoluted flow path.

One potential drawback of flat panelled reactors is the practical scalability of the system, as some level of compromise must be struck between solar collecting surface area, culture depth and areal considerations (Figure 12.8). This problem can be overcome to some extent by connecting a modular array of panels in tandem, although the resultant areal footprint would be likely to be larger than an equivalent volume of other types closed reactors. Other physical drawbacks are associated with temperature control within the system, which can be more energy intensive than other closed systems due to the large surface area to volume ratio. The large surface areas and short light paths within plate reactors can also create problems with photo-damage to the microalgal light harvesting complex if the light levels are too high, although this damaging effect can be lessened by shading or by using turbidostat based cultivation techniques. As with all PBRs, it is also likely that some degree of fouling is unavoidable, especially at the meeting of straight edges, where the liquid velocity is lower. The large surface area to volume ratio also means that the levels of hydrodynamic stress placed on the algae can be higher than in some other systems. Another notable point is that the rapid degassing seen in flat panels could be seen as advantageous in most applications, as it prevents the build-up of excess dissolved oxygen; the converse is true if gas use efficiency is sought, as gas retention is considerably lower in these systems.

Examples of larger-scale plate or panel PBR arrays (Figure 12.8b) are less commonly deployed than for other closed systems. Consequently, to date, there are relatively few examples of commercial-scale algal production using flat panelled configurations. This is most probably due to manufacturing and cost-related issues that still need to be overcome before flat panel PBRs become widely utilised. For example, there have been reports of difficulty with manufacturing at scale, and problems associated with sealing and warping of the materials during temperature stress. As process problems are overcome, however, it is likely that flat panelled systems will increasingly find similar

---

applications to those employed by tubular reactors, although with some distinct process advantages in terms of degassing and solar penetration. At smaller scale, the combination of large surface area and short light path make panelled systems a favourable option, and systems such as the Infors Labfors reactor[46] have found considerable use in growth modelling and screening. In the shorter term, it is likely that higher-value indoor algal cultivation may find considerable use for flat panels, especially if good productivity levels are required. Algenol are one example, using large modular bag-based panels to produce ethanol[47]. In the UK, Algaecytes Ltd has developed its own in-house flat panel reactor for the production of omega-3 oils[48].

12.7.3 Horizontal Tubular Systems

Tubular PBRs encompass a broad range of designs, and can be arranged either horizontally or vertically, in serpentine or in manifold arrangements. Tubular PBRs can also be laid flat on the ground or positioned stacked above one another. Most traditional commercial and research designs tend to favour a single serpentine or horizontal manifold arrangement, with the tubes stacked vertically in order to minimise the footprint of the reactor and maximise areal productivity. Prominent examples include the ground-based, flat horizontal tubular reactor in Cadiz, Spain[49] and the vertically stacked manifold system at the Ben Gurion University, Beer Shevar, Israel[50]. Mixing within these tubular reactors is conventionally achieved with a variety of pumping systems, including centrifugal or diaphragm pumps, as well as airlift driven systems. Popular velocity based pumping systems include centrifugal pumps, in which a rotating impeller increases the pressure and flow rate of a fluid. This type of centrifugal pump is frequently used to move liquids through piping systems and therefore a popular choice for tubular reactors. These pumps operate by allowing fluid to enter the impeller along or near to the rotating axis. The liquid is then accelerated by the impeller flowing outward in either radial or axial directions, where it enters a diffuser or volute chamber upon which it can exit towards the downstream piping system. Centrifugal systems can find particular applications where large discharge through smaller heads is required.

Enclosed, tubular PBR systems have many comparative advantages over open ponds. Most importantly, tubular PBRs have been shown to achieve higher and more consistent levels of areal productivity than can be achieved by algal culture in open ponds (Posten, 2009). This level of consistency is due to the larger illuminated surface to culture volume that are characteristic of tubular PBRs compared to open ponds, with tubes often in the region of 3–10 cm in diameter. The enclosed nature of the system also means better control over both biotic and abiotic factors during cultivation, making it easier to control contamination whilst allowing for a wider repertoire of strains to be cultivated. Another factor that improves productivity within tubular systems is the greater level of turbulence that is achievable within them due to higher liquid velocities. This, in combination with better light penetration, allows for far higher algal growth rates than open ponds, with average reports of biomass productivities ranging from $0.25–1.22\,\mathrm{g\,l^{-1}}$ $\mathrm{day^{-1}}$. Tubular configurations also have the comparative benefit of lower levels of water loss in comparison to open systems.

46 www.infors-ht.co.uk/en/labfors-5-2/
47 ww.algenol.com/
48 algaecytes.com/
49 Molina *et al.* (2001).
50 Richmond *et al.* (1993).

Despite the many process benefits conferred by utilising a tubular configuration, there are also some limitations. One of the major issues with tubular reactors is the fact that the capital and operational costs are considerably higher than that of open ponds, in part due to the relative cost of building materials as well as the considerable energy requirements for operation. As with all closed systems, it is also common to find some degree of wall growth and fouling in tubular PBRs, which can increase the costs associated with the cleaning and maintenance of such a system. Cleaning becomes a particular challenge with larger and more elaborate PBRs. To overcome this problem, several companies have devised ingenious methods of keeping the reactor walls clean, a prominent example being the use of Bio-Beads™ in Varicon Aqua's BioFence™ (Figure 12.9b).

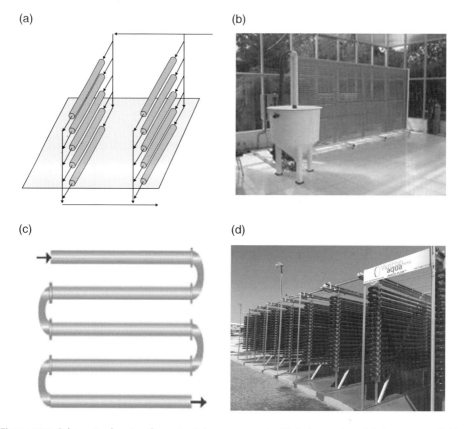

(a)   (b)

(c)   (d)

Figure 12.9 Schematic showing the potential arrangements of tubular reactors. (a) shows a manifold system, illustrating how the manifolds split the main flow[51]. Modified from Chisti (2007). (b) Varicon Aqua's BioFence™, which operates under a manifold system[52] (c) shows a serpentine arrangement[53] (Posten, 2009); panel (d) shows such an arrangement in use in a Varicon Aqua Phyco-Flow in Hawaii[54].

51 Based on a diagram in Chisti (2007).
52 Varicon Aqua Solutions, 2010.
53 Posten (2009).
54 Varicon Aqua, 2015, Hawaii.

In manufacturing terms, the scale-up of tubular reactors can be a reasonably straightforward process, though larger deployments can suffer from a variety of biotic and abiotic problems. One major problem with large-scale tubular systems is that considerable gradients of dissolved oxygen, CO_2 and pH can develop across the system, causing inhibition or underperformance of algal growth[55]. These considerations may place a practical constraint on the run length of many tubular reactors, meaning that very large-scale systems require interruptions in the run length for degassing. It is also worth noting that many of these tubular systems are inappropriate for the more fragile strains of algae, due to the relatively high liquid velocities that cause considerable hydrodynamic shear within the culture (Chisti, 2007). Another important consideration when scaling a tubular system is the multiplication of pumping energy requirements to overcome the frictional forces of multiple tubes and bends. Attempts have been made to minimise this problem by using manifold systems, as shown in Figures 12.9 a and b. As with other enclosed PBRs, the large surface area can make temperature control problematic and costly.

Many horizontal tubular PBRs are already on the market, and may be purchased as specialised equipment from a variety of suppliers. The Varicon Aqua's BioFence™, mentioned above, which has been deployed in over 50 locations worldwide for a variety of uses, including both research and industrial sectors, is a prominent example. The BioFence™ has found widespread use in the production of higher value nutraceutical and cosmetic compounds, and has been employed by Blue Lagoon, Iceland[56] for the production of cyanobacterial metabolites as constituents of cosmetic products. Other examples of commercial production in tubular systems can be found in Ketura, Israel where, as mentioned earlier, *Haematococcus pluvialis* is grown for the production of astaxanthin in a custom-built manifold system. Despite the relatively high levels of productivity in tubular PBRs, current data indicate that around ten times more algal biomass is produced in open pond systems than in closed reactors. This highlights the fact that if the algal biotechnology industry is to reach its envisaged potential, it is somewhat unlikely that closed tubular reactors will be able to cope with the volume and cost requirements necessary for lower value bulk products. It is more likely that closed PBRs will integrate with open-pond systems, either as high rate inoculation platforms or within a two stage production process[57]. It is also likely that tubular PBRs will continue to be used in the production of higher-grade chemicals, where quality control is of the utmost importance and where the use of genetically-modified (GM) microalgae is required[58].

12.7.4 Bubble Columns

Column PBRs have found widespread use within conventional fermentation processes, mainly because the technology can be readily transferred to the growth of many different micro-organisms (Figure 12.10). The mixing regime within column reactors is created *via* air displacement, which is often cheaper than conventional liquid pumping. In fact,

55 Sobczuk *et al.* (2000).
56 www.bluelagoon.com
57 Rodolfi *et al.* (2009).
58 Pulz (2001).

one of the major benefits of a bubble column is that the turbulence created by aeration can be greater than equivalent power inputs with other types of mixing. This aeration results in high levels of mass transfer and maintains low levels of dissolved oxygen[59]. Shear stress is also lower in bubble PBRs than for most velocity driven systems. These features facilitate good rates of biomass production, averaging around $0.24–0.5\,gl^{-1}$ day^{-1}. Other advantages of column reactors are that they are fairly compact whilst being relatively simple to construct in comparison to the larger, horizontal tubular systems. Bubble columns can be made from a variety of materials including PVC and polythene, which can dramatically lower the capital expenditure in many larger-scale operations, whilst the absence of internal parts translates to lower overall maintenance costs.

Bubble columns share many of the drawbacks of other tubular systems in terms of construction materials and scale-up. However, there are also some design specific issues, the foremost being the relatively small illumination to volume ratio, especially when compared to many other types of tubular reactor or flat panels. This reduction in illumination can lead to sub-optimal productivity due to the large dark-zone within the reactor (Figure 12.10a). Another prominent consideration is that although bubble columns compare well in terms of cost with other tubular designs, they are still considerably more expensive than open ponds. Other important considerations for bubble column deployment include practical issues around scale-up; this is because by their very nature they are essentially individual systems (Figure 12.10b). This means that scale-up can only be undertaken by increasing the number of individual units. In this scenario, as each column is an individual batch reactor, this mode of operation incurs increased costs in terms of equipment and labour.

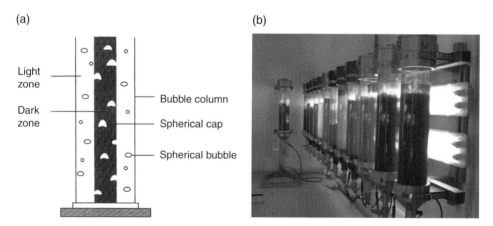

Figure 12.10 Bubble columns. (a) The diagram indicates the relative sections of light and dark zonation within the bubble column[60]. It also shows how air bubbles can deform as they rise through the column. (b) A battery of bubble columns[61].

59 Mirón *et al.* (2000).
60 Modified from Sánchez Mirón *et al.* (1999).
61 Allen (2013).

Bubble columns are a popular choice for intermediate sized systems (5–50 litres) and the relative ease of scale-up using repeating units has led them to be used in a wide array of production processes, ranging from low to high value applications. Bubble columns find particular prominence within the aquaculture industry for the production of algae as feed, or as inoculation vessels for larger PBRs. To date, several commercial reactors are on offer within the marketplace, including one from Varicon Aqua Solutions (Figure 12.10). However, the relative ease of manufacturing bubble columns means that they can be fabricated with limited resources and specialist equipment, resulting in a great variety of in-house designs and configurations. Prominent designs of novel and modular bubble columns can now be found within the literature, and address some of the issues regarding scale-up of bubble columns; this includes vertical systems built by AlgEternal and systems deployed by the University of Texas.

12.7.5 Airlift Reactors

Airlift PBRs encompass a broad family of pneumatic gas-liquid contacting devices, which act to create circulatory motion within a constrained geometry. This type of flow regime differs from that found within a bubble column, being characterised instead by more defined cyclical patterns[62]. The circulatory patterns within airlift PBRs are a function of the geometry and velocity within the system, and are created through interconnecting channels designed specifically for this purpose. The channels are often described as riser and down-comer sections, corresponding to the direction of the liquid travelling within them. The actual motion within the reactor is created by the injection of a mixing gas into the reactor from the bottom of the riser section. The hold-up of the mixing gas within the riser creates a density difference between the riser and down-comer sections. As the gas leaves the fluid by disengaging at the de-gassing zone of the riser, the denser fluid of the riser moves down through the down-comer, and the circulatory motion continues.

ALRs can be categorised as having either an internal or external loop configuration. Internal loop reactors separate their riser and down-comer, either with a draft tube or a split-cylinder arrangement. External loop airlifts have a physically separated riser and down-comer, taking the appearance of two separate interconnected tubes. Bubble size within ALRs is usually of the order of 0.5–5 mm. Figure 12.11 illustrates some of the more common ALR configurations. Airlift PBRs have a variety of operational benefits when compared to other systems. These include relatively high gas and mass transfer, uniform turbulent mixing, low hydrodynamic stress and ease of control, particularly with regards to liquid velocity. ALRs also display lower levels of dissolved oxygen in comparison to horizontal tubular reactors, due to short liquid circulation loops and rapid de-gassing. Many of these characteristics are of particular benefit to certain microbial processes, including algal cultivation. Productivity within ALR types has been reported to reach $0.5–1.5\,\mathrm{g\,l^{-1}\,day^{-1}}$ but despite this, ALRs have still not reached widespread commercial deployment.

Despite the low levels of commercial uptake, there are numerous examples of airlift PBRs in deployment across the globe. One prominent reactor described in the literature is the horizontal serpentine system deployed in Almeria, Spain. The study of this

---

62 Shah *et al.* (1982).

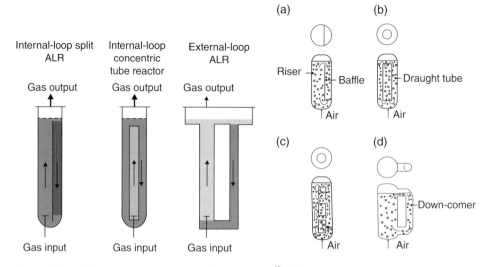

Figure 12.11 Different configurations of airlift bioreactors[63]. (a) Split-cylinder internal-loop; (b) concentric draught-tube internal-loop; (c) draught-tube internal-loop with vertically split draught-tube; and (d) external loop.

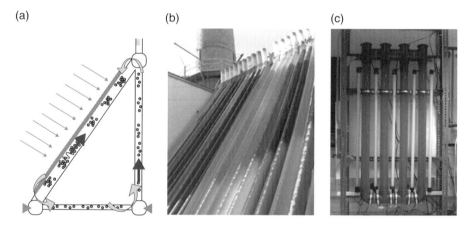

Figure 12.12 Diagram shows the potential arrangements of ALRs. (a) and (b) show how an airlift system can be created by directing the flow of the airlift. Images reproduced, with permission, from Vunjak-Novakovic *et al.* (2005). (c) shows the ALR designed and built by Varicon Aqua Solutions in collaboration with University College London.

system has provided extensive biological and engineering data from a pilot site[64]. Another prominent academic example includes the Massachusetts Institute of Technology (MIT) reactor (Figure 12.12), which was deployed to investigate the potential of flue gas scrubbing with algae. This system takes the form of individual

63 Chisti (1989).
64 Molina *et al.* (2001); Acién (2012).

30-litre triangular modules arranged in an elegant array formation (Figure 12.12a,b[65]). Other airlift configurations include planar designs, an example of which can be seen in Figure 12.12c, which shows a 10 litre ALR produced by Varicon Aqua Solutions[66].

12.8 Concluding Comments

It has been shown to date that there is no optimal way to cultivate microalgae; the choice of individual systems is in fact determined by a multitude of factors which relate to the algal species (and/or product) under consideration as well as the available budget, environmental conditions, local infrastructure and technical expertise available. For commercial production, a robust business plan and life-cycle analysis is essential to steer the user to the most suitable option. To date, open pond systems have clearly been proved to be the most popular systems for mass micro-algal cultivation, driven primarily by favourable operational costs and a few niche products. In this regard, the inherent metabolic diversity of microalgae promises much for their increasing use as photosynthetic platforms for commodity chemicals, as well as nutraceuticals, pharmaceuticals and other high-value compounds. More recent developments and the growth of synthetic biology alongside associated genetic techniques have created considerable interest in the exploitation of GM microalgae. Commercial production of bespoke metabolites from GM microalgae will almost certainly require robust containment measures and will accelerate the global deployment of closed PBR systems in the near future.

Selected References and Suggestions for Further Reading

Acién, F.G., Fernández, J.M., Magán, J.J. and Molina, E. (2012) Production cost of a real microalgae production plant and strategies to reduce it. *Biotechnology Advances*, **30**, 1344–1353.

Allen, M. (2013) http://bubble-columns.com/:

Borowitzka, M. (1992) *Journal of Applied Phycology*, **4**, 267–279.

Borowitzka, M.A. (1999) Commercial production of microalgae: ponds, tanks, tubes and fermenters. *Journal of Biotechnology*, **70**, 313–321.

Brennan, L. and Owende, P. (2010) Biofuels from microalgae – A review of technologies for production, processing, and extractions of biofuels and co-products. *Renewable and Sustainable Energy Reviews*, **14**, 557–577.

Champion Lighting and Supply Co (2005) http://www.championlighting.com/product.php ?productid=17463&cat=1090&bestselle

Chaumont, D. (1993) *Journal of Applied Phycology*, **5**, 593–604.

Chen, C.-Y.,Yeh, K.-L., Aisyah, R., Lee, D.-J. and Chang, J.-S. (2011) Cultivation, photobioreactor design and harvesting of microalgae for biodiesel production: A critical review. *Bioresource Technology*, **102**, 71–81.

---

65 Vunjak-Novakovic *et al.* (2005).
66 variconaqua.com

Chisti, M.Y. (1989) *Airlift Bioreactors*. Elsevier Applied Science, London.

Chisti, Y. (2007) Biodiesel from microalgae. *Biotechnology Advances*, **25**, 294–306.

Christenson, L. and Sims, R. (2011) Production and harvesting of microalgae for wastewater treatment, biofuels and bioproducts. *Biotechnology Advances*, **29**, 686–702.

Franklin *et al.* (2012) US Patent No. 8435767 B2.

GoPixPic (2014) Pin Metal Halide. http://www.proprofs.com/quiz-school/user_upload/ckeditor/metal%20halide_lighting.jpg: www.gopixpic.com

Greenwell, H.C., Laurens, L.M.L., Shields, R.J., Lovitt, R.W. and Flynn, K.J. (2010) Placing microalgae on the biofuels priority list: a review of the technological challenges. *Journal of The Royal Society Interface*, **7**, 703–726.

Hadiyantom, H.S. *et al.* (2013) *Chemical Engineering Journal*, **217**, 231–239.

HarvestKing (2014) Professional 300w LED Grow Lamp. http://growledsolutions.com/wp-content/uploads/harvest-king-main.jpg:

Hu, Q. *et al.* (1996) *Biotechnology and Bioengineering*, **51**, 51–60.

Jiménez, C. *et al.* (2003) *Aquaculture*, **217**, 179–190.

Jorquera, O., Kiperstok, A., Sales, E.A., Embiruçu, M. and Ghirardi, M.L. (2010) Comparative energy life-cycle analyses of microalgal biomass production in open ponds and photobioreactors. *Bioresource Technology*, **101**, 1406–1413.

Joshi, J.B. *et al.* (1996) *The Chemical Engineering Journal and the Biochemical Engineering Journal*, **62**, 121–141.

Lee, C.G. and Palsson, B.Ø. (1996) *Biotechnology Progress*, **12**, 249–256.

Lee, C.G. (2012) http://algaelabsystems.com/algaeconnect/:

Lizzul, A.M.P. *et al.* (2013) *Growth of Chlorella Sorokiniana (UTEX1230) from Exhaust Gases and Wastewater*. London: University College London.

Mrabet, Y. (2009) http://commons.wikimedia.org/wiki/File:Bioreactor_principle.svg:

Mira Images (2015) http://cdn.c.photoshelter.com/img-get/I0000DJkiIsnSI9I/s/860/860/0429B654.jpg:

Mirón, A.S. *et al.* (2000) *AIChE Journal*, **46**, 1872–1887.

Molina, E., Fernández, J., Acién, F.G. and Chisti, Y. (2001) Tubular photobioreactor design for algal cultures. *Journal of Biotechnology*, **92**, 113–131.

Naumann, T., Çebi, Z., Podola, B. and Melkonian, M. (2013) Growing microalgae as aquaculture feeds on twin-layers: a novel solid-state photobioreactor. *Journal of Applied Phycology*, **25**, 1413–1420.

Ojo, E.O., Auta, H., Baganz, F. and Lye, G.J. (2014) Engineering characterisation of a shaken, single-use photobioreactor for early stage microalgae cultivation using Chlorella sorokiniana. *Bioresource Technology*, **173**, 367–375.

Oswald, W.J. (1988) In: *Micro-algal Biotechnology*, Borowitzka, M.A. and Borowitzka, L.J. (eds), Cambridge: Cambridge University Press, pp. 305–328.

Oswald, W.J. (1995) *Water Science and Technology*, **31**, 1–8.

Park, J.B.K. *et al.* (2001) *Applied Microbiology and Biotechnology*, **57**, 287–293.

Park, J.B., Craggs, K.R.J. and Shilton, A.N. (2011) Wastewater treatment high rate algal ponds for biofuel production. *Bioresource Technology*, **102**, 35–42.

Posten, C. (2009) Design principles of photo-bioreactors for cultivation of microalgae. *Engineering in Life Sciences*, **9**, 165–177.

Pulz, O. (2001) *Applied Microbiology and Biotechnology*, **57**, 287–293.

Reef Central Online Community (2013) http://i54.tinypic.com/blkiq.jpg: www.reefcentral.com

Richmond, A. *et al.* (1993) *Journal of Applied Phycology*, **5**, 327–332.

Richmond, A. and Hu, Q. (2013) *Handbook of Microalgal Culture: Biotechnology and Applied Phycology*, 2nd edition. Wiley-Blackwell, Chichester, UK.

Right-light (2014) 70 Watt T8 http://www.right-light.co.uk/media/catalog/product/cache/1/base/700x/9df78eab33525d08d6e5fb8d27136e95/7/0/70-watt-t8-branded-fluorescent-tube-box-of-25-1800mm-6-foot.jpg: www.right-light.co.uk

Rodolfi, L., Chini Zittelli, G., Bassi, N., Padovani, G., Biondi, N. *et al.* (2009) Microalgae for oil: Strain selection, induction of lipid synthesis and outdoor mass cultivation in a low-cost photobioreactor. *Biotechnology and Bioengineering*, **102**, 100–112.

Sánchez Mirón, A. *et al.* (1999) *Journal of Biotechnology*, **70**, 249–270.

Schenk, P., Thomas-Hall, S., Stephens, E., Marx, U., Mussgnug, J. *et al.* (2008) Second generation biofuels: high-efficiency microalgae for biodiesel production. *BioEnergy Research*, **1**, 20–43.

Shah, Y., Kelkar, B.G., Godbole, S. and Deckwer, W.-D. (1982) Design parameters estimations for bubble column reactors. *American Institute for Chemical Engineers Journal*, **28**, 353–379.

Sheehan, J. *et al.* (1998) http://www.osti.gov/scitech/biblio/15003040-tW7nZs/native/

Shi, J., Podola, B. and Melkonian, M. (2014) Application of a prototype-scale twin-layer photobioreactor for effective N and P removal from different process stages of municipal wastewater by immobilized microalgae. *Bioresource Technology*, **154**, 260–266.

Singh, R.N. and Sharma, S. (2012) Development of suitable photobioreactor for algae production – A review. *Renewable and Sustainable Energy Reviews*, **16**, 2347–2353.

Sobczuk, T.M. *et al.* (2000) *Biotechnology and Bioengineering*, **67**, 465–475.

Soratana, K. and Landis, A.E. (2011) Evaluating industrial symbiosis and algae cultivation from a life cycle perspective. *Bioresource Technology*, **102**, 6892–6901.

Stephenson, A.L., Kazamia, E., Dennis, J.C.S., Howe, C.J., Scott, S.A. and Smith, A.G. (2010) Life-cycle assessment of potential algal biodiesel production in the united kingdom: a comparison of raceways and air-lift tubular bioreactors. *Energy & Fuels*, **24**, 4062–4077.

Sunkaier Ltd (2015) http://www.directindustry.com/prod/sunkaier-industrial-technology-co-ltd/mixer-impellers-rushton-turbine-radial-flow-132671-1567049.html:

Terry, K.L. and Raymond, L.P. (1985) *Enzyme and Microbial Technology*, **7**, 474–487.

Tredici, M.R. (2004) Mass production of microalgae: photobioreactors. In: *Handbook of Microalgal Culture: Biotechnology and Applied Phycology*, **1**, 178–214.

Ugwu, C.U., Aoyagi, H. and Uchiyama, H. (2008) Photobioreactors for mass cultivation of algae. *Bioresource Technology*, **99**, 4021–4028.

van Wagenen, J. *et al.* (2014) *Bioresource Technology*, **169**, 566–572.

Vunjak-Novakovic, G., Kim, Y., Wu, X., Berzin, I. and Merchuk. J.C. (2005) Air-lift bioreactors for algal growth on flue gas: mathematical modeling and pilot-plant studies. *Industrial & Engineering Chemistry Research*, **44**, 6154–6163.

Weissman, J.C. *et al.* (1988) *Biotechnology and Bioengineering*, **31**, 336–344.

13

Biofuels from Macroalgal Biomass

Jessica Adams

University of Aberystwyth, Wales, UK

Summary

Previous chapters have presented the issues that make algal biofuels difficult to implement economically, most notably the difficulties of culturing and harvesting large amounts of microalgae sufficiently cheaply, for conversion to biofuel. There exists another algal resource, however, that does not require fresh water, grows rapidly, does not compete for agricultural land and can be used for biofuel, namely seaweed or macroalgae. This chapter discusses the potential and challenges of using macroalgae for biofuel production through biochemical and thermochemical routes including, fermentation to ethanol and anaerobic digestion to methane.

13.1   Macroalgal Resources in the UK

The UK has a coastline in excess of 31,400 km and has an abundance of macroalgae (more commonly known as 'seaweed') growing in the littoral zone for the majority of its length. An estimated 10 million tonnes of seaweed surrounds the Scottish shores alone and it is therefore surprising that macroalgae have rarely been considered as a feedstock for biofuel production. Historically, macroalgae have been used for a number of purposes: as a food for coastal communities, as an animal feed, as a fertiliser on poor soils and as the main constituent in soda production for the glass and soap industries. The amounts of macroalgae harvested for these purposes was not always small – soda production peaked around 1800, when an estimated 400,000 tonnes of wet seaweed per year was burned in Scotland to produce the soda-rich ash, as testified today by the numerous ruins of historic lime-kilns scattered along the coastline. The collection and extraction of compounds from kelps (large, brown algae) has continued in Scotland until the present day, but with the closure of the final alginate extraction plant in 2008 the quantities of kelp processed now are small in comparison to their historic highs.

The term *macroalgae* represents a catch-all term for organisms that are similar in appearance, but not necessarily related. The similarity in structure is due to the selection pressures of living on the coast, and not necessarily to taxonomic relatedness. Generally

Biofuels and Bioenergy, First Edition. Edited by John Love and John A. Bryant.
© 2017 John Wiley & Sons Ltd. Published 2017 by John Wiley & Sons Ltd.

though, macroalgae grow rapidly and are capable of yielding more dry biomass per unit area per year than fast growing land-based crops such as *Saccharimum officinarum* (sugar cane). This extraordinary growth rate is due, in part, to seaweeds' ability to take up nutrients over their entire surface and, thanks to their buoyancy in water, the lack of closely packed cell walls and energy-intensive supporting tissues (*e.g.* cellulose and lignin) which are necessary for land plants. Macroalgae can be sustainably harvested from wild stocks, as is currently routine for commercial businesses collecting *Porphyra umbilicalis* (purple laver) from the west coast of the UK to produce the Welsh delicacy laver bread. However, if macroalgae were harvested for biofuel production, the scale of the required harvest could have a lasting, detrimental impact on the marine environment, especially if not rigorously regulated and monitored. Macroalgal cultivation is therefore a route that would allow the production of large quantities of a particular species or mix of algae. As the larger macroalgal species are benthic (*i.e.* attached to the sea-bed), they are only capable of growing off-shore as far as the light can penetrate through the water. In contrast, cultivated macroalgae can be grown on nets or strings along horizontal ropes attached to buoys, enabling cultivated macroalgae to be grown further off-shore than wild stock.

In addition to wild harvest or cultured macroalgae, a third source of macroalgae for biofuel production is that of cast algae – typically mixed macroalgal species that are washed up on beaches, especially following storms. Though a source of nutrients, food and shelter for a wide range of animal species, when large quantities of macroalgae are washed on to tourist beaches they are often mechanically cleared to provide a more 'tourist friendly' beach. Material cleared in this fashion can be used on fields and allotments as a fertiliser, but may also end up in landfill; alternatively it could be used as a source of biomass for fermentation to biofuel. A final source of macroalgal biomass is the process residues from commodities such as fertilisers or health and beauty products produced from imported macroalgae. Together, the sources of macroalgal biomass described above represent a sizeable biomass resource, though each with its particular merits and disadvantages (summarised in Table 13.1).

13.2   Suitability of Macroalgae for Biofuel Production

For large-scale biofuel production from macroalgae, the main considerations include the quantities produced by proposed species, the composition of the biomass material for conversion to biofuels and the ease of cultivation and harvesting.

The largest type of macroalgae growing off the UK coast are kelps, a generic term for brown, sub-littoral (*i.e.* rarely uncovered at low-tide) seaweeds in the order of the Laminariales. Such algae can often grow to 4 m or more in length. Kelps are characterised by a large flattened blade or frond (split into several strips in some species), a perennial stipe (stem) and a large, multi-branched holdfast attached to a solid base at or below the low-water line. Kelp forests in shallow, subtidal regions are amongst the most productive ecosystems on Earth, generating large amounts of organic carbon. In Nova Scotia, *Laminaria* beds produce 1.75 kg of organic carbon (mainly in the form of carbohydrate) per square metre per year (m^{-2} yr^{-1}). Elsewhere, an average of 1 kg of organic carbon m^{-2} yr^{-1} is more typical of kelp beds.

Table 13.1 Different sources of macroalgae for bioenergy generation.

Macroalgae growth conditions	Advantages	Disadvantages
Wild stock	• High densities of some species in known locations • No cultivation costs or inputs • Rapid growth of macroalgae means fast recovery of the harvested area	• Large-scale removal of particular species may lead to slow recovery of the area and increase the possibility of an invasive species establishment • Manual harvesting is often the only possible method of collection. This is time-consuming and costly
Cultivated macroalgae	• Cultivate desired species only • High densities produced • Can be grown further off-shore than seabed-bound wild stock • Can grow less competitive strains with reduced contamination, *e.g.* those expressing a high concentration of a particular compound	• Large capital costs are required for preparing strings, setting out at sea and harvesting • A monoculture can become rapidly invaded by a destructive species, *e.g.* an invertebrate infestation
Storm- and wave-cast macroalgae	• No cultivation or harvesting costs • In many areas seen as a 'waste' or 'problem', especially on tourist beaches, *e.g.* Blue Flag beaches • If material is to be removed, this typically occurs within a short timescale	• No guaranteed 'harvest' date or quantity thrown up • Mixed species present and possibly a mixed level of degradation • Collection time not guaranteed so further degradation may occur on-shore • Large-scale collection may affect the ecosystem, *e.g.* dune formation, algae-feeding insect populations and insect-feeding bird and mammal populations
Macroalgae product waste streams	• Compounds removed may be high-value extracts, leaving sugars and or alginates suitable for bioenergy generation • Cultivation (if conducted) and harvesting costs are already met • Companies are likely to be interested in increasing their product range, especially those from previous waste streams	• Chemicals added to aid extraction of desired compounds may have toxic or inhibitory effects on any subsequent micro-organisms for biological conversion to bioenergy • Added chemicals may also produce toxic gases or compounds, if thermochemical energy generation used • Companies may have compounds detectable in waste streams, which may reveal I.P. methodology

Kelps are mainly composed of water, typically between 75–90% of the wet weight, with the bulk of the dry matter consisting of the carbohydrates laminarin, mannitol and alginic acid (Figure 13.1). Laminarin is the main storage carbohydrate, a 1,3-linked glucan approximately 25 β-glucose molecules in length with occasional 1,6 linkages causing branches of additional glucose molecules. Using the enzyme

Carbohydrate	Laminarin	Mannitol	Alginic acids
% Dry weight	0–30%	4–25%	17–34%
Structure	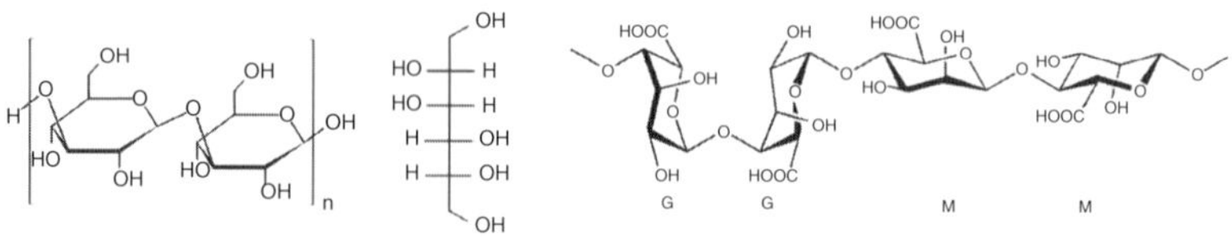		

Figure 13.1 Main carbohydrates in a typical kelp, *Laminaria hyperborean*.

laminarinase, laminarin is readily hydrolysed to single glucose molecules. Mannitol is the alcohol form of the sugar mannose. Mannitol is not readily utilised by the 'standard' brewing and baking yeast, *Saccharomyces cerevisiae*, because the yeast does not possess the enzyme mannitol dehydrogenase, which oxidises the mannitol to another simple sugar, fructose. This process also generates NADH from NAD^+, which needs a transhydrogenase or oxygen through the electron transport chain to regenerate NAD^+.

Alginic acid is a hydrocolloid, composed of D-mannuronic and L-guluronic acids that can occur either as one type only (poly-M or poly-G), or as a mix (poly-MG). The composition and therefore the properties of the alginic acid depend on the macroalgal species and on the location of the alginic acid within the alga itself. Poly-M chains, mainly found within the fronds, are more flexible and linear than the stipe-dominant poly-G, which form a more rigid folded structure. Salts have been formed with both M and G acids; of particular note is the production of a water-insoluble gel through the bonding of poly-G with calcium ions.

As might be expected, the composition of macroalgae differs throughout the year and these changes will have an effect on biofuel production. For example, the 'typical' composition of the kelp *Laminaria hyperborea* may be described as: 77–89% water, with the dried fraction containing 0–30% laminarin, 4–25% mannitol, 17–34% alginic acid, 4–14% protein and 10–11% fibre, with small proportions of fucoidan[1], fats and polyphenols. The peak yields of all components obviously do not all occur simultaneously, something first studied in the 1940s and 1950s by the Scottish Seaweed Research Association examining seasonal variation in the weight and chemical composition of common British kelps. A two-year study of the laminarin, mannitol, dry matter and ash content was determined in the three main UK kelp species – *L. cloustoni* (now *L. hyperborea*), *L. digitata* and *L. saccharina* (now *Saccharina latissima*). In the spring, mannitol was at a low concentration and laminarin was absent, while the ash was at a high concentration. In autumn, this trend was reversed to give high concentrations of laminarin and mannitol, with the mannitol proportion increasing and peaking earlier in the year than the laminarin. The concentration changes in this study are similar to those quoted above in all species examined; for example, the dry weight proportion of laminarin was 1–25% and mannitol concentration between 3% and 21% in *L. digitata*. With these varying concentrations in mind, it is logical to use macroalgae for biofuel production in the summer and autumn months, when the highest concentrations of utilisable sugars are available.

---

1 A generic term covering sulphated polysaccharides, in which the monosaccharide residues are mainly fucose or fucose and glucuronic acid

13.3   Biofuels from Macroalgae

13.3.1   Introduction

To date, the majority of UK-based research into biofuels from macroalgae has been conducted on kelps due to their prevalence, high yields, ease of harvest and/or culture and the fact that kelp is used by few other industries or processes. In other regions of the world, research into biofuels has considered red or green seaweeds, where they can be cultured to produce high yields or are otherwise available, for example as large quantities of cast algae.

Macroalgae can be biologically converted through fermentation or anaerobic digestion (AD) to produce useful, commercial biofuels including ethanol, butanol, methane and hydrogen. In addition, so-called 'platform' organic compounds including lactic and succinic acids can be produced by fermenting macroalgae. Thermochemical conversion methods such as pyrolysis can also be used to produce bio-oil, fractions of which may be used as a direct replacement of fossil fuel-derived diesel. Other energy generation routes could include gasification, combustion and hydrothermal liquefaction.

13.3.2   Ethanol from Laminarin, Mannitol and Alginate

As noted above, macroalgae contains storage carbohydrate molecules that can be degraded to sugars. Using these sugars, a range of products, including ethanol, may be produced through fermentation. Research into the use of macroalgae for bioethanol production does not appear in the literature prior to 2000, possibly due to the fact that the main ethanol-producing yeast, *S. cerevisiae*, does not typically have a transhydrogenase and so cannot ferment mannitol under anaerobic conditions. These conditions are required for ethanol production, so research work in this field began by examining a selection of bacteria and yeasts capable of both ethanol production and utilising mannitol under anaerobic conditions. From this, the yeast *Pichia angophorae* was shown to be the most suitable microorganism screened, as it used both the laminarin and the mannitol, operating under oxygen-limiting conditions to produce 0.43 g ethanol per g of *L. hyperborea*-derived mannitol and laminarin. Subsequent research on ethanol production from kelps included a pre-treatment study and an investigation into the optimisation of several different fermentation parameters, including the dry weight content of milled macroalgae within the fermentation slurry, pH, temperature, novel yeast species in addition to *P. angophorae* and the effect of additional laminarinase. Optimal conditions were then applied to *L. digitata* harvested through the year to study the effect of seasonal variation on bioethanol production. The laminarin and mannitol concentrations in the samples collected followed a similar seasonal trend to that previously published, but interestingly the ethanol followed the laminarin concentrations rather than those of the mannitol, suggesting a higher conversion efficiency or preference for laminarin as the substrate. The highest ethanol yields were seen in July (0.9% v/w), when the laminarin concentrations in the algae were highest, with lower yields occurring in June, despite a similar combined laminarin and mannitol concentration. The ethanol yield was lowest in March, when no laminarin was detected in the samples collected and there was minimal mannitol. Further research has been conducted subsequently to identify marine yeasts for bioethanol production, as species within collections have been shown to tolerate high salinity and operate across a broad pH range. Screening such collections for the ability to produce bioethanol from macroalgae opens the door to a new range of yeasts for bioethanol production.

Using separate saccharification (degradation of the complex polysaccharides to simple sugars) and fermentation (alcohol production) stages has produced higher ethanol yields, especially with the easily hydrolysable brown seaweed, *Alaria crassifolia* and a green species, *Ulva pertusa*. To raise the concentration of glucose released from the biomass, successive saccharifications were conducted on *A. crassifolia* and *U. pertusa* using the commercially available enzyme blend 'Meicelase', which contains a range of endo-glucanases and cellobiohydrogenases. The hydrolysate from the first saccharification was removed and used in the second saccharification, leading to a near two-fold increase in glucose content in the final hydrolysate compared to the initial hydrolysate. This ratio was lower for the subsequent production of ethanol, of × 1.4 (3.4% w/w) and × 1.6 (3.0% w/w) for *A. crassifolia* and *U. pertusa*, respectively, indicating some inhibitory compounds had built up in the hydrolysate. A second technique involved an acid pre-treatment prior to pH adjustment and enzyme saccharification. Following the acid pre-treatment, glucose concentrations were higher initially and produced ethanol yields of 2.6% w/w from the *A. crassifolia* fermentation and 2.8% w/w from *U. pertusa*; however, the successive saccharification method described above gave higher ethanol yields and required less energy input for these macroalgae than the acid pre-treatment, indicating a novel route for improving ethanol yields.

A number of studies have also been conducted on red seaweeds for bioethanol production. For example, work in South Korea considered the properties of 55 species of red seaweeds and identified *Kappaphycus alvarezii (cottonii)* as the most suitable species examined for ethanol production. *K. alvarezii* is one of the largest tropical carrageenophytes, is easily cultured, grows abundantly in aquaculture settings and had the highest proportion of utilisable carbohydrates of the species investigated. The main components in carrageenan are D-galactose-4-sulphate and 3,6-anhydro-D-galactose-2-sulphate, both of which can potentially be fermented. In contrast, the other main storage carbohydrate in red seaweeds is agar. Agar contains both D- and L-type galactose, but the L-type sugars are not usually fermentable and so can become reaction inhibitors if not converted to D-galactose first, reducing the conversion efficiency. Following dilute acid hydrolysis, *K. alvarezii* was fermented and ethanol was produced. The dilute acid used to hydrolyse the carrageenan to galactose also produced inhibitor compounds from the sugars, including furfural[2], hydroxymethylfurfural (HMF) and levulinic acid[3]. As these can reduce microbial processing and can limit ethanol production yields, the authors trialled the addition of activated charcoal to the broth following the acid pre-treatment. This reduced the HMF and levulinic acid by over 40%, but also removed similar proportions of the glucose and galactose present. Despite this, the highest ethanol concentration was 1.7 g L^{-1}, generated following inhibitor removal and was higher than if the inhibitors had remained.

An alternative route to increase the ethanol yield is through an improved conversion of the macroalgal biomass, enabling a larger proportion to be utilised by the ethanol-producing microorganisms. In addition to the laminarin and mannitol, alginic acid is the third major component of kelps, making up approximately 30% of the dry weight.

---

2 Furfural or Furan-2-carbaldehyde is a heterocyclic aldehyde.
3 Levulinic acid or 4-Oxopentanoic acid is a keto-acid that may be generated during the degradation of polysaccharides.

The degradation of alginate is complex due to the different proportions of the uronic acids in different parts of the alga and the changes in the solubility of the alginic acid caused by the binding to inorganic ions such as calcium. Due to the hydrocolloidal properties of the alginate, which increases the viscosity of the fermentation slurry, it is only possible to mix macroalgae fermentations at low percentages of dry solids (5–6% dry solids) compared with other ethanol feedstocks such as *Zea mays* (maize or corn; 25–28% dry solids). The hydrolysis and utilisation of alginic acid therefore provides a double benefit – first, by increasing the dry solids content, thereby raising the initial amount of water-soluble carbohydrate in the fermentation and second, by providing a higher proportion of available carbohydrates for conversion. A further advantage of utilising the alginate is that each unit of alginate fermented to ethanol consumes two units of reducing equivalents (NADH $\rightarrow$ NAD$^+$), whereas mannitol degradation actually generates reducing units (NAD$^+$ $\rightarrow$ NADH). Each of these imbalances of the redox environment can be problematic but when performed together, one can counterbalance the other, allowing ethanol fermentation to occur from both alginate and mannitol simultaneously.

These issues surrounding alginic acid utilisation to improve ethanol production viability are perceived as a major bottleneck in the use of macroalgae for bioethanol production. This restriction was addressed by a group at the Bio-Architecture Laboratories in Berkeley, CA, USA and published in *Science* in 2012[4]. The team identified a 36,000 base pair DNA fragment from the bacterium *Vibrio splendidus*. which contained genes coding for enzymes that control alginate transport and metabolism. The insertion of this DNA fragment with an engineered system for extracellular alginate depolymerisation into *E. coli* led to the production of a bacterial strain that was capable of alginate degradation, uptake and metabolism. A subsequent addition of DNA sequences coding for ethanol synthesis led to a yield of 0.28 g of ethanol per g of dry weight, approximately 80% of the theoretical ethanol yield from the macroalgae. This study demonstrated that all main carbohydrate components of the kelp feedstock, *Saccharina japonica*, were used by the bacteria and a concentration of up to 4.7% w/w ethanol was produced in this way.

A number of advantages may be generated by expressing this suite of alginate-degrading enzymes in *E. coli*, rather than using naturally occurring alginate-metabolising microbes. Limitations in microbial strains with alginate transport systems, such as low production rates or sensitivity in the pH or temperature of the fermentation, can be circumvented through the use of the well-characterised *E. coli* that naturally metabolises glucose and mannitol and has previously been shown to produce a range of different biofuels and platform chemicals. Such organisms may, in future, have a role in biofuel production. In addition, the ability to isolate and identify the function of specific genes can be invaluable in screens of potential alginate-degrading microbes.

13.3.3 Ethanol from Cellulose

A third route to ethanol production from macroalgae, used by a number of researchers, is to extract the most 'standard' (though not the most abundant) feedstock compound found within the macroalgae, namely cellulose. Cellulose is present at low concentrations in a range of seaweeds, potentially up to 10% of the dry weight (*e.g. L. hyperborea*),

---

4 Wargacki *et al.* (2012).

but more typically 3–5% dry weight (*e.g. Ascophyllum nodosum, U. pertusa*). Macroalgal cellulose typically displays a lower crystallinity than the cellulose from terrestrial plants. If extracted and concentrated, macroalgal cellulose could yield high glucose concentrations utilisable by efficient yeast species such as *S. cerevisiae*. Due to the large quantities of macroalgae required to produce cellulose-rich solids, reducing the material weight to approximately 20% of the former weight, research in this area has been conducted on low-cost, bulk sources of macroalgae. This includes waste material from edible seaweed production in Japan and large quantities of an invasive red algae in Hawaii. Water-soluble carbohydrates are removed, and a sodium carbonate (Na_2CO_3) wash is used to convert the insoluble Ca-bound alginate to its soluble Na-alginate form.

To aid non-glucan extraction, an acid pre-treatment is used, though inhibitors to the fermentation occur even when using extremely low acid (ELA) pre-treatments, for example, with 0.06% w/w sulphuric acid combined with a rapid heating and cooling system allowing a *Sacc. japonica* macroalgal solution to reach 170°C in under 1 min. Despite these rapid temperature controls, following retention at 170°C for 15 min, the concentrations of inhibitors generated using this method were such that they caused a significant decrease in enzyme activities. When the pre-treated macroalgae was used unwashed in simultaneous saccharification and fermentation, however, there was no significant reduction in ethanol yield. Despite the loss of glucose hydrolysed from laminarin and other soluble, utilisable sugars through the pre-treatment step, the simultaneous saccharification and fermentation with cellulase, β-glucosidase and *S. cerevisiae* produced 6.65 g ethanol per litre, under optimal conditions. Despite the pre-treatments, this concentration was achieved from material added at 6% dry solids; at higher concentrations, the *S. japonica* preparation became too viscous through the hygroscopic properties of retained alginate and thus the ethanol yield decreased. In a separate study focusing on the cellulose content of the red seaweed *Gelidium*, a stronger acid pre-treatment followed by retention of the hydrolysate meant galactose was released from the agar content in addition to glucose and 5.5% w/w ethanol was produced by the *S. cerevisiae* yeast. An economically viable concentration of ethanol is estimated to be between 4 and 5%, so these two reported studies demonstrate that though cellulose is a small fraction of the dry weight, if it is concentrated and few inhibitors are present, a viable concentration of ethanol is produced; indeed, this is as great or greater than that currently achieved using all the main carbohydrate components in the macroalgae.

13.3.4 Butanol

An alternative alcohol biofuel to ethanol is butanol, produced in conjunction with acetone and ethanol in a process denoted ABE (acetone, butanol, ethanol). Butanol can be used as a transport fuel in a similar manner to ethanol, but with a higher energy content, lower volatility and reduced vehicle corrosion compared to ethanol (see Chapter 2). ABE products are produced by *Clostridium* species, typically *C. acetobutylicum*. Butanol can be produced either directly though a one-step fermentation with acetone and ethanol or in a two-step process involving butyric acid as an intermediate. Continuous and batch fermentations have been used with continuous fermentations producing larger quantities of product than batch fermentations overall, through product removal and new substrate additions. However, the toxicity of butanol and butyric acid to the fermenting organisms is such that cell growth and butanol production are slowed over time, leading to a decreasing concentration of butanol being produced within the period of the continuous fermentation.

C. acetobutylicum has been shown to metabolise glucose, mannitol and water-soluble laminarin, with 70 g glucose or mannitol yielding approximately 10 g of butanol. The butanol to acetic acid ratio changes with the addition of different fermentation feedstocks; for example, the acetic acid concentrations occurring in mannitol fermentations are about three-fold those in glucose fermentations. Use of mannitol also shows an increased concentration of butyric acid. As discussed earlier, mannitol is an alcohol sugar that creates reducing units during its degradation. The conversion of acetic acid to butyric acid is postulated to be a disposal route of the additional hydrogen in nicotinamine adenine dinucleotide (NADH) generated from the mannitol with hydrogen gas production another route.

Butanol has been successfully produced from a range of macroalgal species, including the kelp *Saccharina* and green *Ulva* species. When using *Saccharina* sp. extracts, butanol concentrations were lower than that seen with the sugar equivalent provided as simple sugar feedstocks, suggesting inhibitory effects arising from additional components in the macroalgae.

A pre-treatment study for butanol production from *U. lactuca* included a hot water treatment followed by hydrolysis using cellulase. This caused the solubilisation of 90% sugars in the green algae, without increasing inhibitor concentrations which occurred if using an acid pre-treatment. The hydrolysate was used to produce butanol using *Clostridium acetobutylicum* and *C. beijerinckii*. As with the material from *Saccharina* sp., the *C. acetobutylicum* produced high concentrations of acetic and butyric acids with low butanol and ethanol concentrations. In this study, however, it appeared unable to utilise the rhamnose sugar which, with xylose and uronic acids, is a main constituent of the complex hydrocolloid Ulvan, a sulphated glucuronoxylorhamnan which constitutes 8–29% dry weight in *U. lactuca*. *C. beijerinckii* used half the available rhamnose and produced 3.0 g L^{-1} butanol from the hydrolysate or 0.35 g ABE per g of sugar. In contrast, the *C. acetobutylicum* produced just 0.8 g L^{-1} butanol, or 0.08 g ABE per g of sugar.

Butanol conversion rates and final concentrations remain low compared with ethanol production by *S. cerevisiae*, but as new, improved strains arise through selection or genetic manipulation, *Clostridium* sp. have the potential to improve yields in future, increasing the availability and use of butanol and butanol-derived compounds in the bio-economy.

13.3.5 Anaerobic Digestion

Anaerobic digestion (AD), which is discussed in detail in Chapter 3, occurs when a consortium of bacteria and other micro-organisms hydrolyse and degrade organic material to its constituent parts and ultimately to methane, CO_2 and traces of other gases in an anaerobic environment. (Bio)methane produced by AD can be cleaned or upgraded and used as a replacement for natural gas, providing fuel for cooking, heating, transport and electricity generation (see Chapter 3).

The feedstock used in AD usually has a high moisture content and is predominantly sourced from 'waste' materials, such as human sewage, animal manures and food waste. To increase the dry solids content of the slurry, the feedstock may also contain a proportion of biomass crops. The microbial inoculate for experimental AD is typically taken from an existing AD system (activated sludge), to ensure that an active anaerobic microbial population is present. The stability of AD is therefore based on the balance between the symbiotic growth of the main metabolic groups: the acid forming bacteria, obligate hydrogen-producing acetogens and the methanogens. This microcosmic balance is most easily achieved using a continuous, rather than batch processes, with regular addition of a standard feedstock. In practice it is rarely possible to closely control feedstock composition and availability, but the robustness of the microbial population within the anaerobic digester usually enables proportional adaptations and adjustments to occur. A co-digestion of *Laminaria* and *Ulva* species with milk demonstrated the feasibility of methane production by AD, with the microbial population adjusting to changes in the relative proportions of feedstock and macroalgal composition in the AD.

Much of the research on macroalgal AD has focused on *Ulva* species and in particular algae harvested from 'green tides', proliferations of green algae washed on to shores. An estimated 70,000 m^3 of macroalgae are deposited in this way on the shores of Brittany (northwest France) each summer. These cast algae have an offensive odour when decomposing, due to the high sulphur content of *Ulva* released as hydrogen sulphide, carbon disulphide and thiols, and are therefore an impediment to the tourist industry which forms a key economic sector in that region. Hydrogen sulphide (H_2S) is typically measured at 10,000–20,000 ppm in the biogas generated by AD, whether from *Ulva* alone or in a co-digestion with slurry. These levels of H_2S may present problems: H_2S is toxic, has a corrosive action on gas-processing equipment and following combustion will lead to the production of sulphur dioxide. These problems can be addressed by deploying appropriate 'scrubbing' technologies, but necessarily increase production costs. AD of *Ulva* alone is not highly productive, because of low methane yields and high hydrogen sulphide concentrations. Co-digestion with pig slurry did not lead to a disruption of the process, but neither did it boost methane yields. The hydrogen sulphide content in the co-digestion reached 3.5%, requiring treatment prior to the use of the methane. The addition of a sulphate reduction inhibitor, potassium molybdate, temporarily reduced hydrogen sulphide production, but the concentration rapidly increased in the slurry following each addition.

Other research was conducted on a similar green tide in Denmark, where it was concluded that the process of mechanically clearing the beaches and conducting AD on the material was potentially worthwhile. This was despite issues regarding efficiency and damage caused by a high elemental content in the AD facility from both the macroalgae and from large quantities of sand inadvertently collected during the clearing. There is a high nitrogen load off the Danish coast, so the removal of macroalgae in the study area would reduce the nitrogen yield by approximately 120 tonnes per year and is therefore seen as a beneficial process overall. In lower-nitrogen areas, this would clearly be less beneficial or even detrimental, which reminds us that no one biofuel process suits all situations and environments.

Even without the inclusion of sand, the elemental proportion of macroalgal biomass can be a concern for AD. Macroalgae bio-accumulate heavy metals, so using large proportions of macroalgae in ADs could consequently lead to high concentrations of heavy metals present in the digestate. This in turn could have negative effects when applied to farmland as a fertiliser, a common practice (see Chapter 3). In research on the AD of *Ulva* species, cadmium was identified as being of particular concern in macroalgae from the Baltic Sea. In the Baltic, water exchange is low and the sea contains relatively high concentrations of cadmium. In Sweden, cast algae from the Baltic is classified as a toxic waste due to the high concentrations of cadmium and its use as a bio-fertiliser is subsequently limited. The removal of heavy metals, including cadmium, copper, nickel and zinc from anaerobic digestate has been performed at a research scale, but the increased costs involved means this process is unlikely to be conducted commercially.

While downstream complications may limit the use of macroalgae in AD, the addition of small quantities of macroalgae to the anaerobic slurry could enhance digestion through provision of essential micronutrients to the micro-organisms in the consortium. This improvement in AD was observed following the addition of *Ulva* and the red seaweed *Gracilaria* to the anaerobic sludge from a waste-water treatment plant. Despite the rate-limiting step in the AD being calculated as the hydrolysis of the macroalgae, a synergistic effect was seen with the co-digestion of macroalgae and activated sludge. *Ulva* was found to be a better substrate than *Gracilaria*, as the methane production occurred at a higher rate; the methane potential and percent solubility of the macroalgae were the same. In a separate study, co-digestion of the kelp *S. latissima* with steam-exploded wheat straw improved methane yields, compared with those of the individual feedstocks. Lignocellulosic feedstocks alone are not efficiently digested to methane due to a high C:N ratio. Co-digestion of lignocellulosics with seaweed would therefore be potentially beneficial to the process.

13.3.6 Thermochemical Conversions

If macroalgae are to be considered for thermochemical conversions, two additional considerations must be made in addition to those for biological conversions. The high moisture content needs to be reduced, and the large ash content addressed, as the ash composition would lower melting points leading to the production of molten ash, fouling and corrosion in gas and combustion units and an increase in particle emissions. These challenges, in addition to those regarding the costs associated with harvesting, cultivation and transport, mean that unless high-value compounds can be generated in conjunction with biofuels, it would be unlikely that macroalgae could ever be a viable option for thermochemical conversion.

13.4 Future Prospects

This chapter has focused on the three main biofuel products currently produced from macroalgae, but many more have the potential to be produced commercially. As with micro-algal biofuels (Chapters 11 and 12), the synthesis of high-value or bulk chemicals from macroalgal feedstocks may be a stepping stone to full-scale biofuel production. These platform (bulk) chemicals can be transformed into a wide range of compounds, including food and drink ingredients, plastics and pharmaceuticals. Other chemicals may be generated as by-products from existing pathways. For example, *Clostridium acetobutylicum* does not ferment green macroalgae such as *U. lactuca* to butanol with a high level of conversion. It does, however, anaerobically ferment the deoxysugar rhamnose to the commercially interesting 1,2 propanediol (propylene glycol). 1,2 Propanediol is a commodity chemical widely used in pharmaceutical manufacturing as a vehicle for delivering drugs that are insoluble or unstable in water. It can also be used as a stabilising agent, plasticiser or preservative, predominantly in the food and cosmetics industries. Exploring routes like this one may not lead to the traditional biofuels of ethanol and methane but as the availability of cheap, accessible oil decreases, as it inevitably will, the demand for a range of oil-replacement compounds of a similar quality and cost will increase, providing a wealth of opportunities for biological feedstocks, including macroalgae.

13.5   Conclusion

Current research shows that ethanol can be produced above 5% in macroalgal fermentations, making it economically viable to produce as a biofuel or as an intermediate for a number of other products. AD, too, may be an economically viable option of processing large quantities of cast (*i.e.* free) algae, despite the high hydrogen sulphide production and sand inclusion in the digestion. In addition to these positive reports, lower yielding or high-energy production biofuels may also be viable if a multi-product approach is taken. In order to maximise the value derived from any macroalgal species, future production of biofuels from macroalgae should occur on material cultivated or harvested for higher value compounds within it, whether this be a bioactive molecule or a hydrocolloid such as alginate or agar. Ideally the initial extraction process would take the subsequent production of the biofuels into consideration, so inhibitory compounds included or induced in earlier processes are minimised. In this way, material which would otherwise be perceived as a 'waste' becomes a valuable commodity, used to generate financially viable concentrations of biofuel products. The use of the hydrocolloid as a feedstock for biofuel production is positive, providing alternative routes for conversion if the more valuable hydrocolloid extraction processes are not possible. If 30% of the macroalgal biomass is agar but due to the quality of the material available cannot be removed, then the utilisation for biofuel is worthwhile. It is important to remember, however, that the values of bulk hydrocolloids can be over 20-fold the price of bulk ethanol, with speciality hydrocolloids fetching higher values again, so the intention to convert the hydrocolloids to ethanol should be approached with a solid business plan.

Acknowledgements

The writing of this chapter was supported by the European Regional Development Funding through the Welsh Government (BEACON 80561) and the Biotechnology and Biological Sciences Research Council through the Bioenergy and Biorenewables ISPG (BBS/E/W/10963A01).

Selected References and Suggestions for Further Reading

Adams, J.M.M., Ross, A.B., Anastasakis, K., Hodgson, E.M., Gallagher, J.A. *et al.* (2011a) Seasonal variation in the chemical composition of the bioenergy feedstock *Laminaria digitata* for thermochemical conversion. *Bioresource Technology*, **102**, 226–234.

Adams, J.M.M., Toop, T.A., Donnison, I.S. and Gallagher, J.A. (2011b) Seasonal variation in *Laminaria digitata* and its impact on biochemical conversion routes to biofuels. *Bioresource Technology*, **102**, 9976–9984.

Black, W. (1950) The seasonal variation in weight and chemical composition of the common British laminariaceae. *Journal of the Marine Biological Association of the United Kingdom*, **29**, 45–72.

Horn, S.J., Aasen, I.M. and Ostgaard, K. (2000a) Production of ethanol from mannitol by *Zymobacter palmae*. *Journal of Industrial Microbiology & Biotechnology*, **24**, 51–57.

Horn, S.J., Aasen, I.M. and Ostgaard, K. (2000b) Ethanol production from seaweed extract. *Journal of Industrial Microbiology & Biotechnology*, **25**, 249–254.

Horn, S.J. (2009) *Seaweed Biofuels: Production of Biogas and Bioethanol from Brown Macroalgae*. VDM Verlag, Saarbrucken, Germany.

Hurd, C.L., Harrison, P.J., Bischof, K. and Lobban, C.S. (2014) *Seaweed Ecology and Physiology*, 2nd edition. Cambridge University Press, Cambridge, UK.

van der Wald, H., Sperber, B.L.H.M., Houweling-Tan, B., Bakker, R.R.C., Brandenburg, W. and Lopez-Contreras, A.M. (2013) Production of acetone, butanol and ethanol from biomass of the green seaweed *Ulva lactuca*. *Bioresource Technology*, **128**, 431–437.

Wargacki, A.., Leonard, E., Win, M., Regitsky, D.D., Santos, C.N.S. *et al.* (2012) An engineered microbial platform for direct biofuel production from brown macroalgae. *Science*, **335**, 308–313.

Yanagisawa, M., Nakamura, K., Ariga, O. and Nakasaki, K. (2011) Production of high concentrations of bioethanol from seaweeds that contain easily hydrolyzable polysaccharides. *Process Biochemistry*, **46**, 2111–2116.

14

Lipid-based Biofuels from Oleaginous Microbes

Lisa A. Sargeant, Rhodri W. Jenkins and Christopher J. Chuck

Centre for Sustainable Chemical Technologies, University of Bath, Bath, UK

Summary

Oleaginous microbes, including autotrophic microalgae or heterotrophic yeasts can, under certain conditions, produce copious quantities of oil. This chapter reviews the potential of 'single cell oils' from natural microorganisms as sources of 'third-generation' biofuels. Algae have already been discussed, so this chapter will focus largely on the growth, lipid production and processing of oleaginous yeast for biodiesel.

14.1   Introduction

The current method of lipid production for first-generation biofuels using feedstocks such as oil crops, waste cooking oil and animal fats is insufficient to satisfy the future demand for the fuel and chemical industries (Chapters 1, 6 and 16). Furthermore, first-generation feedstocks from edible oils such as rapeseed or canola (*Brassica napus*), soybean (*Glycine max*) and oil palm (*Elaeis spp.*) have attracted considerable public criticism due to concerns surrounding arable land use and food security (see Chapters 6, 8, 16 and 17). The development of second-generation feedstocks from non-edible plant-based resources such as Jatropha (*Jatropha spp.*) offers potential as a sustainable alternative (Chapter 6), but requires heavy regulation so that their culture does not compete with edible food crops.

To be a suitable feedstock for future fuels, the lipids produced must be economically viable compared to existing fuels and require minimal land use. Lipid production for fuel should also demonstrate superior environmental benefits, whilst maintaining suitable chemical properties for their application. Foremost, the lipids must be capable of being produced in substantial quantities to make a significant contribution to addressing the demand for fuel[1]. To meet these requirements, oleaginous micro-organisms (those that can produce fats and oils) have been considered as potential alternatives to lipids produced by plants. Oleaginous micro-organisms, which produce more than 20%

---

1 Meng *et al.* (2009).

Biofuels and Bioenergy, First Edition. Edited by John Love and John A. Bryant.
© 2017 John Wiley & Sons Ltd. Published 2017 by John Wiley & Sons Ltd.

(w/w) of microbial lipid in their dry cell mass, are also referred to as single-cell oils (SCOs)[2]. Many micro-organisms can produce abundant lipids, including microalgae, bacteria, fungi and yeast. Compared to other vegetable oils, the production of microbial oil has many benefits. The biomass doubling time in exponential growth can be as short as 40 minutes and the oil content in oleaginous microbes can exceed 80% by weight of dry biomass, meaning that oil productivities can be considerably higher than traditional lipid production methods from plants (see Chapter 6). The cultivation of SCOs is also less labour-intensive, less affected by location, season and climate, and is easier to scale up than for terrestrial crops[3]. For example, it has been estimated that the land area needed to produce the same volume of transportation fuel is 180 times greater for soybean and 40 times greater for Jatropha than for microalgae, assuming the microalgae are grown in a photobioreactor (see Chapter 12) to a concentration of 4 kg m^{-3} and an oil content of 50% of dry weight[4]. However, to date, only a restricted number of micro-algal species and a small number of oleaginous yeast species have been investigated as potential sources of biofuel-compatible oils.

The majority of microbial fuel research has been to develop first-generation biodiesel, whereby the glycerides are converted into fatty acid alkyl esters (FAAE, including fatty acid methyl esters, FAME (see Chapters 2 and 6), by a base or acid catalysed transesterification[5]. While biodiesel is compatible with the diesel fuel infrastructure, a low oxidative stability and poor low temperature properties limit its use both in the EU (up to a level of 7%), and in the USA (of up to 5%) (see Chapter 2). The properties of biodiesel are largely reliant on the carbon chain length and saturation of the FAAEs present in the fuel. The common nomenclature used to describe the structure of the fatty acid (FA) chains is to follow the number of carbons in the chain by the number of double bonds present, separated by a colon (as also described in Chapter 6). Therefore, linoleic acid, which consists of a C_{18} chain with one double bond, is commonly referred to as 'C18:1'. Generally, saturates have excellent oxidative stability and combustion quality, but poor low temperature properties, whereas polyunsaturates have very poor oxidative stability but still flow at sub-zero temperatures[6].

Recently, alternative production mechanisms for biodiesels have been developed. For example, hydroprocessing offers a route to alkanes which can be subsequently isomerised, or 'cracked', into suitable hydrocarbon fuels. The process comprises of two stages. In the first step, the triglycerides are deoxygenated under hydrotreating conditions (up to 40 bar H_2, 200°C) before isomerisation, or 'cracking', to improve the cold-flow properties of the fuel[7]. Hydrotreating catalysts are inhibited or poisoned by contaminants such as alkali metals and phosphorous compounds as well as impurities such as waxes, sterols, tocopherols and carotenoids, which are often found in natural oils and fats, so additional pre-treatment of the feedstock may therefore be required. This pre-treatment step is particularly problematic for lipids derived from microbial sources. Hydroprocessing is a more energetically demanding reaction than transesterification,

---

2 Papanikolaou and Aggelis (2011).
3 Subramaniam *et al.* (2010).
4 Allen (2013).
5 Knothe *et al.* (2005).
6 Knothe (2005); Chuck *et al.* (2012).
7 Knothe (2010).

though the resulting fuels are more compatible with current diesel fossil fuels and can therefore be used at far higher blend levels or for aviation.

14.2 Microalgae

The majority of work into microbial lipid production has focused on photoautotrophic microalgae. Microalgae are unicellular, photosynthetic organisms that, when simply viewed through the 'fuel prism', convert CO_2 into fermentable biomass and lipids (see also Chapters 12 and 16). Microalgae also require a nitrogen source as well as a variety of trace elements including P, K, Mg and Fe. Microalgae grow in an aquatic environment, but can thrive under a wide range of conditions. These can include areas of high salinity (*e.g. Dunaliella salina*) or in ponds located on arid land that is unsuitable for agricultural purposes[9]. Algal lipids have mainly been researched for the production of biodiesel, with many reports and articles assessing the viability of biodiesel production in comparison with other available feedstocks.[8]

For autotrophic growth, microalgae require water, nitrogen, phosphorous, CO_2 and light. These nutrients can be obtained from waste streams such as wastewater and flue gases. Wastewater is often high in nitrogen (ammonia, nitrates and nitrites) and phosphorous (phosphates) which, if managed incorrectly, can be the major cause of ecological eutrophication. Therefore, the use of wastewater not only provides an inexpensive growth medium for algal growth, it has dual potential for the treatment of effluent streams[9].

However, producing lipid from CO_2 *via* microalgae is not a simple process. CO_2 fixation in microalgae is achieved through the production of carbohydrate (sugar) molecules, driven by solar energy being converted into ATP. Carbohydrates are then processed into lipid *via* the production of acetyl-CoA, which is channelled into the FA biosynthesis pathway, which we describe below. Enhanced lipid production can be activated by a number of factors, notably metabolic stress, most commonly nitrogen depletion. The transformation of carbohydrate into oil is, however, energetically unfavourable compared to the storage of the carbohydrate as polysaccharide. Many algal species therefore preferentially store energy as polysaccharides rather than as lipids[10].

Because of their ability to fix CO_2 using only the energy from the Sun, a large amount of research has been invested in microalgae, with the belief that oil could be produced cheaply from 'free' resources. However, while the production of algal oils and the subsequent conversion into usable fuels has been demonstrated repeatedly[11], the large-scale commercialisation of this technology for economically viable biofuel production remains elusive (Chapters 2, 6 and 16). As explained in other chapters, it can be difficult to achieve high biomass concentrations and oil productivities from microalgae, because of a number of factors, including light limitation and oxygen accumulation in photoautotrophic cultures[12]. Furthermore, while the CO_2 levels in the atmosphere may be

---

8 Singh and Olsen (2011).
9 Abe *et al.* (2002).
10 Ratledge and Cohen (2008).
11 Chisti and Yan (2011); Singh and Dhar (2011).
12 da Silva *et al.* (2010).

rising, at current atmospheric concentrations of 400 ppm, CO_2 remains a limiting factor for large-scale algal culture. Therefore, the unsustainable demands on fresh water and fertilisers, insufficiency of low-cost concentrated CO_2 and high energy requirements for growing microalgae limit the commercialisation potential for biofuel production[13].

A comparison of 15 published articles based on the life cycle analysis (see Chapters 12 and 16) of oil production from microalgae, led to the conclusion that only high microalgal productivity ($34–50 \, \mathrm{g \, m^{-2} \, day^{-1}}$) combined with extracting the oil from wet biomass would yield an economically favourable process[14]. In order to achieve these yields, heterotrophic growth using organic carbon as both an energy and carbon source has received considerable attention recently. This utilises fermentation technology in the form of stirred tank reactors; mixotrophic growth, using CO_2 and organic carbon in the presence of light is also possible.

Various species of microalgae have been reported to be able to grow heterotrophically (*i.e.* consuming sugar or other organic molecules that are added to the culture, rather than making their own through photosynthesis), including *Chlorella spp., Haematococcus spp.* and *Chlamydomonas reinhardtii*. The carbon sources for heterotrophic growth have included sugars, acetate, glycerol, hydrolysed carbohydrates and molasses[15]. Carbon-rich wastewater streams have also been used so as to remediate polluted water while producing lipid[16]. When grown in a fed-batch culture, the cell density of *Chlorella protothecoides* reached $52 \, \mathrm{g \, l^{-1}}$ in 167 hours[17]. Similarly, fermentations with *Chlorella sorokiniana* using glucose yielded biomass and lipid concentrations of $103.8 \, \mathrm{g \, l^{-1}}$ and $40.2 \, \mathrm{g \, l^{-1}}$, respectively[18]. These yields are significantly higher than those achieved for photoautotrophic growth. For example, the microalga *Chlorella zofingiensis* grown heterotrophically yielded a cell density of $10.1 \, \mathrm{g \, l^{-1}}$ compared to $1.9 \, \mathrm{g \, l^{-1}}$ under photoautotrophic growth conditions[19]. For this reason, heterotrophic growth has been employed as an industrially viable method for the production of lower value compounds. By growing microalgae heterotrophically, the North American biotechnology company, Solazyme, are producing drop-in fuel replacements for the road transport (Soladiesel$_{RD}$®), marine (Soladiesel$_{HRF-76}$®) and aviation (Solajet®) sectors from microalgae[20].

The profile of the lipids produced by microalgae can change when cultured as autotrophs or as heterotrophs. Lower levels of C16:0, C16:3, C18:0 and C18:3, but an increased proportion of C18:1, were observed when *C. zofingiensis* was cultured heterotrophically instead of autotrophically[21]. This shift in lipid content is presumably due to differences in the metabolic flux of the sugars when light is present. In the presence of glucose and under heterotrophic conditions, the glucose is directed into the pentose phosphate pathway (PPP), whereas in the presence of light, the Embden-Meyerhof-Parnas (EMP) pathway serves as the major flux of glucose.

---

13 Klein-Marcuschamer *et al.* (2013).
14 Sills *et al.* (2012).
15 Lui *et al.* (2013).
16 An *et al.* (2003).
17 Xiong *et al.* (2008).
18 Zheng *et al.* (2013).
19 Lui *et al.* (2007).
20 Solazyme.solazyme.com/solutions/fuel
21 Liu *et al.* (2011).

While considerably more productive than photoautotrophic growth, the hetero-trophic growth of microalgae still has several limitations. First, most microalgae are photosynthetic organisms, and thus the number of species capable of heterotrophic metabolism, without genetic modification, is limited. Microalgae also have a broad lipid profile, which while desirable in niche applications, could hinder economic develop-ment due to variations in the feedstock. For algal biodiesel to offer a competitive alter-native, the lipids must cost less than US$ 0.74 l^{-1}, irrespective of the type of fuel to be produced[22]. Although estimates on the cost of algal lipid vary substantially, depending on the method of production, species and productivity, there is a general acceptance that the economics surrounding the cost of the production of algal oil for biofuel alone is currently uneconomical.

14.3   Oleaginous Yeasts

When growing microalgae heterotrophically, the organisms behave similarly to yeast in terms of economics and cultivation methods. However, yeasts exhibit several advan-tages over microalgal feedstocks for lipid production. Yeasts are fully heterotrophic organisms and, as such, all of their energy for cellular respiration and division is acquired from an organic carbon source. Yeast cells do not require light, have shorter doubling times, and can reach much higher cell densities (10–100 g l^{-1} in 3–7 days)[24]. Yeasts are also less susceptible to viral infection than some alternative microbial cultures; it is also possible to control bacterial contamination by using low pH growth conditions. In addi-tion to oils, yeasts produce a range of metabolic co-products, including biologically-derived surfactants and high value carotenoids. This aspect is especially important when proposing to produce relatively low value, commodity products such as fuels, as these co-products are typically rarer and more economically attractive as an adjunct to a potential bio-refinery.

Although there are over 1,600 described yeast species, only about 70 are known to be oleaginous[23]. Oleaginous yeasts are capable of accumulating intracellular lipids of up to 80% of their dry biomass weight in discreet, intracellular lipid bodies. The quantity of oil accumulated by any given species of oleaginous yeast can vary dramatically, depend-ing on the culture conditions. The carbon (typically sugar) to nitrogen (C/N) ratio of the culture broth, temperature, pH, oxygen and concentration of trace elements and inorganic salts all influence lipid productivity[24]. It is therefore crucial to define the culture medium of feedstock that promotes optimal growth and lipid yield.

14.4   Feedstocks for Heterotrophic Microbial Cultivation

Using a heterotroph for SCO production requires a carbohydrate source. Due to the relatively low value of the oil-based product, the carbohydrate feedstock must be as cheap as possible to enable economic viability. Highly refined sugars such as glucose

---

22 Ribeiro and da Silva (2013).
23 Sitepu *et al.* (2014).
24 Li (2008).

cost around US$ 550 per tonne and thus are too expensive for industrial application. The crude sap from sugar-rich plant crops such as sugar cane (*Saccharimum offici-narum*) or sugar beet (*Beta vulgaris* subsp. *altissima*) can be used to support yeast growth, but these uses compete with food-crop growth. Many efforts have therefore focused on using low-cost waste materials as a growth medium for SCO production. Waste substrates include banana (*Musa*) juice, glycerol, molasses, whey and wastewater from monosodium glutamate and olive oil mill processing[25].

More recently, household food waste has also attracted attention as a waste feedstock for lipid production[26] (see also Chapter 3). Second-generation lignocellulosic (plant material) feedstocks such as *Miscanthus*, switchgrass (*Panicum virgatum*), sugarcane bagasse and wheat straw have also attracted interest due to the high carbohydrate content contained within the cellulose and hemicellulose polysaccharides (see Chapters 4, 5 and 7). However, because yeast lack a significant cellulolytic activity (*i.e.* they are unable to degrade lignocellulose), the lignocellulosic material requires pre-treatment and hydrolysis to release significant quantities of monomeric sugars. A number of biological, physical, chemical and physico-chemical pre-treatment methods have been developed to pre-treat lignocellulose (Chapter 5). The pre-treatment regime must be tailored towards the lignocellulose feedstock and the target yeast strain as different products are generated from each regime. Once the pre-treatment has opened up the structure of the polysaccharides and degraded the crystallinity of the cellulose fibrils, enzymes such as cellulase can be added to cleave the polysaccharides into oligo-, di- and mono-saccharides.

One of the most important parameters in the industrial production of lipid from oleaginous yeasts is the lipid co-efficient. This is the percentage of lipid produced per unit of sugar input. For oleaginous yeast cultured on glucose, the maximum lipid yield is 22.4 g of oil from 100 g of glucose. If all the glucose were directed towards lipid biosynthesis, the maximum theoretical yield would be 33% (w/w); however, some of the glucose is also used for cellular metabolism and biomass production. As such, 22% (w/w) is the probable maximum productivity of oil from glucose. Consequently, in the best case scenario, 5 tonnes of sugar is required for the production of 1 tonne of oil. As the carbon source is also used for the synthesis of the cell biomass, there is a fine balance between oil production and the production of oil-free biomass. It has been suggested that a lipid content of 40% (w/w) is probably the optimum value to maximise the overall biomass yield, with anything over this value reflecting a reduction in cellular biomass rather than an increase in cellular lipid[27].

14.5   The Biochemical Process of Lipid Accumulation in Oleaginous Yeast

As briefly mentioned previously, environmental growth conditions play a large part in the amount of lipid obtained from oleaginous yeast. Lipid accumulation generally occurs within a micro-organism when it exhausts a nutrient from the growth medium

---

25 Huang *et al.* (2013).
26 Uckun Kiran *et al.* (2014).
27 Ratledge and Cohen (2008).

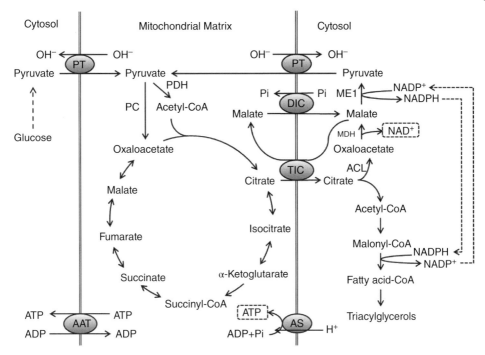

Figure 14.1 Triglyceride biosynthesis pathway from glucose to triglycerides in oleaginous yeast. AAT: ATP-ADP translocase; ACL: ATP-citrate lyase; AS: ATP synthase; DIC: dicarboxylate carrier; MDH: malate dehydrogenase; ME1: malic enzyme; PC: pyruvate carboxylase; PDH: pyruvate dehydrogenase; PT: pyruvate transporter; TIC: tricarboxylate carrier (adapted from Ratledge *et al.*, 1992 and Guay *et al.*, 2007).[28]

needed for sustained growth. This nutrient is often nitrogen, required for DNA replication, RNA synthesis and protein synthesis; thus cell proliferation can no longer occur if it is depleted. Meanwhile, if there is an excess of carbon, such as glucose, in the culture medium, this will be assimilated by the cells and converted into storage lipids, which leads to the accumulation of lipid in suitable vacuoles. Exceptions to this rule-of-thumb exist, for example the yeast *Cryptococcus terricola* is able to accumulate lipid during the exponential growth phase before nitrogen depletion. Factors such as pH and temperature have also been reported to have an effect on the lipid accumulation.

Work undertaken with *Saccharomyces cerevisiae* established the general biosynthesis pathway for FA production in yeasts. However, there are a number of key distinctions between oleaginous and non-oleaginous yeast species. For example, oleaginous micro-organisms produce significantly higher quantities of the key intermediate, acetyl-CoA compared to non-oleaginous micro-organisms (see Figure 14.1).

When glucose is used as the primary energy source, it enters the glycolysis pathway where it is converted into pyruvate through a multistep process yielding two molecules of pyruvate per glucose molecule. The pyruvate molecule can then enter the mitochondrion and into the tricarboxylic acid (TCA) cycle by its carboxylation by pyruvate

28 Ratledge (1992); Guay et al. (2007).

carboxylase (TC) to oxaloactetate or the conversion to acetyl-CoA by pyruvate dehydrogenase (PDH). This leads to the accumulation of citrate within the mitochondrion, which at excess levels is exported from the mitochondria into the cytosol by the tricarboxylate carrier (TIC) in exchange with malate. The citrate is then cleaved by ATP-citrate lyase in the cytosol to yield oxaloacetate and acetyl-CoA.

While some species of oleaginous yeast can produce intracellular lipid continuously, most produce lipid as a stress response induced by environmental change. As mentioned earlier, nutrient limitation has been widely investigated as a route to enhance lipid productivity. Most commonly, nitrogen limitation is used as the trigger for enhanced lipid biosynthesis, though other substrate limitations have been reported. In the case of nitrogen limitation, cell division, as well as protein and nucleic acid synthesis, all cease once the nitrogen has been depleted from the culture medium. The intracellular concentration of adenosine monophosphate (AMP) rapidly decreases upon nitrogen exhaustion, as the AMP is converted into IMP (inosine-mono-phosphate) and NH_4^+ by AMP-desaminase. The release of NH_4^+ ions provides a temporary nitrogen source for DNA replication and protein synthesis[29]. AMP allosterically activates the mitochondrial isocitrate dehydrogenase, which is responsible for the conversion of isocitric acid to α-ketoglutaric acid, and as such, a decrease in AMP leads to intracellular accumulation of isocitric acid. Isocitric acid is found in equilibrium with citrate. These elevated levels of citrate within the mitochondrion result in its export into the cytoplasm in exchange with malate, where it is cleaved into acetyl-CoA.

ATP-citrate lyase (ATP-CL), a multi-enzyme complex only found in oleaginous microbial strains, catalyses the formation of acetyl-CoA and oxaloacetate from citrate. The presence of ATP-CL can be used as a biological marker for oleaginicity. In micro-organisms where ATP-CL is not present, the excess intracellular citric acid caused by nitrogen limitation is generally excreted from the cell into the culture medium. This method of citric acid production has been observed in *Aspergillus niger* and *Candida sp*[30] and has led to yeasts (and other fungi) being classified as either lipid-accumulating or citric acid-producing. In the oleaginous strains, the acetyl-CoA continues in the pathway of lipogenesis, whereas the remaining oxaloacetate in the cytosol is converted to pyruvate, *via* malate by a malic enzyme, which can then re-enter the TCA cycle and generate NADPH. The NADPH is essential for the elongation of the alkyl chain of the FA, and thus malic enzyme controls the level of lipid accumulation[31]. Similarly to ATP-citrate lyase, the malic enzyme is key to providing the oleagenicity of a micro-organism.

Once acetyl-CoA has been generated in the cytoplasm of the cell, it is carboxylated to form malonyl-CoA in an irreversible reaction by the enzyme, acetyl-CoA-carboxylase (ACC), requiring ATP and HCO_3. From this point onwards, the lipid biosynthesis pathway, located in the cytoplasm, is essentially a reduction of malonyl-CoA units to form the alkyl chain of the triglyceride by a multi-enzyme complex known as fatty-acid synthase (FAS). This process yields C16 FA molecules, which are released from FAS by a thioesterase. These free FAs are activated to Coenzyme A to yield palmitoyl-CoA, which are often further elongated to stearoyl-CoA (C18). These are then esterified to glycerol by the α-glycerol phosphate acylation pathway (see also Chapters 6 and 11).

---

29 Evans and Ratledge (1985).
30 Ratledge (1994).
31 Ageitos *et al.* (2011).

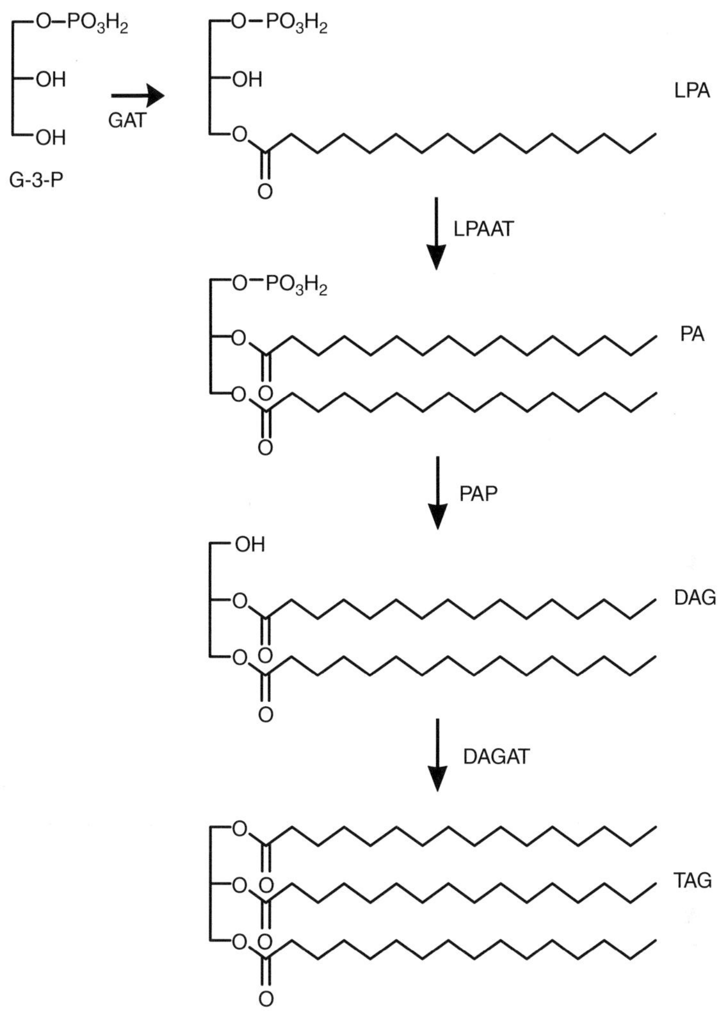

Figure 14.2 Schematic of intracellular triglyceride synthesis *via* the α-glycerol phosphate acylation pathway. DAG: Diacylglycerol; DAGAT: Diacylglycerol acyltransferase; G-3-P: Glycerol-3-phosphate; GAT: G-3-P acyltransferase; lpa: lysophosphatidic acid; LPAAT: Lysophosphatidic acid acyltransferase; PA: Phosphatidic acid; PAP: Phosphatidic acid acyltransferase; TAG: Triacylglycerol[32]. See also Chapters 6 and 14.

In the α-glycerol phosphate acylation pathway (Figure 14.2), the activated acyl-CoA is subsequently used for the acylation of glycerol-3-phosphate (3-G-P). Acylation first occurs at the sn1 position, by G-3-P acyltransferase (GAT), to produce lysophosphatic acid (LA) before a second acylation at the sn2 position to yield phosphatic acid (PA). Following dephosphorylation by phosphatic acid phosphohydrolase (PAP) and subsequent acylation by diacylglycerol acyltransferase (DAGT), triglyceride is produced.

_____

32 Adapted from Papanikolaou and Aggelis (2011).

This final enzymatic step has been established as an important rate limiting step in lipid accumulation, since its over-expression in *S. cerevisiae* led to a three- to nine-fold increase in triglyceride production[33]. Similarly, over-expression of a thioesterase can deregulate the FAs, resulting in the over-accumulation of FAs[34].

While triglycerides are the major component of neutral lipids within SCOs, the intermediates of this pathway may be hydrolysed to produce di- and mono-glycerides, and occasionally free FAs. FA elongation and desaturation occur in the microsomal membrane. Desaturation is initiated by the introduction of a double bond at the Δ9 position of saturated FAs by a Δ9-desaturase. Most yeasts contain this desaturase, which is capable of producing palmitoleic (C16:1) and oleic (C18:1) acids. Some yeasts also contain two further desaturase enzymes, namely Δ12-desaturase and Δ15-desaturase, which can introduce a second and third double bond, respectively; however, the activity of these enzymes is species-specific.

Interestingly, the FA composition of the triglyceride is not randomly distributed on the glycerol backbone, but they are stereospecifically orientated. More than 65% of the FAs at the sn-1 position are C16:0 and C18:1, whereas C18:1, C18:2 and C18:3 are most commonly found at the sn-2 position. The greatest variation in the FA localisation is found at the sn-3 position, which tends to be occupied by C16:0, C18:2 and by the greater proportion of C18:3 in the triglyceride. The positional FA distribution in yeast TAGs tends to the types SUS and SUU, when S and U represent saturated and unsaturated FAs respectively. The lack of saturated FAs in the sn-2 position of the glycerol backbone is also observed in most plant oils.

14.6   Lipid Profile of Oleaginous Microbes

Biodiesel produced from terrestrial crops is relatively simple, containing C16–C20, FAs, with the majority being palmitic acid (C16:0), stearic acid (C18:0), oleic acid (C18:1), linoleic acid (C18:2), linolenic acid (C18:3) and erucic acid (C20:0) (see Chapter 6). Microalgal lipids, however, are more variable and are dependent on the species of algae and the culture conditions. C6–C24 chains are common, as are highly unsaturated esters with up to six double bonds. Due to this variation, it seems likely that the majority of microalgal biodiesel would not meet current biodiesel fuel standards, though it would be suitable for hydrotreatment (treatment with hydrogen under pressure[35].

Similarly to plant storage oils, microbial oils are mainly composed of triglycerides, but can also consist of free FAs, other neutral lipids (such as mono- and di-glycerides and steryl-esters), sterols, polar lipids (*e.g.* phospholipids, sphingolipids, glycolipids) and hydrocarbons. The mono- and di-glycerides along with the polar lipids are generally found in the plasma membrane.

In general, the FA component of yeast lipids is mainly C16 and C18. Palmitic acid (C16:0) constitutes 15–25% (w/w) of the total lipid, while palmitoleic acid (C16:1) is

---

33 BouvierNavé *et al.* (2000).
34 Cho and Cronan (1995).
35 advancedbiofuelsusa.info/wp-content/uploads/2011/03/11-0307-Biodiesel-vs-Renewable_Final-_3_-JJY-formatting-FINAL.pdf

found in concentrations of generally less than 5% (w/w). Commonly oleic acid (C18:1) is the principal lipid accumulated in yeast cells, present in excess of 70% (w/w) in some reported cases, whereas stearic acid (C18:0) and linoleic acid (18:2) are minor components of the oil, found in concentrations of 5–8% (w/w) and 15–25% (w/w), respectively. The relative composition of these lipids can however vary between yeast species (Table 14.1), as well as between different strains. The environmental conditions, growth phase and substrate can all influence the FA profile. Polyunsaturated lipids such as α-linolenic acid (C18:3) are not commonly synthesised in yeast oils. On the whole, yeast oils have a similar profile to plant-derived oils, such as rapeseed and sunflower oil (Chapter 6), making them the suitable replacement for a sustainable feedstock.

14.7   Lipid Extraction and Processing

Following the growth of cells and accumulation of the lipid, the micro-organisms first have to be harvested or separated from the culture medium. As with algae (Chapters 11, 12 and 16), de-watering is often energy-intensive due to the large quantities of water that need to be removed relative to the accumulated biomass, and thus this step can contribute to 20–30% of the total biomass production costs[36]. Common harvesting methods include sedimentation, centrifugation and ultra-filtration. Flocculation can be also used to aggregate cells. Once dewatered, the oil can be extracted from the biomass. Whilst extraction techniques for yeast and algae are very similar, harsher conditions are required for algae due to the relative resilience of the cell. Generally, these include mechanical, organic solvent or supercritical fluid extraction[37].

14.8   Concluding Comments

The potential for microbes to produce biofuels is the basis of first-generation ethanol (fermentation of sugars by yeasts); there is therefore a substantial technical expertise on which to draw for the production of so-called 'third generation' biofuels from naturally oleaginous microbes, notably heterotrophic yeasts. Commercial production of heterotrophic SCO is feasible given an appropriate market and process. However, to date, successful commercial ventures have involved high-value speciality chemicals such as the antioxidant, astaxanthin, from *Phaffia rhodozyma*. The production of fuels from lipid-rich micro-organisms, especially oleaginous yeasts, has great industrial potential, but almost every aspect of the process economics must still be improved. This includes pre-treatment and hydrolysis of feedstocks, improved utilisation of the varied carbohydrates in lignocellulosic hydrolysates by the oleaginous yeasts, faster microbial growth, increased oil accumulation, improved harvesting technologies and upgrading of spent yeast biomass into valuable co-products could all result in an economically-viable process. Nevertheless, with increasing pressures to reduce the use of fossil fuels, oils from oleaginous micro-organisms should have a future.

---

36 Mata *et al.* (2010).
37 Halim *et al.* (2012).

Table 14.1 Lipid content and fatty acid profile for selected oleaginous yeast and algal species.

	Maximum Lipid content (% w/w)	Fatty acid (% of total)													
		14:0	15:0	16:0	16:1	16:2	16:3	18:0	18:1	18:2	18:3	18:4	20:4	20:5	22:6
Yeast Species															
Cryptococcus curvatus	60			32				15	44	8					
Lipomyces starkeyi	65			34				5	51	3					
Rhodosporidium toruloides	66			18				3	66						
Rhodotorula glutinis	72			37				3	47	8					
Waltomyces lipofer	64			37				7	48	3					
Algal species															
Chlorella vulgaris	58	0.5	0.6	23.1	0.2	7.4	5.8	5.2	16.1	20.9	18				
Dunaliella primolecta	54	0.4		21.8	4.5	0.9		1.6	16.3	7	38.7	0.6			
Nannochloropsis sp	68	13.3		17.8					23.9	10.8	28.2	6.1			
Nitzschia ovalis	40	3.2		18.8	28.2	1	4.4	0.3	0.7	0.2	0.4	0.1	2.6	23.7	4
Phaeodactylum tricornutum	31	6.7		14.7	43.6	2			15.8	0.5	0.4	1.1		14.4	0.7

Selected References and Suggestions for Further Reading

Abe, K. *et al.* (2002) *Applied Phycology*, **14**, 129–134.

Ageitos, J.M., Vallejo, J.A., Veiga-Crespo, P. and Villa, T.G. (2011) Oily yeasts as oleaginous cell factories. *Applied Microbiology and Biotechnology*, **90**, 1219–1227.

Allen, M. (2013) bubble-columns.com/:

An, J.-Y. *et al.* (2003) *Journal of Applied Phycology*, **15**, 185–191.

Bouvie-Navé, P., Benveniste, P., Oelkers, P., Sturley, S.L. and Schaller, H. (2000) Expression in yeast and tobacco of plant cDNAs encoding acyl CoA: diacylglycerol acyltransferase. *European Journal of Biochemistry*, **267**, 85–96.

Chisti, Y. and Yan, J. (2011) Energy from algae: Current status and future trends: Algal biofuels – A status report. *Applied Energy*, **88**, 3277–3279.

Cho, H. and Cronan, J.E. (1995) *Journal of Biological Chemistry*, **270**, 4216–4219.

Chuck, C.J., Bannister, C.D., Jenkins, RW., Lowe, J.P. and Davidson, M.G. (2012) A comparison of analytical techniques and the products formed during the decomposition of biodiesel under accelerated conditions. *Fuel*, **96**, 426–433.

Chuck, C.J., Wagner, J.L. and Jenkins, R.W. (2014) Biofuels from microalgae. In: *Chemical Processes for a Sustainable Future* (Letcher, T., Scott, J. and Patterson, D. eds), Royal Society of Chemistry, London, pp. 425–442.

da Silva, T.L. *et al.* (2010) *Applied Biochemistry and Biotechnology*, **162**, 2166–2176.

Evans, C.T. and Ratledge, C. (1985) *Canadian Journal of Microbiology*, **31**, 845–850.

Guay, C. *et al.* (2007) *Journal of Biological Chemistry*, **282**, 35657–35665.

Halim, R., Danquah, M.K. and Webley, P.A. (2012) Extraction of oil from microalgae for biodiesel production: A review. *Biotechnology Advances*, **30**, 709–732.

Huang, C., Chen, X.-F., Xiong, L., Chen, X.-D., Ma, L.-L. and Chen, Y. (2013) Single-cell oil production from low-cost substrates: the possibility and potential of its industrialization. *Biotechnology Advances*, **31**, 129–139.

Klein-Marcuschamer, D., Chisti, Y., Benemann, J.R. and Lewis, D. (2013) A matter of detail: assessing the true potential of microalgal biofuels. *Biotechnology and Bioengineering*, **110**, 2317–2322.

Knothe, G. *et al.* (2005) *The Biodiesel Handbook*. Urbana, IL: AOCS Press.

Knothe, G. (2005) *Fuel Processing Technology*, **86**, 1059–1070.

Knothe, G. (2010) Biodiesel and renewable diesel: A comparison. *Progress in Energy and Combustion Science*, **36**, 364–373.

Li, Q. (2008) *Applied Microbiology and Biotechnology*, **80**, 749–756

Lui, J. *et al.* (2007) In: *Biodiesel-Feedstocks and Processing Technologies* (Stoytcheva, M. *et al.* eds), InTech, Rijeka, Croatia, pp. 58–78.

Liu, J., Huang, J., Sun, Z., Zhong, Y., Jiang, Y. and Chen, F. (2011) Differential lipid and fatty acid profiles of photoautotrophic and heterotrophic *Chlorella zofingiensis*: Assessment of algal oils for biodiesel production. *Bioresource Technology*, **102**, 106–110.

Lui, J. *et al.* (2013) In: *Biofuels from Algae* (Pandey, A. *et al.* eds), Elsevier, San Diego, pp. 111–142.

Mata, T.M., Martins, A.A. and Caetano, N.S. (2010) Microalgae for biodiesel production and other applications: A review. *Renewable and Sustainable Energy Reviews*, **14**, 217–232.

Meng, X., Yang, J., Xu, X., Zhang, L., Nie, Q. and Xian, M. (2009) Biodiesel production from oleaginous microorganisms. *Renewable Energy*, **34**, 1–5.

Pandey, A., Lee, D.-J., Chisti, Y. and Soccol, C. (2014) *Biofuels from Algae*. Elsevier, San Diego and Burlington, USA.

Papanikolaou, S. and Aggelis, G. (2011) Lipids of oleaginous yeasts. Part I: Biochemistry of single-cell oil production. *European Journal of Lipid Science and Technology*, **113**, 1031–1051.

Ratledge, C. (1992) In: *Industrial Applications of Single Cell Oils* (yle, D.J. and Ratledge, C. eds), Taylor & Francis, Abingdon.

Ratledge, C. (1994) In: *Technological Advances in Improved and Alternative Sources of Lipids* (Kamel, B.S. ed.), Heidelberg: Springer, pp. 235–291

Ratledge, C. and Cohen, Z. (2008) Microbial and algal oils: Do they have a future for biodiesel or as commodity oils? *Lipid Technology*, **20**, 155–160.

Ribeiro, L.A. and de Silva, P.P. (2013) Surveying techno-economic indicators of microalgae biofuel technologies. *Renewable and Sustainable Energy Reviews*, **25**, 89–96.

Ruan, Z., Zanotti, M., Zhong, Y., Liao, W., Ducey, C. and Liu, Y. (2013) Cohydrolysis of lignocellulosic biomass for microbial lipid accumulation. *Biotechnology and Bioengineering*, **110**, 1039–1049.

Sills, D.L., Paramita, V., Franke, M.J., Johnson, M.C., Akabas, T.M. *et al.* (2012) Quantitative uncertainty analysis of life cycle assessment for algal biofuel production. *Environmental Science & Technology*, **47**, 687–694

Singh, A. and Olsen, S.I. (2011) A critical review of biochemical conversion, sustainability and life cycle assessment of algal biofuels. *Applied Energy*, **88**, 3548–3555.

Singh, N.K. and Dhar, D.W. (2011) Microalgae as second-generation biofuel. *A review. Agronomy for Sustainable Development*, **31**, 605–629.

Sitepu, I.R., Garay, L.A., Sestric, R., Levin, D., Block, D.E. *et al.* (2014) Oleaginous yeasts for biodiesel: Current and future trends in biology and production. *Biotechnology Advances*, **32**, 1336–1360.

Subramaniam, R., Dufreche, S., Zappi, M. and Bajpai, R. (2010) Microbial lipids from renewable resources: production and characterization. *Journal of Industrial Microbiology & Biotechnology*, **37**, 1271–1287.

Uckun Kiran, E., Trzcinski, A.P., Ng, W.J. and Liu, Y. (2014) Bioconversion of food waste to energy: A review. *Fuel*, **134**, 389–399.

Xiong, W. *et al.* (2008) *Applied Microbiology and Biotechnology*, **78**, 29–36.

Zheng, Y. *et al.* (2013) *Applied Energy*, **108**, 281–287.

15

Engineering Microbial Metabolism for Biofuel Production

Thomas P. Howard

School of Biology, Faculty of Science, Agriculture and Engineering, Newcastle University, UK

Summary

While first-generation biofuels are a market reality, research into new, drop-in biofuels is advancing rapidly. Metabolic engineering and synthetic biology are enabling the production of compounds that are more suited to the existing infrastructure and engines. Some of these compounds are already at pilot-scale production. Other compounds, that actually possess identical chemical properties to fossil fuels, are at proof-of-concept. This chapter will explore these new biological tools and how microbial platforms and products are being designed to produce the biofuels of the future.

15.1   Introduction

Biofuels are a market reality. In 2001, ethanol in the USA constituted a little over 1% of gasoline (petrol) consumption, but by 2011 this had risen to nearly 10%. During this time, biodiesel production increased by more than 100-fold. Mandates for a minimum percentage biofuel to petroleum-distillate blends enacted in Europe[1], the USA, China, Brazil and India are driving this shift. These mandates stem from concerns surrounding the increasing demand for transport fuels (due to rising global population, increasing industrialisation and prosperity), combined with concerns about fuel security and the transport sector's impact as the second largest source of global greenhouse gas emissions. Of the current generation of biofuels, ethanol is produced predominantly (but not entirely: see Chapter 5) by yeast from the fermentation of sugar cane or corn (maize) starch and is mixed with gasoline, while biodiesel (Chapters 2 and 6), comprising fatty acid ethyl esters (FAEEs) derived from animal fats or plant oils, is mixed with diesel (automotive gas oil).

Current biofuels are not without problems. Ethanol is hygroscopic, meaning that it has a high tendency to absorb water from the air. This property makes ethanol corrosive to transport infrastructure and reduces the heat of fuel combustion. Moreover, ethanol

---

1 Although at the end of April 2015, the European Parliament lowered the required percentage (by 2020) from 8.6% to 7%.

Biofuels and Bioenergy, First Edition. Edited by John Love and John A. Bryant.
© 2017 John Wiley & Sons Ltd. Published 2017 by John Wiley & Sons Ltd.

has only 70% the energy density of petrol, meaning less distance travelled per unit volume. Ethanol may also contain soluble and insoluble contaminants such as halide and chloride ions, increasing the corrosive nature of the fuels. In many countries, ethanol is restricted to a 10–15% maximum blend with petroleum distillate (the blend-wall) – Brazil is the notable exception. Without re-engineering our petroleum-focused transport infrastructure, this is unlikely to rise. In the USA, ethanol production has slowed as saturation of the 10% ethanol blend (E10) market has been reached and in 2010/11 the USA became a net exporter of ethanol for the first time. Whilst the US Environmental Protection Agency has recently approved the use of 15% ethanol blends (E15), retail outlet liability issues and automobile warranty concerns have held back uptake of E15.

Unlike ethanol, biodiesels have more than 90% of the energy content of the petroleum diesels they supplement. Although there may be less functional constraint with biodiesel on the maximum blends that can be accommodated in petroleum-distillate, biodiesels do also have problems associated with their use. Solid waxes may form in biodiesel at low temperatures (see Chapter 6), making them difficult to transport and handle, and restricting their use in cold climates and at high altitude. Biodiesel is moderately hygroscopic, carrying some of the same issues as outlined for ethanol. Like ethanol, there is a restriction on the ratio of biodiesels to petroleum distillate. However, in 2011, biodiesel blends in the USA peaked at 2.2%, less than half of the upper blend limit of 5%. In order to increase the amount of biofuel in retail fuel, modifications to engines and infrastructure are required. The first part of this chapter addresses our ability to design biofuels that fit our existing transport infrastructure.

The use of the feedstock to generate biofuels is perhaps the most contentious issue surrounding biofuel production (Chapters 8, 16 and 17). The current generation of biofuels are derived from 'first-generation' carbon sources, for example, sugar cane (*Saccharum officinarum*; Brazil), corn (*Zea mays*) starch (USA), or animal fats and plant oils, for example palm (*Elaeis spp.*) oil, soybean (*Glycine max*) oil, and rape-seed/canola (*Brassica napus*) oil. The use of such feedstock has the effect of further cementing the already strong link between food and fuel prices. This is a matter for concern, as the demand for both food and transportation fuel is predicted to double by 2050. There are also concerns that demand for palm oil, in part for biodiesel production, is linked to deforestation in Southeast Asia (see Chapters 8 and 17). These issues pose serious questions for the environmental credentials of biofuels. These concerns however, can be part of the solution in the move towards industrialisation, because the organic feedstock for microbial growth is the biggest cost factor in biofuel viability. By exploiting cheaper, non-food feedstock, it may be possible to improve both the economics of biofuel production and their social impact. Finally, whilst rising petroleum prices will make biofuels more attractive, efficient carbon conversion and high yields are essential for scale-up of bulk chemicals such as fuels. The second part of this chapter addresses possible metabolic engineering approaches to maximise biomass conversion.

15.2   Designer Biofuels

15.2.1   Introduction

The chemical nature of fuel is incredibly important and not all petroleum-based fuel blends are alike. The retail diesel sold in Canada is not the same as the retail diesel sold in Mexico, though you are unlikely to be aware of this at the pump (Chapter 6). However, whilst it may be possible to run a car in Mexico on Canadian diesel blends, you will run into problems if you try the reverse experiment. This is because the blends sold in Canada are designed to operate at lower temperatures and as such have a higher proportion of branched and cyclic hydrocarbons than those sold in warmer climes. The octane rating, a measure of gasoline and jet fuel performance and the cetane number, a measure of diesel fuel performance are also important indicators of fuel operation. The use of ethanol as a substitute petrol distillate has been driven primarily by its high level of natural production by yeast, not for its desirability as a fuel chemical (although the Model-T Ford was originally designed to run on an ethanol-petrol mix: Chapter 5). Metabolic engineering can, through the rational design of artificial metabolic pathways, produce biofuel chemicals to the standards and with the properties that both industry and legislation require. These biofuels, sometimes referred to as advanced biofuels, are considered to fall into three categories: isoprenoid-derived biofuels (*e.g.* farnesane), short chain alcohols (*aka* higher alcohols such as butanol) and petroleum-replica hydrocarbons (*e.g.* alkanes and alkenes). A fourth category, polyketide-derived fuels, are relatively unexplored and shall not be considered here, while fatty acid-derived fuels (FAEEs and fatty alcohols – currently in use, but derived from chemical processing) will be considered, because of the examples they provide for microbial metabolic engineering and their relevance to petroleum-replica hydrocarbons.

15.2.2 Isoprenoid-Derived Biofuels

On the 1 February 2013, Amyris Biotechnologies, Inc., a California-based company, announced the first commercial shipment of Biofene® from its 50 million litre fermentation plant in São Paolo, Brazil. Biofene®, a long chain branched hydrocarbon molecule called farnasane, is chemically derived from farnesene. Farnesene is a highly branched C_{15} isoprenoid (or terpene) molecule (Figure 15.1a). Whilst farnesene itself can be used as a fuel, it is highly unsaturated. This means it has a low cetane number and low oxidative stability. Chemical hydrogenation to farnesane markedly improves the cetane number and cold flow properties. Biofene® is currently in use fuelling the buses of São Paolo and Rio de Janeiro.

Farnesene is a product of the isoprenoid metabolic pathway and both *Escherichia coli* and yeast (*Saccharomyces cerevisiae*) have been engineered to produce farnesene, as well as other potential isoprenoid-fuel molecules. The isoprenoids show potential for use in fuels due to the fact that they represent a large class of hydrocarbons with a wide range of branched and ring structures. Such features are important features for fuel blends. Short, branched isoprenoids may find use in gasoline supplementation, whereas longer molecules may be added to diesel. Part of the attractiveness of isoprenoids comes from the fact that, despite a large diversity of chemical structures, they are all synthesised from two universal C_5 precursors: isopentenyl diphosphate (IPP) and dimethylallyl diphosphate (DMAPP). IPP and DMAPP are generated from one of two pathways that use either acetyl CoA or pyruvate and glyceraldehyde as the starting compounds (Figure 15.1a); these are the mevalonate (MEV) pathway and the 2-C-methyl-D-erythritol-4-phosphate (MEP) pathway. In order to generate isoprenoids, IPP and DMAPP molecules are combined 'head-to-tail' in a reaction catalysed by prenyltransferase enzymes (Figure 15.1b). This results in the production of geranyl pyrophosphate (GPP). GPP has a C_8 backbone with two branched methyl groups (making a C_{10} molecule). Sequential addition of further IPP molecules to GPP increments the chain length by four carbons to produce farnesyl pyrophosphate (FPP), and then geranylgeranyl

(a)

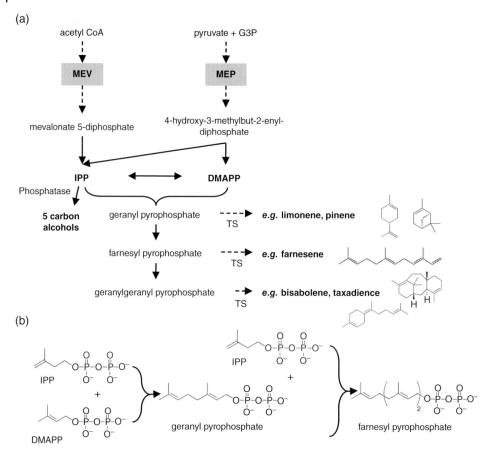

(b)

Figure 15.1 Metabolic pathways to isoprenoid-derived fuels. (a) Two alternative pathways generate the starter molecules IPP (isopenetyl diphosphate) and DMAPP (dimethylallyl diphosphate): *via* mevanolate (the MEV pathway; cytosolic) or *via* 2-C-methyl-D-erythritol-4-phosphate (the MEP pathway; plastidial). IPP and DMAPP are condensed to form geranyl pyrophosphate, which may then be converted to C_{10} hydrocarbons, or may undergo further rounds of extension with IPP prior to conversion to C_{15} and C_{20} hydrocarbons, *via* the activities of terpene synthases (RS). (b) Condensation of IPP and DMAPP to form geranyl pyrophosphate and subsequent elongation with another molecule of IPP to form farnesyn pyrophosphate. At each step, a pyrophosphate molecule is released.

pyrophosphate (GGPP) (Figure 15.1b). Each molecule is highly branched. A methyl group is found on every fourth carbon atom followed by a carbon–carbon double bond.

Terpene synthases are largely responsible for the subsequent variety of isoprenoid structural diversity, though further modifications may be made. Terpene synthases convert the intermediates (GPP, FPP and GGPP) into C_{10} monoterpenes, C_{15} sesquiterpenes and C_{20} diterpenes. Terpene cyclases are responsible for the conversion into cyclic and even polycyclic isoprenoids. GPP, FPP and GGPP intermediates may also be joined in a 'head-to-head' fashion to create very long chain molecules such as squalene (C_{30}). A variety of potential isoprenoid fuels (some of which are shown in Figure 15.1) have been produced following deregulation and/or over-expression of the MEV and MEP pathways in *E. coli* and *S. cerevisiae* to supply the appropriate precursor substrates.

These include five carbon alcohols, formed from dephosphorylation of IPP in *E. coli* using pyrophosphatase activity from *Bacillus subtilis*:

1) *pinene*, produced following heterologous expression of pinene synthase from the loblolly pine (*Pinus taeda*) in *E. coli* and monoterpene synthases from two species of sage (*Salvia fruticosa* and *S. pomifera*) in yeast;
2) *limonene*, following expression of limonene synthase from spearmint (*Mentha spicata*);
3) *farnesene* (Biofene® precursor) following expression of farnesene synthase from *Artemisia annua*;
4) *bisabolene* following expression of a bisabolane synthase from the fir tree *Abies grandis*; and
5) the cyclic alkene *taxadiene* by expression of a taxadiene synthase from *Taxus*.

The latter is of medical interest, as it is a precursor-molecule for the anti-cancer drug paclitaxel (taxol). The range of structures that can be derived from the same metabolic pathways and the success of Biofene® means that isoprenoids will remain an attractive target for designer biofuel production.

15.2.3 Higher Alcohols

The production of higher alcohols (C_4 and C_5 alcohols) is an attempt to overcome some of the limitations of ethanol (a C_2 alcohol) biofuel. Higher alcohols possess a higher energy density and lower hygroscopicity than ethanol. Metabolic engineering strategies have been employed, not only to increase carbon chain length but also to introduce branches to the molecules. *Iso*propanol and 1-butanol are perhaps the furthest progressed of these potential fuel molecules and are naturally produced *via* fermentative processes (Figure 15.2a). *Iso*propanol biosynthesis has been achieved in *E. coli* using heterologous expression of a fermentative pathway using acetyl-CoA starting molecules. This pathway uses three genes derived from *Clostridium acetobutylicum* and one from *Clostridium beijerinckii* with the over-expression of a fifth *E. coli* gene. Expression of these enzymes can yield almost 14 g/L from *E. coli* shake flasks, depending upon growth conditions, representing a higher titre than that achieved from the native pathway in *Clostridium*, which is typically 2 g/L. At these concentrations, *iso*propanol is toxic to *E. coli*, but continuous removal of *iso*propanol from the media dramatically improves titres to 143 g/L over 240 h. This value represents 67% of the theoretical yield.

1-butanol is a four-carbon alcohol that contains 84% of the energy density of gasoline, limited miscibility with water and 100% miscibility with gasoline. Production of industrial butanol occurs *via* catalytic conversion of petrochemical-derived propylene. For biofuel use, microbial fermentation of renewable feedstock is required. 1-butanol is naturally produced by *C. acetobutylicum* and other Clostridial species. *C. acetobutylicum* itself also possesses the advantage that it can utilise both traditional and novel feedstock: monomeric sugars, complex carbohydrates, industrial wastes (whey and glycerol) and short-chain fatty acids (FAs, *i.e.* acetic, butyric and lactic acids). However, industrial fermentation typically yields between 7 and 16 g/L, below that required for industrial-scale production. Genetic tools for the study and manipulation of *C. acetobutylicum* are limited in comparison to yeast and *E. coli*, but they are improving. A recent report demonstrated that several parallel strands of metabolic manipulation resulted in approximately 19 g/L of 1-butanol.

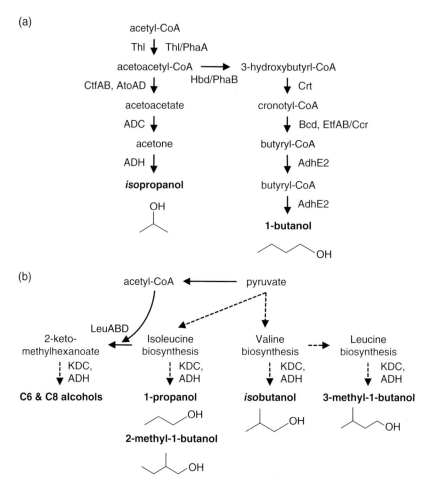

Figure 15.2 Metabolic pathways to higher alcohols. (a) Fermentative routes to higher alcohol (*iso*propanol) production introduced into *E. coli*. ThI (thiolase), CtfAB (coenzyme A transferase, A and B subunits) and ADC (acetoacetate decarboxylase) from *C. acetobutylicum*, ADH (butanol dehydrogenase) from *Clostridium beijerinckii* and over-expression of *E. coli* AtoAD (acetoacetyl-CoA transferase). *C. beijerinckii* 1-butanol biosynthesis pathway. ThI (thiolase), Hbd (3-hydroxybutyrl-CoA dehydrogenase), Crt (cronotase), Bcd (butyrl-CoA dehydrogenase), AdhE2 (butaraldehyde/butanol dehydrogenase). In the synthetic butanol pathway PhaA (thiolase from *Ralstonia eutropha*) replaced ThI, PhaB (acetoacetyl-CoA reductase from *R. eutropha*) replaced Hbd, Ccr (Cronotyl-CoA reductase (encoded by *Ter*) from *Treponema denticola*) replaced Bcd and EefA & EtfB. (b) Non-fermentative routes in higher alcohols. 2-ketoacid-derived short chain alcohol biosynthesis utilises substrates from amino acid biosynthesis (2-ketoacids) derived from a common pyruvate pool. Ketoacid decarboxylases (KDC) and alcohol dehydrogenases (ADH) catalyse the 2-step conversion form ketoacid to aldehyde, and from aldehyde to corresponding alcohol.

Butanol yields from *C. acetobutylicum* are hampered by a slow growth rate and an intolerance to 1-butanol over 2%. Attempts have been made to overcome these issues by engineering 1-butanol biosynthetic pathways into more tractable hosts. This has been demonstrated in the bacterial hosts *E. coli, Pseudomonas putida, B. subtilis* and

Lactobacillus brevis, the cyanobacterial host *Synechococcus elongatus* and in the yeast *S. cerevisiae*. Each of these presents a particular advantage over the use of *C. acetobutylicum* but while metabolic engineering has improved yields of 1-butanol within these hosts, yields remain stubbornly low. The greatest success in heterologous 1-butanol production has been through the development of artificial metabolic pathways, created in *E. coli*, independent of the *Clostridium* 1-butanol pathway (Figure 15.2a, green). In this approach, enzymes that were considered reversible were replaced by those from other species that were deemed to catalyse alternative, irreversible reactions. For example, the activity of butyryl-CoA dehydrogenase and its two redox partners were replaced with cronotyl-CoA reductase from *Treponema denticola* (*ter*). Cronotyl-CoA reductase catalyses the irreversible conversion of cronotyl-CoA to butyryl-CoA. This creates a one-way route from which substrate can only move towards 1-butanol production. The authors of this study also over-expressed the native pyruvate dehydrogenase complex to increase the availability of the starting acetyl-CoA and boosting nicotinamine adenine dinucleotide (NADH). This yielded titres of approximately 5 g/L 1-butanol. Using a similar approach, in which the cronotyl-CoA reductase replacement was also made, another group over-expressed formate dehydrogenase (to relieve pyruvate build-up and provide increase reducing power), acetyl-CoA acetyltransferase (to encourage better acetoacetyl-CoA formation), and removed a phosphate acetyltransferase (to minimise acetate synthesis and cause a build-up of acetyl-CoA). This allowed 1-butanol production at between 15 and 30 g/L, depending on growth conditions (70–88% maximum theoretical conversion).

Application of synthetic metabolism has also been applied to the production of branched-chain alcohols (*iso*butanol, 2-methyl-1-butanol, 1-propanol and 3-methyl-1-butanol). As has been described, branched hydrocarbons are desirable for improved cold-flow properties. The native branched amino acid (isoleucine, leucine and valine) biosynthesis pathway intermediate keto-acids are exploited (Figure 15.2b). These routes are non-fermentative. 2-ketoacids are converted to their corresponding higher alcohol through a two-step reaction catalysed by a heterologous ketoacid decarboxylase (KDC), to form an aldehyde intermediate, and an endogenous alcohol dehydrogenase (ADH), to covert this to an alcohol. Longer chain alcohols have also been produced though the introduction of an elongation step catalysed by LeuABD. This latter feat is particularly noteworthy, as it involves the rational redesign of the enzyme catalyst itself. LeuABD elongates 2-ketoisovalerate (C_5) by two carbon atoms from acetyl CoA and then decarboxylates it to 2-ketoisocaproate (C_6). LeuABD, however, is promiscuous enough to use 2-keto-3-methyl valereate, resulting in the production of a 2-keto-4-methylhexanoate (C_7) intermediate that is decarboxylated to a C_6 alcohol. The researchers used previously resolved structures of LeuA, KDC and ADH to rationally design a new LeuA with an improved binding site to enhance the yield of C_6 alcohols. This not only improved yield but also led to the production of longer alcohols, up to C_8. Further manipulation and exploitation of these pathways is likely to yield improvements in yield as well as providing a wider palette of alcohol biofuel molecules.

15.2.4 Fatty Acid-Derived Biofuels

Fatty acids (FAs; phosphoglycerides and triacylclycerides (TAGs)) are the primary components of cell membranes. TAGs are also widely laid down in seeds and fruit as storage compounds (Chapter 6). These fats and oils, both from plant and animal sources,

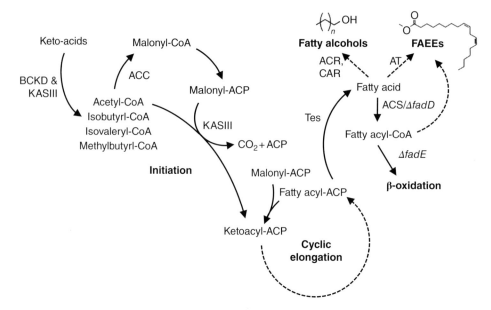

Figure 15.3 Metabolic pathways of FA metabolism. FA biosynthesis commences through the condensation of acetyl-CoA with malonyl-ACP. This creates a ketoacyl-ACP molecule that can be cyclic-elongated through the sequential addition of further malonyl-moieties at 2-carbon increments. Free FAs may be liberated through the action of thioesterases (Te) with different chain length specificities, a modification which also serves to increase the FA pool size. FA titres have also been increased by over-expression of ACC (acetyl-CoA carboxylase) and removal of flux into the ß-oxidation pathway using either *fadD* (ACS; acyl-Co-A synthetase) or *fadE* (acyl-CoA dehydrogenase) knockouts. Conversely, over-expression of ACS with ΔfadE serves to increase fatty acyl-CoA pools for FAEE production. Expression of branched chain ketoacld dehydrogense (BCKD) and KASIII from *Bacillus subtilis* can be used for the production of branches chain FAs in *E. coli* as branched CoA primer units are incorporated in FA biosynthesis.

are currently exploited as biofuels (biodiesel: see Chapter 6). However, FAs are not suitable for direct use and are chemically converted to fatty acid methyl esters and fatty acid ethyl esters (FAMEs and FAEEs). These chemicals are mixed with diesel. Globally, diesel demand outstrips the demand for petrol, and is growing at three times the rate. There are several strategies for the microbial production of FAs and their derivatives.

FA biosynthesis in prokaryotes and eukaryotes differs somewhat in the nature and arrangement of the enzymes. Moreover, different species utilise slightly different precursor molecules, meaning that the final FA products are different. Broadly speaking however, their biosynthesis follows a similar cyclic elongation scheme (Figure 15.3; see also Chapters 6 and 14). In *E. coli* this occurs through the cyclic elongation of a growing acyl chain with malonyl-acyl carrier protein (-ACP) following an initial condensation of acetyl CoA with malonyl-ACP. The ketoacyl-ACP molecule is elongated in a four-step process, the final step of which is the incorporation of another C_2 moiety from malonyl-ACP and the cycle continues. This is terminated by a thioesterase (Te) catalysed reaction in which the -ACP molecule is removed leaving a FA of predominantly C_{16} or C_{18} chain length. FA yields can be increased in several ways: first, introduction or over-expression of thioesterases increases the FA pool, with the added benefit of providing some control over the C_n of the FAs; second, over-expression of acetyl-CoA carboxylase (ACC), to increase production of malonyl-CoA (and hence malonyl-ACP) from acetyl-CoA; and third, removal of the FA breakdown pathway (the β-oxidation pathway) by gene knock outs, *fadD* (fatty acyl-CoA synthetase) and *fadE* (acyl-CoA dehydrogenase) being the most common. In *E. coli*, FAs can also be increased utilising the hosts endogenous mechanism for regulating FA biosynthesis and degradation, the transcription factor FadR. FadR plays an important role in regulating the expression of FA biosynthesis and FA degradation genes. When the FA content of the cell is low, FadR binds to regulatory elements within certain gene promoter regions, increasing the expression of FA biosynthesis genes and repressing key genes of the β-oxidation pathway. As FA content rises, FadR preferentially binds FAs and fatty acyl-CoAs, rather than DNA, removing inhibition of β-oxidation pathway gene expression and removing the stimulatory effect on FA biosynthesis gene expression. All these strategies can yield FAs at approximately 2 g/L.

These FAs provide the starting molecule for subsequent conversion to biofuel such as FAEEs, alcohols, methyl ketones or the alkanes and alkenes discussed in the next section. To engineer *E. coli* to produce C_{12} to C_{18}, FAEEs over-expression of thioesterases and removal of β-oxidation pathway, as described above, was combined with the over-expression of fatty acyl-CoA synthetase (*fadD*) to increase fatty acyl CoA production, along with the introduction of an ethanol pathway and expression of a non-specific acyltransferase. Researchers recently described a carboxylic acid reductase (CAR) that, when expressed heterologously in *E. coli*, was able to produce a range of fatty alcohols directly from FAs. It has also been demonstrated that FA biosynthesis in *E. coli* can be engineered to produce branched FAs: the presence of branched-chain molecules within fuel blends is necessary to maintain performance at low temperature and high altitude. Wild-type *E. coli* is unable to produce branched FAs endogenously because the native β-ketoacyl-ACP synthase III (KASIII) enzyme, which catalyses the initial step in the FA elongation cycle, accepts only linear acetyl-CoA or propionyl-CoA substrates. The substrates required by KASIII for branched FAs are also not present within *E. coli*. Branched-chain FAs can be produced *in vitro* by the FA elongation enzymes from *E. coli* if an alternative KASIII enzyme and suitable precursor molecules are present however. Expression in *E. coli* of KASIII (FabH2) and the branched-chain α-keto acid dehydrogenase (BCKD) complex from *Bacillus subtilis* overcame these obstacles, providing branched molecules as starter substrates for the endogenous FA elongation cycle.

15.2.5 Petroleum Replica Hydrocarbons

Alkanes and alkenes are the primary constituents of retail fuel. Petroleum-replica hydrocarbons (bio-alkanes and -alkenes) therefore possess the distinct advantage over other biofuel molecules in that they are chemically and structurally identical to the fuel molecules they seek to replace. There are no issues surrounding their compatibility with current transport infrastructure. Because of this, they possess the potential for much greater penetration of the fuel market than any other biofuel molecule. Biogenic sources of alkane have been known for some time, but the recent discovery and elucidation of a two-step cyanobacterial alkane biosynthetic pathway has opened up the possibility of production of direct replacement fuel molecules (Figure 15.4). In this scheme, the intermediate of the FA elongation cycle, the growing fatty acyl-ACP chain is reduced to a fatty aldehyde by fatty acyl-ACP reductase, and then the fatty aldehyde is deformylated

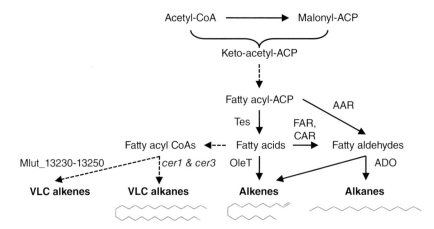

Figure 15.4 Metabolic pathways to petroleum-replica biofuels. Natural and synthetic alkane biosynthesis pathways have been recreated in *E. coli* and in yeast. The initial elucidation of a two-step alkane route to alkanes, *via* (AAR) acyl-ACP reductase and ADO (aldehyde deformylase) was demonstrated by expressing these genes in *E. coli*. Likewise, bacterial production of alpha-olefins *via* Ole T from *Jeotgalicoccus* spp. Exploitation of FAs, rather than fatty acyl-ACP using FAR (FA reductase complex from *Photorahbus luminescens*), allowed alkane output to be determined by the predominant FA species. A similar approach of producing alkanes directly from FAs was reported for CAR (carboxylic acid reductase from *Mycobacterium marinum*) though this report was principally aimed at tailoring fatty alcohol content. A three gene cluster (Mlut_ 13230-133250) from *Micrococcus luteus*, produces VLC alkenes when expressed in *E. coli*. The *cer1/cer3* pathway has been genetically identified from *Arabidopsis thaliana* and expressed in yeast to produce VLC alkanes (>C_{30}). Alkanes and alkenes of this chain length are not useful as fuel molecules *per se*, but knowledge of these pathways may help with engineering tailored synthetic alkanes routes.

to an alkane by a fatty aldehyde deformylase (ADO; originally described as fatty aldehyde decarbonylase, the mechanism of action of this enzyme is still under investigation). It is now clear, however, that formate, rather than CO, is released during alkane biosynthesis. This reaction produces C_{17} and C_{15} alkanes in cyanobacteria, with a similar outcome observed when the pathway is heterologously expressed in *E. coli*. Alkane production in other species is also close to full elucidation, notably the production of very long chain alkanes in plants and the production of terminal alkenes by the cytochrome P450 OleT. Expression of *oleT*$_{JE}$ from *Jeotgalicoccus* led to the production of terminal alkenes in *E. coli*, while the expression of the *Arabidopsis* genes *cer1* and *cer3* in yeast lead to the production of very-long chain alkanes (Figure 15.4). Three genes from *Micrococcus luteus* have also been heterologously expressed in *E. coli* to produce VLC alkenes.

These metabolic pathways all produce the hydrocarbons that the original host organism produces. In order to supply alkanes and alkenes that are required by industry, several recent reports have demonstrated different strategies by which alkane biosynthetic pathways can be adapted to increase the type of hydrocarbon output. A team of undergraduate students from the University of Washington, competing in the Internationally Genetically Engineered Machines (iGEM) competition in 2011, expressed the two-step cyanobacterial pathway for alkane biosynthesis in *E. coli*. They additionally expressed the *fabHB* gene (coding for β-ketoacyl-ACP synthase III) from *B. subtilis* and grew their *E. coli* in the presence of the three-carbon propanoate. The presence of propanoate in *E. coli* growth media is known to increase the pool of propionyl-CoA. Unlike the native FabH activity, *B. subtilis* FabH is less fussy about the substrate it uses. Use of the propionyl-CoA molecule in the KASIII catalysed reaction meant that the three-carbon compound was incorporated into the FA elongation cycle as well as the normal two-carbon acetyl-CoA (Section 2.3, Figure 15.3). The result was that some of the FAs in the modified *E. coli* are one carbon longer than in the original strain. Given that the cyanobacterial pathway results in the loss of one carbon during alkane biosynthesis, this modification resulted in the production of even- as well as odd-chain length alkanes.

In a further example, researchers replaced the cyanobacterial gene for fatty acyl-ACP reductase with the aldehyde generating fatty acid reductase (FAR) complex from *Photorhabdens luminescens*. The FAR complex is normally responsible for providing fatty aldehydes for bioluminescence in several luminescent bacteria, including *P. luminescens*. This change had the advantage that alkanes could be produced directly from FA, rather than fatty acyl-ACP molecules. It was therefore possible to tailor alkanes by manipulating the FA pool, a strategy that had been employed before to modify the fatty alcohol output of the cells. This new route, unlike the cyanobacterial fatty acyl-ACP reductase, was compatible with expression of a thioesterase to modify chain length. Moreover, expression of a 5-gene operon encoding the branched chain keto-dehydrogenase complex and FabH from *B. subtilis* (Section 2.3, Figure 15.3) also enable branched FAs to be produced in *E. coli* and, when expressed in conjunction with the novel pathway, branched alkanes were detected.

Taken together, these reports demonstrate the ability to tailor the alkane and alkene output of microbial cells to industrially-relevant molecules. Yields of alkanes vary between 2 and 40 mg/L over 24 h. Production therefore remains low. The ability to structurally tailor the chemical output of the cells to those fuels demanded by the transport sector is a big advantage and direct-replacement biofuels will therefore remain a target for further research.

15.3 Towards Industrialisation

15.3.1 Introduction

For successful (*i.e.* economically viable) industrialisation of advanced biofuels, there are three areas where metabolic engineering can be deployed: the production of industry-relevant fuel molecules (discussed above); high productivity and yield; and feedstock cost. Ethanol has a 90–92% theoretical maximum conversion and the aim for advanced biofuels is to attain conversion in the region of 85% theoretical maximum. However, it has taken 7,000 years to develop *S. cerevisiae* into the high-ethanol producing organism it is today, and the time-frame available for developing new designer microbes is much shorter. The tools available are sophisticated and growing more so by the day; data-rich, high-throughput automation of strain development, genome sequencing, DNA synthesis and the rapid increase in fundamental biological knowledge is providing a great deal of optimism that these goals may be achieved.

15.3.2 Bioconsolidation

The identity of the carbon feedstock is perhaps the most contentious issue currently surrounding biofuels. In the USA, in 2011, ethanol represented 10% of all transport fuel, but consumed 40% of the country's corn. Corn prices have risen as a result and North

American farmers have switched to varieties of corn preferred for biofuel, rather than food production (see Chapters 8, 16 and 17). As global demand for both food and fuel rises, this further linking of food and fuel prices is unwanted. Correct exploitation of biofuel feedstock could help this situation and may make biofuels more economically viable. Developing microbial platforms that can process cheaper, non-food feedstock to advanced biofuels is known as consolidated bioprocessing, or bioconsolidation.

The cheapest and most abundant source material on the planet is plant biomass (Chapters 5, 7, 16 and 17). This biomass may be cobs or straw left over after harvest, or it may be material grown expressly for this purpose; switchgrass (*Panicum*) and elephant grass (*Miscanthus*) have received a lot of attention in this regard (Chapter 7). Although plant biomass consists of approximately 70% sugars, these are not readily accessible for microbial growth. The lignocellulosic material that forms the bulk of such material is recalcitrant to digestion. Heat and chemical treatment, as well as biochemical processing, are required before it can be fed to microbes (Chapter 5), pushing up costs. One solution may be to modify the plants, prior to microbial fermentation to enable easier access to lignocellulosic sugars. Another approach is one in which the microbial biofuel producer is also capable of digesting recalcitrant biomass (Figure 15.5). As with tailoring biofuel output. it has also been possible to some extent to modify the input pathways too. FAEEs have been produced in *E. coli* that secrete xylanases capable of digesting hemicellulose. Other strains have been developed that express cellulase, xylanase, β-glucosidase and xylobiosidase, making it possible to produce a range of biofuel components (pinene, limonene) from pretreated switchgrass biomass.

Harvesting photosynthesis directly for biofuel production is conceptually very appealing: CO_2 may be fixed and converted to biofuels by the same host (Figure 15.5). Two groups of microbes are under investigation in this regard: cyanobacteria (Chapter 9) and microalgae (Chapter 11). Cyanobacteria can grow much faster than terrestrial crops (potentially generating 20 times more biomass per day than soybeans). They are also genetically tractable: *Synechococcus elongatus* has been engineered to produce *iso*butanol through heterologous expression of KDC and ADH-genes (Section 15.2.3), and FA production has been boosted through gene knockout of the FA-recycling acyl-ACP synthetase and expression of a thioesterase (Section 15.2.4). Algae have high lipid productivity and can be cultivated on marginal land in poor-quality water, but are traditionally more genetically difficult to work with. However, economic models for algal biofuels are not so promising, largely due to the large, upfront investment that is required in infrastructure.

Current biofuel schemes seek to maximise production of carbohydrates and lipids. One group at UCLA have illustrated the benefits of looking beyond traditional feedstock. In the first instance, they asked whether protein hydrosylates could provide the carbon skeletons required for biofuel production (Figure 15.6a). Unlike recalcitrant lignocellulosic sugars, proteins can be readily hydrolysed by proteases to short peptides and amino acids. The use of proteins provides several benefits: under current and advanced biofuel generation reduced nitrogen by-products, in the form of proteins, are not recycled. This results in a net loss of nitrogen and an increase in nitrous oxide production (a greenhouse gas 300 more potent than CO_2). Moreover, input of nitrogen fertiliser is required at the agricultural stage of the process; fertiliser that is currently supplied through the energy-intensive Haber-Bosch process.

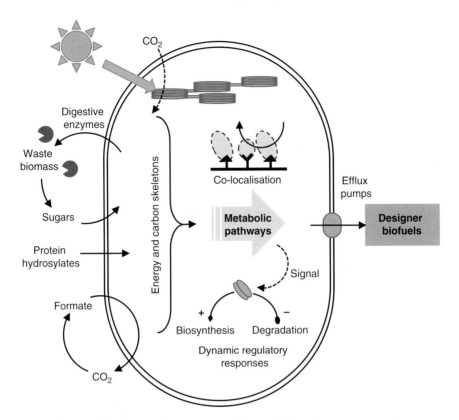

Figure 15.5 Strategies for improving the chances of industrialisation. Alternative sources for energy and carbon skeletons for biofuel synthesis include photosynthetic carbon fixation by algae and cynaobacteria; secretion of digestive enzymes for the digestion of waste biomass (typically *E. coli*) or electrochemical formate production (*Ralstonia eutropha*). Within the cell enhanced or synthetic metabolic pathways for designer biofuels may be co-localised using scaffolding techniques (*E. coli*) or through sub-cellular compartmentalisation of different organelles (*S. cerevisiae*) and may be placed under dynamic regulatory responses such as the FadR transcription factor (*E. coli*) that respond to the activity of the biofuel pathway. Recovery of the fuel molecule, removal of toxic compounds and alleviation of feedback inhibition may be aided by the use of efflux pumps to export the end product from the cell (*E. coli*).

A microbial platform, as part of a biorefinery process, that allows recycling of usable nitrogenous compounds could be highly advantageous. In order to use amino acids and peptides as a carbon source, the carbon skeleton must be enzymatically liberated from attached amino groups. Thermodynamics and cellular regulation though tend to favour the creation of amino acids, not their degradation. With this in mind, *E. coli* was first engineered to accept a wider variety of amino acids as the sole carbon source rather than wild-type *E. coli*. Next, key genes involved in carbon:nitrogen regulation (including quorum sensing *luxS* and *lsA*) were removed, whilst others for the deamination of certain amino acids were introduced. Key to success was the removal of enzymes responsible for recycling the ammonia released from deamination back to amino acids – these deletions ensured that the ammonia release from amino acids was a one-way process (Figure 15.6a). At the same time they introduced pathways for the production of *iso*butanol. The carbon skeletons released from amino acid deamination could then be used in the production of

(a)

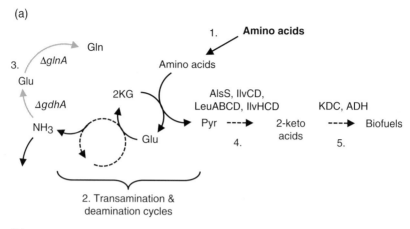

(b)

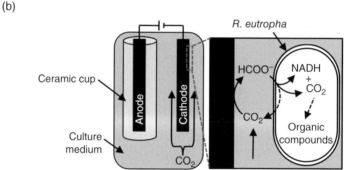

Figure 15.6 Novel sources of energy and carbon skeletons. Researchers at UCLA have reported two alternative strategies for generating the energy and carbon skeletons required for biofuel molecules. (a) *E. coli* was engineered to produce higher alcohol biofuels from protein hydrolysates. 1) *E. coli* was engineered by chemical mutagenesis to accept a wider range of amino acids as the sole carbon source rather than wild type *E. coli*. 2) Three transamination and deamination cycles are introduced as well as 3) the removal of GdhA and GlnA to prevent re-assimilation of ammonia. 4) Pyruvate is converted to the various 2-ketoacids by endogenous LeuABCD and IlvHCD activity, or through over-expression of AlsS and IlvCD. 5). Final conversion to higher alcohols follows the reactions shown in Figure 15.2 catalysed by introduced ketoacid decarboxylase (KDC) and alcohol dehydrogenase (ADH). The authors report 56% theoretical maximum yield and the ability to utilise a range of different sources, including microalgae, yeast and bacterial feedstock.
(b) Chemolithotrophic bacteria fix CO_2 using NADH, indepentent of the light reactions found in plant and cyanobacteria. One such bacteria, *R, eutropha*, was engineering to produce *iso*butanol, while endogenous pathways to polyhydroxybutyrate were removed. Genetically-modified *R. eutropha* were also used to identify O_2- and no stress as a limiting factor on cell growth, which was attributed to anode chemistry upon application of electrical currents to the media. Consequently, an inexpensive ceramic shield was used to quench O_2- and NO diffusion from the anode, and cell growth was no longer inhibited. *Iso*butanol was produced at 140 mg/L from electricity and CO_2.

alcohols. Another alternative energy source explored by the researchers at UCLA was the use of electrochemically generated H_2 as a feedstock. Microbes already display a dazzling array of different feedstock and chemolithotrophic bacteria, such as *R. eutropha* are a prime example. *R. eutropha* is capable of using H_2 as an energy source to fix CO_2. Combining engineering of the culture set-up along with metabolic engineering of *R. eutropha*, the researchers were able to generate *iso*butanol from electricity and CO_2 alone (Figure 15.6b). One of the limitations of large-scale renewable energy generation has been the ability to store the electricity generated 'off-peak'. The ability to use this to fix CO_2 directly into liquid transportation fuel has obvious appeal.

15.3.3 Molecular and Cellular Redesign

The optimisation of metabolic pathways in suitable microbes can be undertaken to improve yield under a given set of conditions, but these conditions can change as the process in scaled-up. Scaling without losing performance is a major challenge. As a general rule of thumb, for every 1,000-fold scale-up (*i.e.* from 50 mL to 50 L), new problems ('bugs') may be encountered that need to be debugged and the final scale of the reactors may be as large as 600,000 litres. One interesting route to solving this is to build dynamic regulatory mechanisms that respond to cellular conditions into the microbial biofuel platform. This has been achieved by placing the heterologous biofuel producing genes under control of endogenous promoters that are under control of the transcription factor, FadR (Section 15.2.4 and Figure 15.5). FadR binds to the promoter or medium chain FAs/fatty acyl-CoAs. As the FAs/fatty acyl-CoA content rises, FadR releases its grip on the promoter DNA with the effect that β-oxidation pathway genes under control of FadR sensitive promoters are turned on, whilst FA biosynthetic genes under control of such promoters are turned down. The effect is to balance, at the transcriptional level, FA biosynthesis and degradation. By placing genes for biofuel biosynthesis (FAEEs in this case) downstream of FadR sensitive promoters, yield was increased 3-fold (28% theoretical maximum) as the heterologous biosynthetic pathway was now more sensitive to changes in the cellular composition.

As has been described in previous sections, it is possible to engineer synthetic pathways to tailor the output to industrially-relevant molecules. It is also possible to re-engineer entire pathways to run in the reverse direction. Possibly one of the most remarkable feats of biological engineering is the reversal of the FA β-oxidation pathway to create, rather than degrade, long chain fatty acyl molecules. Reversal of this pathway requires the absence the β-oxidation pathway's natural FA substrate and the presence of an alternative source (*e.g.* glucose) for generating acetyl-CoA precursors. In such a situation, and with endogenous β-oxidation control mechanisms removed, degradative thiolases operate in a synthetic direction. This generates acetoacetyl-CoA from acetyl-CoA and these molecules are elongated by two carbon units each turn of the cycle. The introduction of various 'terminating' steps allows production of the desired biofuel output molecule. The activation of acetyl-CoA to malonyl-CoA, required in FA biosynthesis (Figure 15.3), is unnecessary in this scheme, saving one ATP molecule. The scheme also utilises acyl-CoA rather than acyl-ACP intermediates, thus avoiding the usual requirement for the conversion of acyl-ACP, *via* free FA, to fatty acyl-CoA for biofuel production, which uses one ATP molecule. The authors report on the production of linear alcohols ($C_{\geq 4}$) and extra long FAs ($C_{\geq 10}$) at greater efficiency than many previous studies: production of *n*-butanol and FAs were at higher titres (≈ 14 g/L and ≈ 7 g/L, respectively) and yields (33% and 28% $^w/_w$ product/glucose consumed, respectively) than previously reported (4.65 g/L and 28% $^w/_w$ for *n*-butanol and ≈ 2 g/L and 6% $^w/_w$ for FAs). Only the production of 1-butanol using a modified, irreversible Clostridial pathway, driven by NADH-accumulation. exceeded this level of production (Section 15.2.3).

Co-localisation of metabolic enzymes is another way in which pathway optimisation can be undertaken and scaffolding is one way in which enzyme activities can be co-localised (Figure 15.5). This has been achieved using protein scaffolds (for the MEV pathway and glucaric acid pathway), RNA scaffolds (for H_2 production), DNA scaffolds (for the MEV pathway again) and membrane-associated scaffolds (for FA biosynthesis). Notably, some of the work involving RNA, DNA and membrane scaffolds was undertaken by undergraduate students as part of the annual iGEM competition, further evidence of the catalytic role iGEM is having stimulating scientific research in this area. One of the advantages of scaffolds, as demonstrated by the use of a proteinaceous scaffold, is that it enabled the researchers to empirically test the effect of recruiting different enzyme stoichiometry to the complex; for example, optimal flux through the MEV pathway was achieved by recruiting more of the enzyme that catalysed the final step in the pathway. This resulted in a more efficient flux through the pathway, less toxic intermediate and a 77-fold increase in final titre. In eukaryotic systems, co-localisation of the enzymes required for the biosynthesis of branched chain alcohols (Figure 15.2b) to the mitochondria led to almost 500 mg/L *iso*butanol compared with 150 mg/L for cytosolic over-expression of the same enzymes.

15.3.4 Biofuel Pumps

Many of the biofuel molecules discussed above have known antimicrobial properties. In order to be industrially relevant, the yields of these molecules must often exceed the native tolerance levels. This has already been required to maximise and maintain ethanol yields of yeasts. In addition to developing pathways for advanced biofuels, strains must therefore be biofuel-tolerant. One of the means of achieving such tolerance is through the activity of efflux pumps (Figure 15.5). Efflux pumps are a class of membrane transporters in which compounds are exported from the cell using proton motive force. Expulsion of the biofuel compound has the added effect of removing the pathway endpoint, and potentially increasing pathway flux by alleviating feedback inhibition.

Efflux pumps in Gram-negative bacteria, such as *E. coli*, are composed of an inner-membrane protein responsible for target recognition and proton exchange (to drive export), a periplasmic linker and an outer membrane channel. Such efflux pumps can either be specific to a particular solvent or may be broad-range pumps. The number of identified pumps is low however, and does not cover the range of biofuel molecules required. To overcome this limitation, data-mining of published bacterial genomes allowed researchers to identify new pumps with similarity to a few well characterised pumps, and to clone 43 uncharacterised pumps for testing against advanced biofuel molecules. A competition-based strategy testing biofuel tolerance was first implemented: strains expressing different pumps were grown in isolation and then pooled in media with or without a particular biofuel and then allowed to undergo a series of growth/dilution cycles. The researchers followed the success of each strain and identified two pumps that consistently outperformed all others when cells were grown in the presence of limonene. One of these two pumps was transferred to a strain of *E. coli* that produced limonene with the result that limonene production from these cells rose. Given that limonene production in the cells with and without the efflux pump was below toxic levels, this presumably indicates the important role product inhibition can have on product biosynthesis. Similar results have been obtained using ABC transporters (better known as lipid and drug exporters), which were used to excrete isoprenoids synthesised within the cell, while further work has demonstrated strategies for improving transporter function and efficiency. This is important, for simply increasing the number of pumps to increase efflux can have a detrimental effect on cell viability by saturating the cell membrane with pores.

15.3.5 Synthetic Biology and Systems Engineering

Synthetic biology should be seen as a tool to enable improved metabolic engineering and understanding of metabolic flux. The tools of the trade may have their roots in the molecular biology advances of the past four decades, but many of the over-arching principles take their cue from engineering. Synthetic biology relies on modular, standard and well-characterised parts, computer-aided modelling and computer-aided data analysis; it also depends on key enabling technologies that can be automated in a high-throughput manner. These technologies include DNA sequencing and synthesis, genome sequencing (and the large amount of generated data), rapid assembly of genetic constructs (based on standard parts) and systems-wide data sampling (*e.g.* transcriptomics and metabolomics). One of the drivers of synthetic biology is to make biological systems easier (and more predictable) to engineer. The application of synthetic biology to metabolic engineering therefore allows us to both understand and engineer metabolic complexity. For example, standard DNA parts, assembled using automated platforms for the generation of large numbers of constructs, allows empirical testing and rapid combinatorial analysis of different strain genotypes (potentially under different fermentation conditions too). This allows a much wider interrogation of the design space than was possible a few years ago. Built upon this foundation, traditional biochemical and flux analysis can be combined with functional genomics to identify pathway components with high flux control coefficients. The data generated is used to propose new genotypes and the cycle is repeated. If the goal of microbial biofuel production is to produce an economically viable product, then such an approach will be invaluable to optimise the microbial system.

Generating strain genotypes to test metabolic flux control coefficients can be done at many levels. The amount of enzyme that is present in the cell can be tested by integrating different numbers of genes into the genome, employing different copy number vectors, varying transcriptional control mechanisms (promoters), translational control mechanisms (ribosome binding sites, riboswitches) and post-translational control mechanisms (protein stability). Ensuring full enzyme activity is a common problem: allosteric regulation, cofactors, protein–protein interactions and correct protein folding can influence this, and should also be considered. Protein engineering may be employed to alter substrate specificity and/or reaction kinetics of the enzyme, though this approach is not yet reliably predictable. Pathway engineering involves looking beyond a single enzyme-catalysed reaction, but at multiple steps along a biosynthetic route, and potentially at pathways which co-exist within the cell, competing for, or supplying substrates to, the target biochemistry.

Above this, productivity and flux may be influenced by host choice and physiology, as well as the fermentation conditions the cell finds itself living in. Given the number of parameters that may influence metabolic behaviour of the culture, it can be seen why iterative, high throughput systems, aided by computer modelling and correct statistical analysis, are crucial for success. Design of experiment (DoE), commonplace in optimising engineering and manufacturing processes (and with its roots in agricultural science), has been used to empirically test optimum codon-usage for heterologously expressed genes in several species and is the basis for the codon-optimising algorithms employed by synthetic biology company DNA2.0. Such approaches can be employed in pathway and system optimisation. Complementary to modelling, machine learning has been combined with DoE to successfully optimise protein expression in an *in vitro* setting. This approach (combined with automated liquid handling and yield phenotyping) provided a large degree of automation to the whole optimisation cycle.

Global transcriptional machinery engineering (gTME) is a methodology for producing complex metabolic phenotypes that has been applied to the production of various chemical compounds, including biofuels. The driver for the implementation of gTME is the recognition that the desired molecules are often produced by a number of metabolic pathways acting together in ways that are not yet fully understood. The technique relies on error-prone PCR to create mutant libraries of general (basal) transcription factors. In bacteria for example, σ factors regulate the promoter preferences of RNA polymerase. The main σ factor in *E. coli*, σ^{70}, is encoded by *rpoD*. Mutated *rpoD* genes are expressed at low levels in a wild-type background. The wild-type *rpoD* carries out the functions necessary for cell viability, while the expression of the mutant *rpoD* changes the way in which RNA polymerase activates its target promoters. The effect is therefore a dominant effect. The result is a change in the transcriptome at a global level. The mutant library is screened for cells with cellular metabolism altered to suit the desired phenotype and the process can be iterated in a directed evolutionary fashion. gTME has been applied to both prokaryotic and eukaryotic systems. Yeast and *E. coli* have been successfully engineered for improved tolerance to high glucose and ethanol concentrations, and an increased conversion efficiency from one to the other. Importantly, the sum of the transcriptional changes required for the desired phenotype was greater than the individual parts. gTME may therefore allow top-down changes complimentary to the bottom-up approaches described previously.

15.4 Conclusion

The expansion of first-generation ethanol biofuel into the retail market has demonstrated that microbial production of biofuels can be performed at a large scale. However, the global demand for both food and transportation fuel is expected to more than double by 2050. Future biofuel production will therefore be guided not just by the inevitability that it must be economically competitive and produced in very large quantities, but it will also be required to do so with minimal impact on food supplies. Furthermore, future biofuels should demonstrate a strong, positive net energy yield. For example, for every 1 unit of input energy, US corn ethanol produces 1.3 units of output energy, compared to 8 units of output energy for Brazilian sugar cane ethanol. They should also do so with minimal environmental harm, and preferably with environmental benefit. Achieving this is beyond the scope of microbial metabolic engineering alone, but designing microbial platforms to fit the desired processing pipeline and feedstock will be needed if this is to be achieved.

Selected References and Suggestions for Further Reading

Akhtar, M.K., Turner, N.J. and Jones, P.R. (2013) Carboxylic acid reductase is a versatile enzyme for the conversion of fatty acids into fuels and chemical commodities. *Proceedings of the National Academy of Sciences, USA*, **110**, 87–92.

Freemont, P. and Kitney, R. (2012) *Synthetic Biology – A Primer*. Imperial College Press, London.

Harger, M., Zheng, L., Moon, A., Ager, C., Han, J.H. *et al.* (2013) Expanding the product profile of a microbial alkane biosynthetic pathway. *ACS Synthetic Biology*, **2**, 59–62.

Howard, T.P., Middlehaufe, S., Moore, K., Edner, C., Kolak, D.M. *et al.* (2013) Synthesis of customized petroleum-replica fuel molecules by targeted modification of free fatty acid pools in *Escherichia coli*. *Proceedings of the National Academy of Sciences, USA*, **110**, 7636–7641.

iGEM (2015) Synthetic biology: based on standard parts. Available at: https://www.igem.org/Main_Page

Kung, Y., Runguphan, W. and Keasling, J.D. (2012) From fields to fuels: recent advances in the microbial production of biofuels. *ACS Synthetic Biology*, **1**, 498–513.

Peralta-Yahya, P.P., Zhang, F., del Cardayre, S.B. and Keasling, J.D. (2012) Microbial engineering for the production of advanced biofuels. *Nature*, **488**, 320–328.

Wen, M., Bond-Watts, B.B. and Chang, M.C.Y. (2013) Production of advanced biofuels in engineered *E. coli*. *Current Opinion in Chemical Biology*, **17**, 1–8.

16

The Sustainability of Biofuels

J.M. Lynch

Centre for Environment and Sustainability, University of Surrey, Guildford, UK

Summary

The term 'sustainability' embodies several different concepts, emphasis on which depends on the context in which the term is used. In this chapter, economic sustainability is discussed briefly before going on to look at greater length at environmental sustainability. The latter involves several sub-topics, including mitigation of climate change, food *vs* fuel and change of land use. Life cycle assessment is a key component of discussions on sustainability: several examples are presented.

16.1   Introduction

Traditionally, biomass such as wood has been used for cooking and heating. This is particularly still true today in Africa where wood is the base for about 70% of heat and power generated. Forestry is generally regarded as a highly sustainable industry, but with the global increase in deforestation and forest degradation, that sustainability is being questioned, especially in tropical countries where very high proportions of the populations are dependent on forests for their livelihoods. The illegal logging trade is estimated to be worth between US\$ 30 billion and US\$ 100 billion annually, with governments losing US\$ 10 billion annually in tax income[1]. Stolen wood is estimated to depress world timber prices by 16%. Some of the deforestation is caused by a move from tropical forest to energy crops such as palm oil, which fetch high prices in the short term but which are not sustainable (see Chapter 17). In the UK today, some commercial heat and power generation facilities produce energy from wood chips that replaces in part fossil sources, but all that wood is taken from public and private forests with sustainable management.

In recent years, biomass is increasingly used for the production of liquid biofuels used in transport vehicles. Various food commodities including vegetable oils, sugar, cereals and

---

1 Lynch *et al.* (2013).

Biofuels and Bioenergy, First Edition. Edited by John Love and John A. Bryant.
© 2017 John Wiley & Sons Ltd. Published 2017 by John Wiley & Sons Ltd.

other starchy crops are used to produce so-called first-generation biofuels: oils are converted to biodiesel *via* transesterification, while sugars – either directly from sugar cane and sugar beet or enzymatically derived from starches – are converted to bioethanol *via* fermentation (see Chapters 1, 5 and 6). Second-generation bioethanol using the cellulose from all kinds of plant cells (*e.g.* straw, wood) are under development (see Chapter 5), as are synthetic fuels from any type of biomass *via* gasification and the Fischer Tropsch process. Mostly the feedstock is food crops, which could therefore potentially compromise global food security (see Chapter 17). However, there has also been an interest in algae as feedstocks (Chapters 11, 12 and 13). The sustainability of all these systems is critical as to whether public and private investments can be made into their future development. Scientific achievements have been made in wood fuel, heat and light generation, fermentation and distillation, esterification of vegetable oils, but significant developments go on in biofuel quality, biogas, biomass gasification and biomass combustion systems.

As many of these developments in biofuel technology are driven by strong policy support, a proper analysis of the implications with respect to both stated objectives and unintended side effects is needed to provide policy-makers with the evidence to make decisions. In January 2007, the OECD Directorate for Trade and Agriculture, together with the Co-operative Research Programme on Biological Resource Management for Sustainable Agricultural Systems, organised a workshop on Bioenergy Policy Analysis in Umea, Sweden. The account which follows is in part a summary of the conclusions made there[2], together with an analysis from the follow-up workshop on modelling environmental, economic and social aspects in the assessment of biofuels of the Co-operative Programme, which was held in Copenhagen in 2008. It will also be set in the context of the opportunities and problems of bioenergy in relation to energy demand[3].

16.2   Bioenergy Policies

Support for bioenergy in general, and biofuels in particular, is provided in almost all producing countries. The objectives and their relative priorities are heterogeneous (Chapter 1) and include the reduction of fossil energy use in times of high crude oil prices and finite reserves, the reduction of greenhouse gas (GHG) emissions in the light of the evidence on climate change, the generation of new outlets for agricultural produce given relatively low farm prices, the development of rural areas (both in developed and developing countries), and others. Bioenergy is often considered to be capable of solving all these problems. Of particular significance is the EU Renewable Energy Directive (RED) 2009/28/EC, which promotes the use of biofuels in order to help Europe meet its strategy and commitments on reducing GHG emissions, improving the security of energy supply and increasing the use of renewable energy sources that would reduce the dependence on oil imports and make greater use of indigenous sources.

There are numerous policies in several countries largely covering two main areas. On the one hand, tax incentives, guaranteed prices and direct support for investment and production are given to bridge the gaps between production costs and market prices.

---

2 Lynch and von Lampe (2011).
3 Lynch and Harvey (2011).

On the other hand, the use of biofuels is directly increased by blending requirements and mandates for public fleets. In addition to these, a wide set of other policies directly or indirectly support the production and use of bioenergy, such as fuel standards, public research and trade measures.

There are several links to policies in other sectors. Direct links to agricultural policies exist through energy crop payments (see Chapter 7) and, in the EU, through granting of permission for non-food crop production on 'set-aside' land, while larger use of commodities for the production of bioenergy can lead to higher agricultural prices (see Chapter 17) – an objective that is common to agricultural policies. Air quality objectives are behind both bioenergy and environmental policies, while a more intensive production in agriculture can have negative environmental implications. Similarly, while the objective of increased energy security is joint to energy policies, higher fuel prices due to blending obligations could conflict with the objective of low energy prices.

16.3 Economics of Bioenergy Markets

Strong growth in bioenergy markets can be stimulated from attempts to reduce climate change or by lack of support for policies for increased oil prices in real terms. Increased use of renewables can play a role in reducing fossil energy use and GHG emissions. Even though the contribution of a number of bioenergy options are limited and almost all options only work with public support, there are some small-scale power and heat generation systems working under feed-in tariff regimes. Strong growth in bioenergy production will result from high crude oil prices and strong public support. This will result in high profitability. However, this will inevitably result in higher and more volatile prices in agricultural markets, but whether the effect of more volatile prices is inevitable has been questioned and depends on the responsiveness of the bioenergy industry's demand for agricultural products. The requirements of land for biofuel production are shown in Figure 16.1.

In most countries, costs of biofuel production at present exceed the energetic value of the resulting product, making biofuel supply dependent on public intervention. At the same time, strongly growing energy consumption around the world, and in particular the growth in transport fuel consumption, raise concerns about the future availability of fossil fuel sources as well as the developments in GHG emissions. While some smaller-scale facilities for power and heat production exist that benefit from specific feed-in tariff regimes, most of the biofuel industry is in the hands of large-scale enterprises, benefiting from lower production costs. Cross-border investments have been relatively rare to date, but are likely to become more important, and individual activities of foreign direct investments tend to be of significant magnitude.

With the current high prices for crude oil and hence for fossil fuels, and given the strong public support, many of the biofuel plants enjoy substantial profitability, which explains the explosive growth in production capacity and biofuel output in a number of countries. While much of the current fluctuation in agricultural commodity prices is caused by other factors, such as weather-related low harvests in Australia and other major producing countries and reduced public stocks, the growth in biofuel production and in the associated demand for feedstock commodities clearly further supports prices for cereals, oilseeds and sugar crops. At the same time and to the degree that profit margins remain

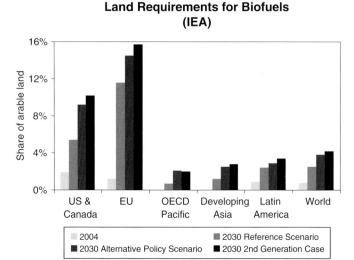

A range of scenarios were to have been considered by IEA on the land required to produce biofuel in the year 2030. Figures for 2004 are given for comparison.

Figure 16.1 Land requirements for biofuels.

important for biofuel producers, the low responsiveness with respect to feedstock prices together with the globally low stock levels may render international commodity markets more volatile than in the past, resulting in an increased risk of extreme situations.

16.4 Environmental Issues

The environmental and energy performance of bioenergy production systems varies significantly across forms of bioenergy, feedstocks, conversion processes and production regions. But in addition to these factors, life-cycle assessments of bioenergy production systems vary considerably, due to differences in data used and methods applied. Most studies indicate that the use of biomass for heat and power generation, preferably combined, tends to be more efficient in terms of fossil energy savings and GHG emission reductions than most forms of first-generation biofuels, at least when the fuels are produced from food commodities within the northern hemisphere. Second-generation biofuels offer higher potentials on both fronts, but are not yet commercially readily available. The implications for the local environment are ambiguous. On the one hand, higher intensification of agricultural production systems can reduce biodiversity and increase soil erosion and pollution of water and soil. On the other hand, there are potential co-benefits between energy use and nature protection. The use of innovative bioenergy cropping systems can result in a high energy yield but also reduce environmental pressures compared to some food cropping (see for example, Chapter 8). These reductions in environmental pressure can be made from less nutrient input, enhanced crop diversity and less use of heavy machines. The use of forest residues can support fire prevention measures in otherwise unmanaged forests in Southern Europe. The use of cuttings from grassland can maintain biodiversity-rich grassland and landscape diversity and provide a limited amount of bioenergy.

Table 16.1 The highest risks (C) where most environmental damage to occur comes from traditional agriculture. Lower risks (A) are achieved through innovative systems or perennial cropping.

Environmental risks in different cropping systems.

	'Innovative' double cropping	Wheat	Maize	Sugar beet	Poplar perennial - for comparison
Erosion	A	A	C	C	A
Soil compaction	A	A	B	C	A
Nutrient inputs into surface and groundwater	A	A	C	B/C	A
Pesticide pollution of soils and water	A	B	C	C	A
Water abstraction	A/B	B	A/B	A/C	B
Link to farmland biodiversity	B	B/C	B/C	B	A/B

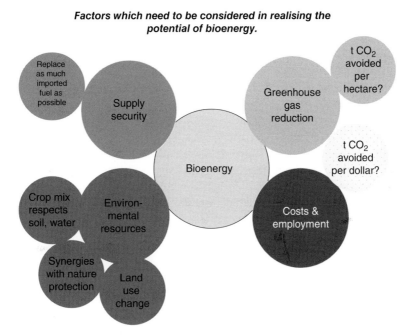

Factors which need to be considered in realising the potential of bioenergy.

Figure 16.2 Factors which need to be considered in realising the potential of bioenergy.

More research is needed to fully understand the complicated links between bioenergy production and use and the consequent net changes in fossil energy use, emissions of carbon dioxide and other GHGs, land use, soil carbon stocks, soil conservation, water quality and quantity, and biodiversity and landscape impacts.

Table 16.1 gives examples of low to high environmental risks for a broad range of annual crops when subjected to Life Cycle Assessment. Figure 16.2 gives some illustrations of how best the potential of bioenergy might be used while protecting the environment.

16.5 Life Cycle Assessment

16.5.1 General Features

Life Cycle Assessment (LCA) is a methodology which is able to reveal environmental and energy performances in relation to the reduction of GHG emissions and dependence on fossil fuel. A detailed analysis was provided by Cherubini *et al.* in 2009[4]. In order to investigate environmental impacts of bioenergy and biofuels, it is essential to account for other problems such as acidification, nitrification, land occupation, water use and toxicological effects of fertilizers and pesticides. LCA may give different results for apparently similar bioenergy systems. Policy-makers can sometimes be shy of LCA by seeing it as a complex and irrelevant mathematical process. In practice the opposite is true as it is constructed by listing everything that is relevant to the process and then set that on paper as a word model from which questions can be asked, and simple equations constructed to generate answers to the questions. However, LCA can also be valid in comparing just two or three components. For example, the energy and GHG savings per hectare per year (with replacement of inefficient coal or efficient natural gas) can be tabulated and that can be analysed in relation to lignocellulosic crop yields. Another useful analysis would be the effect of nitrogen fertilisation of a bioenergy crop on global warming potential (GWP) and how this affects soil quality in terms of soil organic carbon. Amongst other considerations are the type of management of raw materials, conversion technologies, end-use technologies, system boundaries and the reference energy system with which the bioenergy chain is compared.

A full LCA should:

- recognise the biomass carbon cycle, including carbon stock changes in biomass and soil over time;
- include nitrous oxide and methane emissions from agricultural activities;
- select the appropriate fossil reference system;
- evaluate homogeneity of the input parameters in Life Cycle inventories;
- recognise the influence of the allocation procedure when multiple products are involved; and
- estimate future trends in bioenergy, such as second-generation biofuels and biorefineries.

The conclusions and recommendations of Cherubini *et al.* (2009)[5] were:

- Each bioenergy system (especially those based on dedicated energy crops) should avoid the depletion of carbon stocks or, at least, any decline in carbon stock of any pool should be taken into consideration in calculating the GHG mitigation benefits of the system.
- Perennial grasses like switchgrass (*Panicum virgatum*) and *Miscanthus* can enhance carbon sequestration in soils if established in set-aside and annual row crop land, thus increasing the GHG savings of bioenergy systems (see also Chapter 7).
- LCA results of bioenergy from dedicated crops should be expressed on a per hectare basis, since the available land for the production of biomass raw materials is the biggest bottleneck (see also Chapter 17).

---

4 Cherubini *et al.* (2009).
5 Cherubini *et al.* (2009).

- LCA results of bioenergy systems based on biomass residues should be expressed on a per unit output basis, if there is a need to be independent from the kind of biomass feedstock, or per unit input basis, in order to be independent from the kind of biomass feedstock, or per unit input basis, in order to be independent from the final products and conversion processes.
- The production of liquid biofuels usually requires more fossil energy inputs than the generation of electricity and heat from biomass.
- As a consequence, electricity or heat generation from biomass may achieve larger GHG and fossil energy savings per hectare devoted to biomass production, than the production of transportation biofuels.
- Bioenergy chains which have wastes and residues as raw materials show the best LCA performances, since they avoid both the high impacts of dedicated crop production and the emissions from waste management (see also Chapter 3).
- Given the constraints in land resources and competition with food, feed and fibre production, high biomass yields are extremely important in achieving high GHG savings, although the use of chemical fertilizers to enhance production can reduce the savings.
- Fossil energy savings and GHG mitigation will be increased if agricultural co-products such as bagasse and straw, and process residues such as lignin, are also used for energy production to run the biomass conversion plants (as discussed in Chapter 5).
- When agricultural residues are collected from fields and used for bioenergy production, the effects of removal on that particular soil type cannot be neglected, and the GHG implications such as lower yields, N_2O emissions from land and decline in soil carbon pools, should be accounted when compiling the overall the overall energy balance of the bioenergy system.
- High biomass conversion efficiency to energy products is fundamental to maximising GHG emission savings.
- A lower degree of savings is achieved when power from natural gas or cogeneration sources are displaced; a high emission savings rate is achieved when coal-generated power, especially with low efficiency, is displaced.
- The initial use of products, followed by the use of energy ('cascading'), especially in the case of wood, can enhance GHG savings given scarce resources of biomass and/or land.

16.5.2   OECD Copenhagen Workshop, 2008

An OECD workshop held in Copenhagen in 2008 examined several aspects, including life-cycle assessments of development of alternative fuels. Several of the papers arising from the workshop were published in an issue of *Biomass and Bioenergy*[6].

The following are a selection of the workshops discussions and findings:

16.5.2.1   Comparison of Fossil Fuels with Biofuels Generated During Land Use Change (LUC)

A question was posed[7]: 'What would a specific area of land be used for if not for the production of biofuel?' It was concluded that LCA generated highly variable results depending on the type of biofuel and that geography is important, especially temperate *vs* tropical

---

6 OECD Copenhagen workshop (2008).
7 Reinhardt and von Falkenstein (2011).

conditions. The LCA of oil-seed rape (*Brassica napus*), *Miscanthus*, short-rotation coppice willow and forest residues were considered as land use systems for energy crops[8]. The focus was the impact on soil organic carbon as well as GHG emissions, acidification and eutrophication. Oil-seed rape had the worst LCA. As forest residues are by-products, their environmental impact was small.

The conceptual issues of LUC in relation to GHG emissions from soil and biomass were extensively considered by Delucci (2011)[9] citing 158 references. He concluded that the impacts of GHG emissions from the initial LUC can at least partially be offset by the impacts of carbon sequestration from the reversion (see also Chapter 16).

16.5.2.2 Biofuel LCA in the Regional Context

Reinhard and Zah (2011)[10] specifically considered the consequential LCA of production of methyl esters of rapeseed (canola) fatty acids (see Chapter 6) in Switzerland. This depends more on the environmental scores of marginal replacement products on the world market than on local production factors. Displacing food by RME production can reduce GHG emissions when GHG-intense soy meal from Brazil is substituted by rape and sunflower meal, which is a co-product of the vegetable oil production.

For passenger vehicles in Spain, the use of ethanol and fatty acid methyl ester (FAME: see Chapters 2 and 6) from sunflower has been considered for domestic and imported raw materials in relation to integrated land use change (ILUC)[11]. This has been done in order to realise the 10% substitution under the conditions of the Renewable Energy Certificate (REC) by 2020. It was shown that ethanol production would not displace land, whereas FAME will displace some amounts of land. This has been considered more widely in the global economy for agricultural production and will impact imports into the EU[12].

Iowa is a particular 'hot spot' for biofuels in the USA[13]. The Environmental Policy Integrated Climate (EPIC) model has been used to predict environmental impacts of biofuels in that state. In particular, the impact in relation to land in the Conservation Reserve Program (CRP) for land set aside and increased rotation to corn against soybean in intensive and extensive margins has been evaluated. Returning CRP land into production has a vastly different environmental impact as non-cropped land showing much higher negative environmental effects when brought back into row crop production.

The methodology to explore the future spatial distribution of biofuel crops in Europe has been made by Hellman and Verburg (2011)[14]. Both biodiesel and bioethanol, as well as second-generation biofuel crops were considered in the macroeconomic model. Areas which stand out for production have a combination of well-developed infrastructural and industrial facilities, alongside large areas of suitable arable land.

---

8 Brandão *et al.* (2011).
9 Delucci (2011).
10 Reinhard and Zah (2011).
11 Lechon *et al.* (2011).
12 Banse *et al.* (2011).
13 Secchi *et al.* (2011).
14 Hellman and Verburg (2011).

16.5.2.3 Social Aspects

The country with most experience of biofuels is Brazil (see Chapters 1 and 5). The social sustainability of Brazilian bioethanol has been considered by Lehtonen (2011)[15]. Policy intervention can transform power structures. The key questions asked were sustainability of what, how and for who can only be answered through fine-grained analysis of prevailing power structures.

With biofuels becoming 'mainstream', the social impacts such as increased food prices and 'land grab' by plantation developers need to be considered[16] (see also Chapter 17). Climate change policies are driven by energy security, but they may not yield high social or environmental benefits. The public reaction to this can sometimes be analysed in focus groups such as has been used to address the impact of genetically-modified food crops, which can then influence policy approach and implementation. Some of the social impacts to consider are:

- Why targeting marginal land comes at the expense of marginal communities;
- Why extraction of commodities by large corporations tends to come at the expense of local communities; and
- Why the switch to cash crops, and the creation of paid jobs, may not be a net gain for rural communities.

The production and use of biofuels is never carbon-neutral and the social debate needs to analyse the motivation for production, the motivation for consumption and the choice of scale.

16.5.2.4 Algae

The potential production of algal biodiesel has tended to follow land cropping for bioenergy and was not addressed in the Copenhagen debates, but is included here to present the full picture. It is probably even more important to perform LCA on algal biodiesel, as the processes for production are more complex (see Chapters 11, 12 and 13). Sander and Murthy (2010) performed a 'well-to-pump' LCA to investigate the overall sustainability and net energy balance of an algal biodiesel[17]. It quantified a major obstacle to algal technology as the need to efficiently process the algae into useable components. Thermal dewatering of algae requires high amounts of energy, currently provided mostly from fossil fuels; this produces a very negative input to LCA. They concluded that new technologies are needed to make algal biofuel a commercial reality, which is a conclusion reached by many other analysts.

Chlorella vulgaris has often been the alga of choice for biofuel in raceway or air-lift bioreactors (see Chapter 12 for discussion of algal culture systems). In an LCA of these two systems it was shown that cultivation in typical raceways would be significantly more sustainable than in closed air-lift bioreactors[18]. The GWP and fossil-energy requirements in this operation were found particularly sensitive to:

1) the yield of oil achieved during the cultivation;
2) the velocity of circulation of algae in the cultivation facility;

---

15 Lehtonen (2011).
16 van der Horst and Vermeylen (2011).
17 Sander and Murth (2010).
18 Stephenson *et al.* (2010).

3) whether the culture media could be recycled or not; and
4) the concentration of carbon dioxide in the flue gas.

A comparative LCA study was made for microalgae grown in ponds in Australia in relation to canola (oil-seed rape) and ultra-low sulphur (ULS) diesel. Comparison of GHG emissions and costs were investigated. Algae GHG emissions compared very favourably with canola and ULS diesel. However, the costs were not so favourable: algal biofuels were the most expensive, highlighting the need for high production rates to make the biodiesel economically attractive. In this respect, a combinatorial LCA was made to inform process design of industrial production of algal biodiesel[19]. The system was divided into five distinct steps:

1) microalgae cultivation;
2) harvesting and/or dewatering;
3) lipid extraction;
4) conversion (transesterification) into biodiesel; and
5) by-product management.

An LCA was done on various technology options for each process. The optimal option was found to be a flat panel enclosed photo-bioreactor and direct transesterification of algal cells with supercritical methanol.

Algal biodiesel production has also been analysed using the CA-GREET[20] model[21]. Current commercial data were used in a 'pond-to-wheels' LCA by Passell *et al.* (2013).[22] The environmental impacts of the base-line case (commercial data) were an order of magnitude higher than soy biodiesel and petroleum diesel. However, a future case with improved efficiencies provided lower impacts, again highlighting the need for more technical breakthroughs before algal biodiesel can become a commercial reality.

The source of the algae may be important. For example, in the salt flats of Namibia, aerial views reveal extensive red mats of the halo-tolerant alga *Duniella* in which the salt induces glycerol production to over 85% of its biomass[23]. Glycerol can be burnt in modified diesel engines as a result of the novel McNeil combustion cycle[24] and the algal mats can probably be harvested and dried easily. No LCA has been done for the process at this stage, but this unusual system might be more favourable for rural communities in southern Africa than many of the processes described above.

16.6 Conclusions

A clearer understanding of the individual objectives and their measurement as well as of the functioning and interrelationships of policy measures is important. A proper impact analysis will need to include both implications with respect to the different policy objectives, and of any side effects relative to agricultural markets

---

19 Brentner *et al.* (2011).
20 California-Modified Greenhouse Gases, Regulated Emissions, and Energy use in Transportation
California-Modified Greenhouse Gases, Regulated Emissions, and Energy use in Transportation
21 Woertz *et al.* (2013).
22 Passell *et al.* (2013).
23 Lynch and Harvey (2011); Harvey *et al.* (2012).
24 McNeil (2010).

and production systems. This all needs to be considered in relation to developing patterns of integrated land use.

To tackle issues like energy security and climate change, the reduction of energy consumption deserves more attention. Within the supply side, the direct use of biomass for heat and power generation tends to perform better in terms of net energy use and GHG emissions than their transformation to liquid biofuels for transport vehicles. When promoting biofuels, emphasis should be given to the technological development of promising second-generation fuels.

Policy support for energy cropping should build on an assessment of environmental impacts. Some bioenergy crops are different from food crops and open up new possibilities for optimising production systems. An important bioenergy potential comes from waste and residues and can in many cases be used at low cost. LCAs of the net emissions need to be analysed and this should not only cover CO_2. For example, wetlands can generate much methane, and cattle and cows are also an important contributor to this gas pool, which has a GHG effect about 20 times that of CO_2.

We need to develop renewable energy sources and, conditional on the policy objectives, bioenergy is a promising route if there is environmental benefit. On a large scale there will be serious competition with increasing food demands in the near future. LUC due to energy cropping could have major environmental impacts, which would negate potential carbon benefits for decades (or longer). For example, one of the greatest risks is the decline in soil structure if cropping is not managed critically. There is a need for cross-cutting research into technological options and approaches to biomass production and utilisation. In helping to generate evidence for policy, but also to implement policy, Earth observation by remote sensing from planes or satellites are likely to be useful tools. A variety of satellite systems are now available which can get resolutions of up to one metre and through clouds using radar[25]. This is particularly useful to protect the main source of bioenergy as wood fuel under the UN REDD+ (Reduction of Emissions Due to Deforestation and Forest Degradation) Programme, but in COP 19[26] in Warsaw in December 2013, the critical decision was taken that this can only be fully evaluated in an LUC scenario, where the move to biofuel is one important component.

Selected References and Suggestions for Further Reading

Banse, M., van Meiji, H., Tabeau, A., Woltjer, G., Hellman, F. and Verburg, P.H. (2011) Impact of EU biofuel policies on world agricultural production and land use. *Biomass and Bioenergy*, **35**, 2385–2390.

Brandão, M. *et al.* (2011) *Biomass and Bioenergy*, **35**, 2323–2336.

Brentner, L.B. *et al.* (2011). *Environmental Science and Technology*, **45**, 7060–7067.

Cherubini, F., Bird, N.D., Cowie, A., Jungmeier, G., Schlamadinger, B. *et al.* (2009) Energy- and greenhouse gas-based LCA of biofuel and bioenergy systems: key issues, ranges and recommendations. *Resources, Conservation and Recycling*, **53**, 434–447.

---

25 Lynch *et al.* (2013).
26 19th session of the Conference of the Parties to the United Nations Framework Convention on Climate Change, Warsaw, Poland, November 2013.

Delucchi, M. (2011) A conceptual framework for estimating the climate impacts of land-use change due to energy crop programs. *Biomass and Bioenergy*, **35**, 2337–2360.

Harvey, P.J. *et al.* (2012) *Proceedings of 20th European Biomass Conference*, European Union, Brussels, pp. 85–90.

Hellman, F. and Verburg, P.H. (2011) Spatially explicit modelling of biofuel crops in Europe. *Biomass and Bioenergy*, **35**, 2411–2424.

Lechon, Y. *et al.* (2011) *Biomass and Bioenergy*, **35**, 2374–2384.

Lehtonen, M. (2011) *Biomass and Bioenergy*, **35**, 2425–2434.

Lynch, J.M. *et al.* (2013) *Nature*, **496**, 293–294.

Lynch, J. and Harvey, P.J. (2011) Opportunities and problems of bioenergy. *Biochemist*, **33**, 39–43.

Lynch, J.M. and von Lampe, M. (2011) The need for bioenergy policy analysis. *Biomass and Bioenergy*, **35**, 2311–2314.

McNeil, J. (2010) *Combustion method*. GB Patent No 2460996B.

OECD Copenhagen workshop (2008) Modelling environmental, economic and social aspects in the Assessment of Biofuels. *Biomass and Bioenergy*, **35**(6), June 2011.

Passell, H. *et al*, (2013) *Journal of Environmental Management*, **129**, 103–111.

Reinhardt, G.A. and von Falkenstein, E. (2011) Environmental assessment of biofuels for transport and the aspects of land use competition. *Biomass and Bioenergy*, **35**, 2315–2322.

Reinhard, J. and Zah, R, (2011) *Biomass and Bioenergy*, **35**, 2361–2373.

Sander, K. and Murthy, G.S. (2011) Life cycle analysis of algae biodiesel. *Biomass and Bioenergy*, **15**, 707–714.

Secchi, S. *et al.* (2011) *Biomass and Bioenergy*, **35**, 2391–2400.

Stephenson, A. *et al.* (2010) *Energy Fuels*, **24**, 4062–4077.

van der Horst, D. and Vermeylen, S. (2011) Spatial scale and social impacts of biofuel production. *Biomass and Bioenergy*, **35**, 2435–2443.

Woertz, I.C. *et al.* (2013) *Environmental Science and Technology*, **48**, 6060–6068.

17

Biofuels and Bioenergy – Ethical Aspects

John A. Bryant[1] and Steve Hughes[2]

[1] *College of Life and Environmental Sciences, University of Exeter, Exeter, UK*
[2] *Egenis, University of Exeter, Exeter, UK*

Summary

Ethical decisions are made in several different ways, according to which ethical 'framework' is being used. The relationship between these ethical frameworks and environmental ethics is not straightforward but nevertheless, a number of different approaches to environmental ethics can be identified. When these are applied to the topic of biofuels and bioenergy, different approaches place different emphases on particular aspects of the debate. This is illustrated by our analysis of three particular issues, namely food *vs* fuel, environmental damage and biodiversity, land-grab and the rights of indigenous people.

17.1 Introduction to Ethics

17.1.1 How Do We Make Ethical or Moral Decisions?

Ethics may be defined as the study of or classification of the ways in which people make moral decisions, that is decisions relating to the philosophical concepts of right and wrong. However, in wider, non-academic contexts, the distinction between *ethics* and *morals/morality* is often lost so that they are used interchangeably. Thus, the *unethical* is often used to mean *immoral.*

So, how do people decide between right and wrong? The most straightforward system or framework is one that assumes that there are actions that are always wrong and actions that are always right. Because we know what is right and what is wrong, it is our duty to carry out the right action, giving us the term **deontological ethics** (the Greek word *deon* means duty). The Ten Commandments in the Old Testament of the Bible are often regarded as examples of deontological ethics, but as one of us has discussed elsewhere[1], several of them are actually more complex than this. We are on safer ground with Immanuel Kant (1724–1804) who developed the idea that moral decision-making could be classified as a series of moral imperatives and that it is our duty to obey these

---

1 Bryant (2013).

Biofuels and Bioenergy, First Edition. Edited by John Love and John A. Bryant.
© 2017 John Wiley & Sons Ltd. Published 2017 by John Wiley & Sons Ltd.

imperatives. For this reason, deontological ethics is sometimes known as **Kantian ethics.** One of Kant's major ('categorical') imperatives was (putting it in Modern English) that no-one should use another human being as a means of fulfilling their own wishes.

A modern variant of deontological ethics is **rights-and-duties** ethics. The concept here is that we can define a set of rights (as in the UN's Universal Declaration of Human Rights and the European Convention of Human Rights) that should be available to all and that it is a duty not to infringe anyone's rights. Thus, in exercising my rights I have a duty to take your rights into consideration. This may seem rather far from the topic of biofuels but as will become apparent, it is very relevant.

Some ethicists are very critical of the rights-and-duties version of deontology. Mary Warnock for example, believes that it removes concepts of right and wrong from our dealings with each other. Furthermore, it has led to some very tenuous claims about moral rights and wrongs. For example, in 2014, the billionaire Donald Trump made the (legally unsupportable) claim that building a wind farm near a golf resort that he owns in Aberdeenshire, Scotland infringes his human rights as defined by the European Convention on Human Rights. Nevertheless, despite its short-comings, rights-and-duties have been very useful at the international level, especially in dealing with despotic regimes.

Returning briefly to Kantian ethics, another imperative was that no-one should ever tell a lie. This brings us to another ethical system. Many of us can think of situations in which telling a lie is at least 'more right' than telling the truth, for example, telling a lie in order to save someone's life. Saving a life thus 'trumps' the imperative to tell the truth, because the outcome of telling a lie is good. This is an example of **consequentialism**, an ethical system in which the rightness or wrongness of an action is decided on the basis of its consequences. A particular form of consequentialism, **utilitarianism**, judges between two actions on the basis of which of them promotes more happiness or satisfaction. Consequentialism in its various forms is the most widely used way of making moral decisions in developed 'Western' countries, albeit that it is tempered by some deontology (*e.g.* obeying the law) and at a personal level is often combined with a strong sense of moral relativism, the idea that there are very few universally applicable standards of right and wrong and that morals are mostly a matter of individual choice and preference[2].

For two other ethical systems we need to go back to the Ancient Greeks, especially Plato and Aristotle. They believed that right moral decisions were made by people of good character who exhibited and practised the classical virtues, especially courage, wisdom and temperance (moderation). This is known, perhaps obviously, as **virtue ethics.** We should note that the Greek list omits self-giving love, in Greek, *agape;* sometimes known as charity, but the modern use of this word does not convey the original meaning. This is very much associated with the Christian era, as we have discussed elsewhere[3]. Over the centuries, and especially since the 18th century, virtue ethics fell out of favour. However, in the late 20th century, it experienced a revival which still continues, in both religious (especially Christian) and secular contexts.

We owe one more ethical influence to the Greeks and that is **natural law.** Aristotle suggested that all components of the world, whether living or non-living, had a purpose in existence, their *telos.* Exercise of true virtue would enable any component, whether a rock or a plant or an animal, to fulfil its *telos.* A link between virtue and natural law was

---

2 For further discussion of this, see Bryant and la Velle (2017).
3 Bryant (2013); Bryant and la Velle (2017).

also made by the Christian theologian, Thomas Aquinas, who believed that natural law actually expressed God's purposes for the world and that virtue would lead us to perceive and observe these. Vestiges of natural law are still seen in the Roman Catholic approaches to reproductive ethics and, relevant for this chapter, in some of the arguments presented by some environmental campaigners.

It is clear that none of these ethical frameworks is perfect. Deontology is at times too rigid. Consequentialism may lead to us thinking that the ends always justify the means and in its utilitarian guise may lead to legitimate interests of minorities being ignored. Virtue ethics may lapse into a vague 'all you need is love' (we describe this to our students as 'Beatles ethics'), while natural law can so easily fall into the naturalistic fallacy ('what is should be', which we might translate as 'if it's natural, it's good or right'). Nevertheless, in discussing any ethical issue, we need to be aware that a person making an ethical decision may employ any of these frameworks or a mixture of two or more[4].

17.1.2 Environmental Ethics

Many ethical decisions will involve more than one stakeholder: more than one person or group has an interest in the outcome of the decision and sometimes it is very difficult to work out, between the competing sets of interests, where the best decision lies. In dealing with a topic such as biofuels, ethical consideration will involve people, both as individuals and as larger groups, and 'the environment'. The environment becomes one of the stakeholders in the decision-making. We thus need a brief consideration of environmental ethics.

There are two broad approaches to valuing the environment and thus in the way it is regarded as a 'stakeholder'. The first assigns to it **intrinsic value**, that is we value it just because 'it is there'. This approach is often linked with particular ways of thinking of the position of humankind within nature. Thus, those holding biocentric or ecocentric views, regarding humans as just 'plain members and citizens'[5] of the 'land community', are likely to assign intrinsic value to the environment. For some, there may be a religious element in that the environment is the work of a creator-God (as in the monotheistic religions) or because nature itself is regarded as divine (as in pantheistic religions).

The second approach assigns **extrinsic value** to the environment and generally that means we value it for what it provides to humankind, which can include, along with the obvious material provision ('ecosystem services'), things like leisure and aesthetic enjoyment. Human needs are thus taken into account. In practice though, many people who assign intrinsic value to the environment also recognise its extrinsic value[6]. It is actually very difficult, in the face of human needs, to maintain a totally intrinsic, eco/bio-centric approach (and it is probably unrealistic to try). Thus, in discussions of climate change and of biofuel production and of a whole range of environmental issues, we see a mixed approach to ethical decision-making.

Based on what has been written here, we are now set up with an understanding of ethical frameworks and an appreciation of how 'the environment' can be a stakeholder in ethical decision-making. We are thus in a position to introduce an ethical discussion of biofuels. As we do so, we encourage readers to think about which combination of ethical framework and valuation of the environment they would use in their own analysis.

---

4 See, for example, the discussion on virtue ethics in *Beyond Human?*
5 Quoting Aldo Leopold (1949).
6 Readers who want a fuller treatment of this are referred to Chapter 14 in Bryant and la Velle (2017).

17.2   Biofuels and Bioenergy – Ethical Background

As we saw in Chapter 1, one of the main drivers for use of alternative fuels is the need to move away from fossil fuels. This need arises from the increase in atmospheric CO_2 caused by the burning of such fuels. Unless we can reduce significantly the use of fossil fuels, the Earth's temperature is set to rise to levels which are unprecedented in human history. The planet has of course been here before, several times during geological history. However, the changes are taking place so fast that there are concerns about the ability of organisms and indeed whole ecosystems to adapt, let alone any of the problems raised for human society.

Biofuels are amongst the mix of alternative energy sources that are contributing to and will continue to contribute to reduction in the use of fossil fuels. In early 2014, they comprised only a small fraction of the energy budget and a great deal of what they do contribute is in the form of biomass for direct combustion. Furthermore, much of this comes from 'traditional' sources, as was discussed in Chapter 1. However, there has already been some diversification. New sources of biomass, for example giant grasses like *Miscanthus,* are grown specifically for this purpose (Chapter 7), food and agricultural waste is 'digested' to produce 'biogas' (Chapter 3), sucrose and waste plant material are fermented to ethanol (Chapters 1, 5 and 8), plant oils have been added to the mix of combustible fuels (Chapter 6), and so on.

We regard the ethical case for reducing very significantly our use of fossil fuels as incontrovertible. Furthermore, we believe that biofuels can make a major contribution to our ability to do this. However, as the contribution made by biofuels to the energy budget increases, so the range of ethical issues arising from biofuel production also grows. We can summarise the main issues in a series of questions:

- Can biofuels be produced on an adequate scale, without reducing significantly our capacity to grow food?
- Can biofuels be produced sustainably, without inflicting serious damage on the environment?
- Can biofuels be produced without infringing the rights and damaging the livelihoods of traditional farmers and indigenous peoples?

We now deal with these questions in turn.

17.3   The Key Ethical Issues

17.3.1   Biofuel Production and the Growth of Food Crops

The amount of agricultural land on the planet is decreasing because of the need to house the increasing population of the world. Climate change is having, on average, deleterious effects on crop yields and will also lead to reductions in land area because of rising sea levels. At the same time, about 1 billion people are severely undernourished. While it is true that a significant proportion of these are hungry because of poverty – they cannot afford adequate food – it is also true that increases in crop productivity are not keeping up with increases in the population. All this suggests that we cannot afford to switch land use from food to biofuels and yet, as we have seen in Chapter 1, there are pressing reasons for reducing very significantly our use of fossil fuels.

However, the magnitude of the food *vs* fuel problem varies significantly from one part of the world to another. Three examples illustrate this. First, as we noted in Chapter 1, Brazil is the world's second largest producer of and the largest exporter of fuel-ethanol. The feedstock for Brazil's ethanol production is sucrose from sugar cane and despite the scale of production, sugar cane grown for fuel occupies only 1% of the country's arable land. Indeed, it is claimed that biofuel production does not reduce the amount of land dedicated to growth of food crops. This claim was supported by the World Bank in its report on the effects of biofuel production on global food prices.

The second example is the USA. It overtook Brazil as the world's largest fuel-ethanol producer in 2005, but unlike Brazil, US fuel-ethanol production is from fermentation of carbohydrates from corn (maize). The current rate of production is about 40 billion litres per year; this represents a 6.5-fold increase on the rate of production in the year 2000, with the increase between 2007 and 2012 being especially rapid. There has also been a large increase in bio-diesel production, mainly from oil-seed rape and soybean but with some contribution from sunflower oil. Current production stands at about 5 billion litres per year, meeting the targets set by the Federal Renewable Fuel Standard. However, this amount of fuel oil is just 0.6% of the amount of gasoline used annually in the transport sector in the USA, showing just how difficult it is to move from fossil to renewable fuels.

The increased production of fuel ethanol in the USA, noted above, has been achieved first by diverting land use from other crops to growth of corn and especially by an increase in the proportion of harvested corn that is used for ethanol production. The latter reached 25% in 2007 and 40% in 2012 when, for the first time, the amount of corn used for ethanol production exceeded the amount used in animal feed. Furthermore, this diversion of a large proportion of the corn yield from human and animal nutrition to fuel production has led to a 20% reduction in the amount of corn exported from the USA.

The third example is the UK. Current use of oil is about 93 billion litres per year. As with much of the rest of the EU, the main liquid biofuel is biodiesel; the EU produces about 55% of the world's biodiesel. Yield of biodiesel from a typical UK oil crop, oil-seed rape, is about 954 litres per hectare. The UK has about 6.1 million hectares of agricultural land and if all this was used for growing oil-seed rape for biodiesel, the yield would be about 5.82 billion litres or about 6.26% of the total annual usage. This clearly is a non-starter, especially as it is predicted that by 2030, the amount of agricultural land will be 2 million hectares short of what is needed to grow food. The UK cannot produce, using standard agricultural practice, anything like enough liquid biofuel to meet its needs. In fact, the current proportion of total liquid fuel use made up by biofuels (biodiesel and ethanol) in the UK is just over 3%. While these data are specific to the UK, the situation will be similar in any relatively small, relatively densely populated industrial countries in which agricultural land is at a premium. Making significant inroads into fossil fuel use is again seen to be very difficult.

But has the push towards biofuels affected food prices? Prices of staple foods certainly rose sharply in 2008, bringing yet more people into food poverty. A report undertaken by a consortium of global organisations, including the UN's Food and Agriculture Organisation, the World Trade Organisation and the World Bank, concluded that increases in biofuel and especially bioethanol production reduced the supply of food and caused price increases. For many people in wealthy developed countries, the price increases were bearable, but even in those countries, the poorest citizens found the increases to be difficult or impossible to meet. The situation was even worse in poorer countries in which food poverty undoubtedly increased. The latter was not helped by the fact that land use in some poor countries is diverted from growth of food for the native population to growth of staples and cash crops for export. Furthermore, it is particularly noteworthy that in at least one of these poorer countries, Somalia, the exported staple crop, cassava, is used for fuel ethanol production in wealthier countries.

Making ethical decisions is often not easy and this is especially so in the area of environmental ethics. The case for reducing the use of fossil fuels is obvious; using biofuels looks like an ideal component of the reduction process. Equally obvious is the need to feed people, a problem that grows by the minute (world population is currently increasing at a rate of approximately 150 per minute). Of these competing demands, most people choose food production and this has led to strong demand for '*Food not Fuel*'. A campaign under this banner has been conducted by a consortium of relief agencies and the ethical tension becomes very apparent in that several members of the consortium also campaign vigorously about the need to stem the tide of climate change.

Both the USA and the EU have lowered their targets for the adoption of liquid biofuels, in order to reduce effects on growth of crops for food. In the meantime, research goes on, first into potential biofuel crops that can grow on land not suitable for agriculture (but see next section) and second, into the production of fuel ethanol from waste plant biomass such as corn and wheat straw (see Chapter 5) and from crops currently grown for biomass (such as *Miscanthus* – see Chapter 7). For biomass crops, it is again important that they do not occupy good agricultural land[7]. Third, there has been extensive investment into research on production of biofuels from algae, fungi and bacteria (Chapters 9 to 15), culture of which does not use agricultural land. There has been some success in this area and it has been claimed that micro-algae are capable of producing 300% more fuel per unit area than oil-seed rape. Even so, it will be a long time before these sources meet a significant proportion of our liquid fuel needs. It is estimated that by 2022, between 35 and 40% of the world's biofuels will come from algae and a further 5% coming from bacteria. However, this is still a long way from meeting global needs for liquid fuel. If current targets are met (see above), then in the EU for example, biofuels will only contribute 10% of the total of liquid fuels.

17.3.2 Is Growth of Biofuel Crops Sustainable?

Sustainability and *sustainable* have become 'buzz words' over the past 10 to 15 years, but what do they actually mean? A general definition of *sustainable* is capable of being maintained at the current rate or intensity. This definition is very familiar to one of us from his days as a competitive long-distance runner: was a particular pace sustainable throughout the race? However, in environmental ethics, it has acquired a more specific meaning: can an activity (which might be farming, or something more nebulous such as economic growth) be continued without (further) damage to the environment or without depletion of natural resources.

This question of whether biofuel production is sustainable, both in a general sense and in relation to specific environmental factors, is dealt with in Chapter 16. Here we

---

7 We are not talking here of woody biomass crops grown to fuel power plants. According to a recent report, using wood as a biomass fuel may not be any 'cleaner' than using a fossil fuel.

want to point out the main areas of debate. First, we have already seen that at present, production of liquid biofuels cannot be regarded as sustainable in respect of food production. It may be possible to avoid that problem if, as we suggested above, non-agricultural land could be used for biofuel crops. However, that brings its own problems. If a natural habitat is cleared (and, most likely, the plant material is burned), then the CO_2 locked up in the habitat is released to the atmosphere. Because of this, the saving of CO_2 emissions from fossil fuels is outweighed by emissions from the cleared habitat and it may be many years before carbon neutrality is achieved. Cultivation of palm oil represents an obvious example of this. It is grown as a source of two different mixtures of lipids (from the fruit mesocarp and kernel: Chapter 6), which have extensive uses in the manufacture of cosmetics and in the food industry. Indeed, the use of palm oil for biofuel is relatively recent. A very large proportion of the world's palm oil crop is grown on land cleared from long-established forests, especially tropical rain forests in Southeast Asia and Africa. These forests are a major repository of fixed CO_2. Even if all the palm oil from these plantations was used as fuel, it would take many years before this resulted in a net reduction in emissions. Indeed, one estimate suggests that it could take as long as 220 years for a plantation to become carbon-neutral[8]. This has led one of the world's major oil producers to state that palm oil is not a sustainable source of biofuel.

Use of previously uncultivated land for growth of biofuel crops also causes a decrease in biodiversity, because of loss of the original habitat and the replacement of a diverse, established plant community with a monoculture. Tropical rain forest is again a dramatic example. As has been described elsewhere[9], this important habitat has been cleared at an alarming rate, not just for palm oil but for several other uses including logging, ranching and growth of more 'conventional' crops such as soybean. Tropical rain forests currently occupy about 6% of the world's land area but contain nearly 50% of the known animal and higher plant species, some of which are 'iconic figures' for the conservation movement. Destruction of rain forests in parts of Southeast Asia for example, has led to the loss of up to 80% of the habitat for orang-utans. Both orang-utan species are now endangered[10].

17.3.3   Biofuel Production, Land Allocation and Human Rights

The final aspect of sustainability that we wish to consider relates to land allocation and its implications on human rights, especially the rights of indigenous people and local farmers. The amount of land devoted to biofuel/bioenergy crops nearly tripled between 2005 and 2010. The rate of increase has slowed since 2010 but even so, many thousands of hectares are newly allocated to biofuel/bioenergy crops each year. In relation to this, the Nuffield Council on Bioethics was clear in its consultative report on biofuels[11], that since global climate change is largely attributable to overconsumption of non-renewable hydrocarbons, there is a moral imperative to develop alternative fuels from renewable bioresources. However, the Council was equally clear that the endeavour should be governed

---

8 Achten and Verchot (2011).
9 Bryant and la Velle (2017).
10 http://www.orangutan.org.au/palm-oil
11 Nuffield Council on Bioethics (2011).

by a set of ethically argued performance yardsticks. Most obvious among these was the sustainability requirement in relation to greenhouse gas output: biofuel production should not result in a net increase of the associated carbon footprint relative to current practice. This may either be expressed in terms of greenhouse gas emissions or fossil energy consumed as a collateral of the overall chain of activity from production to consumption. Either way, the yardstick, while as much a simple economic as an ethical requirement, may not be simple to apply. In the context of fuels derived from agricultural crops, it may seem that an objective linear arithmetical calculation of the carbon units consumed in planting, fertilising, irrigating, harvesting, transport (including worker transport), processing, and process reagent and byproduct disposal is a straightforward matter. However, this approach overlooks other indirect, less easily estimated consequences, such as loss of sequestered carbon during changes of land allocation, for example from forest to arable or from smallholder cultivation to mono-cropping. It also discounts the spatially remote knock-on effects such as the intensification of carbon inputs in other regions, which might happen as a consequence of the need to compensate for the diversion of land away from food production.

As soon as we enter this realm of argument, uncertainties regarding chains of causality and indirect consequences throw up a challenge to reliance on consequentialist ethics alone. Land allocation and the issue of food *vs* fuel became a contested area (see Section 17.3.1) since the introduction in the USA of Renewable Fuel Standards (RFS) associated with the so-called first generation of biofuels and the international trading of replacement carburants such as carbohydrate-derived ethanol. On the one hand, there are lobbies representing those who have invested in the production chain, from land acquisition, to crop production, to process plant and transport, who are keen to articulate the global environmental benefits and seek to raise the minimum provisions of the RFS. Against this are those who were alarmed by the rise in world food commodity prices which accompanied the first wave of corn-derived ethanol production, which was widely attributed to shift of land allocation from food to fuel. The latter argue for a suspension of plans to elevate the RFS so as to moderate the environmental pressures of land allocation to fuel production. Each evoke consequentialist arguments to support their position but as highlighted above, uncertainties over causal relationships and local heterogeneity in the production circumstances undermine the ethical postures.

This kind of conflict is further complicated by the potential intervention of biomass (cellulosic)-derived ethanol transformed from waste product streams. Corn stover (the residual plant material left in the field after harvesting of the grain) is a popular example, though to date there has been little commercial production contributing to the RFS, largely due to the costs and complexities of rendering lignocellulose down to fermentable substrates. However, cellulosic ethanol plants are now running in Brazil (based on sugar cane bagasse) and the USA (based on corn stover) and there is potential to expand these activities. Nevertheless, while the advent of so-called advanced biofuels of this nature seems to sidestep the food *vs* fuel confrontation, since land reallocation is not required (waste is from crops that have been used in the normal way) they, like grain ethanol, carry some potential direct and indirect consequences for which we must account. As a first instance, biomass ploughed into the land post-harvest (rather than being used for ethanol production) can contribute significantly to its net capacity for carbon sequestration. Then, for many farmers, especially in less developed countries, crop by-products constitute valuable resources for shelter, insulation or husbandry which may be traded in the local community, contributing both to individual livelihoods and social capital. Projections for the global expansion of cellulosic ethanol based on the assumption of zero-value by-products diverted from farming communities to opportunist capital- or state-driven centralized processing facilities are therefore problematic. This is not only because those non-edible feed-stocks may already actually be off-setting carbon elsewhere, but because community economics and livelihoods may be put at risk. These possibilities raise concerns about distributive justice and procedural justice, as was recognized by the principles established for biofuel standards in the Nuffield Report (see above).

Distributive justice resides in the rights of farmers and farming communities to pursue their livelihoods and to draw sustainably on the natural resources on which they depend. It is the duty of responsible state officials or entrepreneurs to protect this right and to give it due consideration in the zoning of biofuel production. The same principle applies to land reallocation for production of oil crops such as palm oil and of cereals and sugar cane for ethanol production (food *vs* fuel again), just as also does the principle of procedural justice. Procedural justice recognises the rights of stakeholders to be consulted and to participate in decision-making, which is especially significant in the case of farming communities and the reallocation of their land or crop by-products to fuel production. Indicators of conformation with these principles in pursuance of an ethical fuel standard could be straightforward to monitor, in which case it would be required of the producer and trader to demonstrate that an established procedure had been followed. This would lead us to lean more heavily on deontological (rights and duties) than consequentialist ethics. However, at a local level, where indirect effects may be easier to assess, it may be appropriate to take a consequentialist view. For instance, some argue that despite the reallocation of land from food to fuel, the improved infrastructures and resource management protocols developed to support fuel production may be expected to lead to augmented rather than depleted food availability. Clearly, the extent to which this beneficial consequence might be achieved would be very much dependent upon local circumstances and upon how assessments are made, but they could underpin a checklist for ethically predicated good practice.

However, despite any over-arching motivation for good practice, there have been many claims that land has been re-allocated with little or no consultation. GRAIN (an NGO) estimated that between 2003 and 2013, there had been at least 300 examples of 'land-grabs' for biofuel production (especially oil-palm)[12], totalling over 17 million hectares, mainly in Africa (Figure 17.1), but also in Asia, South America and Eastern Europe. In many instances, the new owners of the land are investors from other countries, especially China. Now, we might be inclined to think that an NGO such as GRAIN would be biased in its reporting. However, the World Bank itself is largely in agreement with GRAIN's findings. Furthermore, the World Bank has reported that;

> …these projects are not providing benefits to local communities. Environmental impact assessments are rarely carried out, and people are routinely booted off their land, without consultation or compensation … Investors are deliberately targeting areas where there is 'weak land governance'.[13]

---

12 http://www.grain.org/article/entries/4653-land-grabbing-for-biofuels-must-stop. Note that millions of hectares have also been grabbed for large-scale food crop production.

13 http://www.grain.org/article/entries/4026-the-world-bank-in-the-hot-seat

Figure 17.1 Aerial photo of the lands taken by Addax Bioenergy for its sugar cane plantation in Sierra Leone. Photo: Le Temps, Switzerland/public domain.

These land-grabs lead to a loss of livelihood and clear infringement of human rights for those who previously occupied the land. Focusing on just one region of Ethiopia (Gambella), the Oakland Institute in California concluded[14] that the national government had: 'perpetrated human rights abuses in resettling indigenous communities ... to allow for land investment deals to move forward.' Furthermore, the Institute reports that they '... did not find any instances of government compensation being paid to indigenous populations evicted from their lands.'

It is thus very clear that the ethical standards set out by the Nuffield Council and also by the World Bank, in respect of land re-allocation, are in many instances not being adhered to. While there is certainly great urgency in the need to produce non-fossil fuels, highlighted by the October 2014 report of the IPCC (stating that fossil fuels need to be phased out completely by the end of the 21st century), this does not justify human rights abuses, even under the most pragmatically entrenched utilitarian ethical systems.

Finally, in this section we consider that a frequently acknowledged additional driver of the development of biofuels is national security, which translates into an ambition to secure a fuel supply, independent of imported oil, for military and 'defence' purposes. Beyond this, threats to the supply of fossil fuels themselves may arise from international and regional conflicts (as currently exemplified by the Middle East), terrorism and the fragility of transport networks and stockpiling provisions. Threats to fuel security are seen as something which governments would have

14 http://www.oaklandinstitute.org/oakland-institute-exposed-human-right-impact-%E2%80%9Cland-grabbing%E2%80%9D-ethiopia

us individually and collectively prioritise and respond to in the name of patriotic virtue. It would be easy to see how under such circumstances, the ethical safeguards discussed above might be 'trumped' and set aside. However, patriotism is a problematic virtue for the ethicist. On the one hand, we can appreciate how there might be a duty to defend the state, which has provided us with security and opportunity for personal growth and development. On the other, the assumption of a right to assert the interests of one's state above the basic human rights of remote and disempowered peoples to distributive and participative justice is certainly brought into question by the Kantian imperative, namely that we must not use another person (or persons) merely as a means of fulfilling our own needs or wishes. We must therefore be rather cautious of the way in which biofuel lobbyists may appeal to national security to support their ambitions.

17.4   Concluding Comment

Production of biofuels has been regarded as a social and environmental good. Their use reduces our dependency on fossil fuels and is seen as a key strategy in the 'battle' against climate change. However, as we have seen, the actual situation is very much less clear-cut than that. Biofuels raise issues in both social and environmental ethics: there is an ethical 'downside' to several aspects of biofuel production. This does not mean that biofuel production should be stopped. Many applications of science raise social and ethical issues and thus biofuel production is not unique in this. Awareness of the problems leads to a search for a means to solve them. Some solutions have been briefly mentioned in this chapter; several of them are discussed in greater length in other chapters. Thus, we and Jim Lynch (in the previous chapter, Sustainability of Biofuels), emphasise that biofuel production *can* be a social and environmental good, but it will take deep wisdom and good governance to ensure that it is so.

Selected References and Suggestions for Further Reading

Achten, W.M.J. and Verchot, L.V. (2011) Implications of biodiesel-induces land-use changes for CO_2 emissions: Case studies in Tropical America, Africa and Southeast Asia. *Ecology and Society*, **16**, 14.

Bryant, J. (2013) *Beyond Human?* Oxford, Lion-Hudson.

Bryant, J. and la Velle, L. (2017) *Introduction to Bioethics*, 2nd edition. Wiley-Blackwell, Chichester, UK.

GRAIN (2013) Available at: http://www.grain.org/article/entries/4653-land-grabbing-for-biofuels-must-stop

Leopold, Aldo (1949) *The Sand County Almanac*. Oxford and New York, Oxford University Press.

Moran, E.F. (2006) *People and Nature*. Blackwell, Malden, MA/Oxford UK.

Nuffield Council on Bioethics (2011) *Biofuels: Ethical Issues*.

The Orang-Utan Project (2013) *Palm Oil*. Available at: http://www.orangutan.org.au/palm-oil

18

Postscript

John Love and John A. Bryant

College of Life and Environmental Sciences, University of Exeter, Exeter, UK

Since the Industrial Revolution, societies throughout the world have become increasingly reliant on cheap, energy-dense fossil fuels. In transport and heating, our need for fossil fuels is obvious; our reliance on fossil fuels for materials, food production or medicines is perhaps less obvious. Even less apparent, though, is the manner in which fossil fuels underpin our modern cultures; the (relatively) easy availability of a (again, relatively) cheap form of energy has enabled mass mobility and a mixing of humanity on a previously unprecedented scale. Our reliance on fossil fuels is a truly global phenomenon.

Even if we disregard the environmental cost of pollution and (as some do) the threat of global climate change for our planet, from a supply perspective alone, continuing consumption of fossil fuels is not sustainable in the long term. Biofuels may not be the answer to the energy problems of the future, but they are the current answer to reducing fossil fuel use within the technological, economic and cultural constraints that prevail today. The scientists that contributed their time and expertise to write this volume are an eclectic group and we have the fortune of knowing most of them personally. All are passionate about their research, but perhaps more importantly, all are aware of the urgency of the problem. As explained in Chapters 2 and 15, it may take 20 years to move from research to commercialisation and we have, as yet, no way of picking a 'winner'. In fact, there will probably be more than one winner; after all, crude oil is a complex mixture of many different hydrocarbons that must be separated and distilled before they can be used. Biofuels research, therefore, must be as wide as possible, and it is this breadth of opportunity that we hope to have represented in this volume. From first-generation ethanol and biodiesel (not forgetting methane, produced by anaerobic digestion[1]), to second-generation fermentation of lignocellulose and on to the advanced biofuels derived from natural or synthetically engineered microbes, replacing petroleum with a socially responsible, technically suitable and sustainable platform is challenging.

---

1 As mentioned in Chapter 3, methane may be used as a transport fuel, in addition to its more widespread use in being burned to generate electricity.

Biofuels and Bioenergy, First Edition. Edited by John Love and John A. Bryant.
© 2017 John Wiley & Sons Ltd. Published 2017 by John Wiley & Sons Ltd.

Many people throughout the globe are, in their own way, contributing to that effort. Unfortunately, we were unable, in the restricted space of this volume, to outline all of the possible bioenergy solutions that are actively researched; notable omissions are bio-hydrogen, the generation of electricity from bacterial fuel cells or the manner in which future biofuel production might be integrated with recycling and water purification. We have only briefly mentioned the economic, societal and ethical dimensions of biofuels, but hope that this volume has represented with honesty, the potential these new technologies offer. Ultimately, we hope that this volume will have provided a broad perspective on the recent exciting and innovative developments in biofuels research.

As we hinted above, there are those who ignore or dismiss the evidence that the Earth is warming because of the burning of fossil fuels and yet the scientific evidence for this is very strong indeed. 'Going green' is not simply a lifestyle choice for the affluent, but is absolutely essential if we wish to maintain global prosperity now, and for the future. Continuing to act as if climate change 'has nothing to do with us' is not an option if we are to keep the average temperature increase to 2 °C above pre-Industrial Revolution levels. Indeed, recent research[2] has shown that large proportions of our known reserves of fossil fuels must remain unused 'in the ground' if there is to be a realistic chance of achieving this target. These calculations indicate that 88% of the world's known coal reserves, 52% of known gas reserves and 35% of known oil reserves should not be touched. Those figures are stark reminders that 'something must be done.'

As indicated in the first chapter of this book, it is relatively straightforward to generate electricity without reliance on fossil fuels, although there remain problems associated with scale-up and with energy storage. It is thus encouraging to note that China is committed to reducing the use of coal, which has led researchers at the London School of Economics to suggest that CO_2 output from that vast country will peak by 2025 (much earlier than previously thought) and then decline[3]. Furthermore, at a recent meeting of chief ministers of the G7 nations, there was a commitment to the total phasing out of fossil fuels by the end of this century[4]. In the context of preventing run-away global warming, this is encouraging but provision of fuels for transport, a key part of modern life at all levels, is still a problem. Currently, we are a long way from providing transport fuels without relying on coal, oil and gas; thus the research and associated developments described in this book are very important. Nevertheless, we cannot know what form the energy of the future will take or how diverse, centralised or dispersed our energy sources will be. To a certain extent, even if biofuels are only used to reduce the exploitation of current fossil fuel reserves, their use buys us time to make that transition.

---

2 McGlade and Ekins (2015).

3 http://www.lse.ac.uk/GranthamInstitute/wp-content/uploads/2015/06/Chinas_new_normal_green_stern_June_2015.pdf

4 http://www.theguardian.com/environment/2015/jun/08/g7-fossil-fuel-pledge-is-a-diplomatic-coup-for-germanys-climate-chancellor

Selected References and Suggestions for Further Reading

Bhardwaj, A.K., Zenone, T. and Chen, J. eds (2015) *Sustainable Biofuels: an Ecological Assessment of the Future Energy.* Higher Education Press, Beijing, China and De Gruyter, Berlin, Germany and Boston, MA.

IPCC (2014) *Fifth Assessment Report: Climate Change 2014.* Available at: https://www.ipcc.ch/report/ar5/syr/

IPCC (2014) *Climate Change 2014: Mitigation of Climate Change. Contribution of Working Group III to the Fifth Assessment Report of the Intergovernmental Panel on Climate Change* (Edenhofer, O., Pichs-Madruga, R., Sokona, Y., Farahani, E., Kadner, S. *et al.*, eds), Cambridge University Press, Cambridge, UK/New York.

McGlade, C. and Ekins, P. (2015) The geographical distribution of fossil fuels unused when limiting global warming to 2C. *Nature*, **517**, 187–190.

Index

Page numbers in *italics* refer to illustrations; those in **bold** refer to tables.

Biofuels and Bioenergy, First Edition. Edited by John Love and John A. Bryant.
© 2017 John Wiley & Sons Ltd. Published 2017 by John Wiley & Sons Ltd.